Advances in

VIRUS RESEARCH

VOLUME 58

Advances in
VIRUS RESEARCH

Edited by

KARL MARAMOROSCH
Department of Entomology
Rutgers University
New Brunswick, New Jersey

FREDERICK A. MURPHY
School of Veterinary Medicine
University of California, Davis
Davis, California

AARON J. SHATKIN
Center for Advanced Biotechnology and Medicine
Piscataway, New Jersey

VOLUME 58

Amsterdam Boston London New York Oxford Paris
San Diego San Francisco Singapore Sydney Tokyo

This book is printed on acid-free paper. ∞

Academic Press
An imprint of Elsevier Science.
525 B Street, Suite 1900, San Diego, California 92101-4495, USA
http://www.academicpress.com

Academic Press
84 Theobolds Road, London WC1X 8RR, UK
http://www.academicpress.com

International Standard Book Number: 0-12-039858-3

PRINTED IN THE UNITED STATES OF AMERICA
02 03 04 05 06 07 MM 9 8 7 6 5 4 3 2 1

CONTENTS

Role of Lipid Rafts in Virus Assembly and Budding

DEBI P. NAYAK AND SUBRATA BARMAN

I.	Introduction	1
II.	Lipid Rafts	2
III.	Functions of Lipid Rafts	5
IV.	Assembly of Influenza Virus	19
V.	Conclusion	22
	References	23

Novel Vaccine Strategies

LORNE A. BABIUK, SHAWN L. BABIUK, AND MARIA E. BACA-ESTRADA

I.	Introduction	29
II.	The Ideal Vaccine	30
III.	Principles of Vaccination	32
IV.	Conventional Vaccines	35
V.	Genetically Engineered Vaccines	39
VI.	Marker Vaccines	52
VII.	Vaccine Formulation and Delivery	54
VIII.	Epilogue	65
	References	66

The Potential of Plant Viral Vectors and Transgenic Plants for Subunit Vaccine Production

PETER AWRAM, RICHARD C. GARDNER, RICHARD L. FORSTER, AND A. RICHARD BELLAMY

I.	Current Trends in Vaccine Development	81
II.	An Overview: The Use of Plants for Recombinant Protein Expression	83
III.	Plant Protein-Expression Systems	86
IV.	Stable Transgenic Expression Systems	87
V.	Plant Virus Expression Systems	89
VI.	Enhancing Protein Yield	98
VII.	Expression of Immunogenic Molecules	100
VIII.	Immunogenicity and Vaccine Responses	106
IX.	Edible Vaccines and Human Health	112

X. Conclusion 113
References 114

Treatment of Arenavirus Infections: From Basic Studies to the Challenge of Antiviral Therapy

Elsa B. Damonte and Celia E. Coto

I. Introduction 125
II. Classification 126
III. Arenaviruses as Agents of Emerging Diseases 128
IV. The Virus 131
V. Virus Infection and the Host Cell 132
VI. The Replicative Cycle and Possible Targets for Therapeutic Agents . 133
VII. Present Treatment of Human Disease 143
VIII. Concluding Remarks 145
References 147

Evaluation of Drug Resistance in HIV Infection

Benedikt Weissbrich, Martin Heinkelein, and Christian Jassoy

I. Introduction 157
II. Targets of Antiviral Drug Therapy in HIV Infection 159
III. Mechanisms of Antiviral Drug Resistance 168
IV. Technologies for Measuring Drug Sensitivity 173
V. Clinical Implications 183
VI. Conclusions 187
References 188

Perspectives on Polydnavirus Origins and Evolution

Matthew Turnbull and Bruce Webb

I. Introduction 203
II. Classification of Polydnaviruses as Viruses 206
III. Ichnoviruses 208
IV. Bracoviruses 228
V. Origin of the Polydnaviridae 239
VI. Conclusion 244
References 245

Bacteriophage ϕ29 DNA Packaging

Shelley Grimes, Paul J. Jardine, and Dwight Anderson

I. Introduction 255
II. Overview of ϕ29 DNA Packaging 256
III. Components of the Defined ϕ29 DNA Packaging System 260
IV. The DNA Packaging Process 272

V. Aims and Prospects 289
References 290

The Interaction of Orthopoxviruses and Interferon-Treated Cultured Cells

C. JUNGWIRTH

I. Introduction 295
II. Interaction of Poxviruses with the IFN Type I-Treated Host Cell ... 296
III. The Posttranscriptional Inhibition of Poxvirus-Specific Gene Expression in IFN Type I-Treated Chick Embryo Fibroblasts 299
IV. Cytotoxicity Enhancement of IFN Type I-Treated and Vaccinia Virus-Infected Cells 301
V. Inhibition of Vaccinia Virus Replication by Nitric Oxide Synthase Induced by IFN Type II 303
VI. Metabolic Activities Are Required for the Response of the IFN Type I-Treated Host Cell to Vaccinia Virus Infection 303
VII. Double-Stranded RNA Molecules Detectable during Vaccinia Virus Replication and Their Role in the Interaction of Vaccinia Virus with the IFN Type I-Treated Host Cell 308
VIII. Modulation of the Cellular IFN Type I and Type II Responses by Poxvirus-Specific Gene Products 309
IX. Concluding Remarks 315
References 317

Index 327

ADVANCES IN VIRUS RESEARCH, VOL. 58

ROLE OF LIPID RAFTS IN VIRUS ASSEMBLY AND BUDDING

Debi P. Nayak and Subrata Barman

Department of Microbiology, Immunology and Molecular Genetics
UCLA School of Medicine
Molecular Biology Institute
Los Angeles, California 90095-1747

I. Introduction
II. Lipid Rafts
III. Functions of Lipid Rafts
A. Protein Transport, Protein Sorting, and Protein Targeting
B. Membrane Signaling and Endocytosis
C. Virus Assembly and Budding
IV. Assembly of Influenza Virus
V. Conclusion
References

I. Introduction

Different enveloped viruses assemble and bud from the membrane of specific intracellular organelles such as the endoplasmic reticulum (ER), different Golgi compartments (*cis*-, mid-, or *trans*-Golgi or *trans*-Golgi network, TGN), and the plasma membrane (Garoff *et al.,* 1998). Furthermore, viruses that assemble and bud from the plasma membrane do not bud randomly, but bud asymmetrically from the surface of polarized epithelial cells. Some viruses bud from the apical surface, whereas others bud from the basolateral plasma membrane of polarized epithelial cells (Compans, 1995; Nayak, 2000). Moreover, it is known that the protein composition of the viral envelope is distinctly different from that of the budding host membrane and the lipid composition does not fully represent that of the host cell membrane, although all viral lipids are derived from host. In general, the viral envelope excludes the majority of the host proteins or in some cases may selectively include a few specific host proteins which may be functionally important for virus replication. Mechanisms by which viruses exclude host proteins are multiple. Many viruses inhibit or shut off host protein synthesis in a number of ways, thereby preventing incorporation of newly synthesized host proteins into the viral envelope. However, this does not explain the exclusion of preexisting host membrane proteins from the

0065-3527/02 $35.00

viral envelope. In general, it is believed that interaction of the viral envelope proteins with internal components (e.g., matrix proteins and/or nucleocapsids) and interaction among the matrix proteins, which may form a sheet under the lipid bilayer, will exclude most of the host membrane proteins from the budding site and the viral envelope. Lipids of the viral envelope are derived from host cells and vary in composition depending on the cells in which viruses are grown and the membranes from which the viruses bud. However, the lipid composition of the viral envelope does not quantitatively represent that of the budding organelle or plasma membrane. In other words, the lipid composition of viruses budding from the plasma membrane such as apical or basolateral membranes of polarized epithelial cells is not same as the average lipid composition of these membranes. This would suggest that lipid bilayers of plasma membranes are not homogeneous or uniform, but are mosaic in nature and contain heterogeneous microdomains and that viruses bud from the specific microdomains present within these membranes. The apical membranes possess mosaic microdomains, which are called lipid rafts, and some viruses selectively bud from these lipid rafts. In this review we will briefly discuss the nature of these lipid rafts and their role in the assembly and budding of some viruses. The structure, origin, and functions of these lipid rafts will not be discussed in detail. Readers are referred to some of the excellent reviews published recently (Brown and London, 1998a, b; Rietveld and Simons, 1998; Simons and Ikonen, 1997).

II. Lipid Rafts

Membrane lipids exist as lipid bilayers and form an amphipathic barrier between the cell and its environment and thereby prevent free leakage of its internal content to the outside environment as well as unregulated entry of external elements into the cell. It is generally believed that lipids of the plasma membrane function mainly as a solvent for membrane proteins and exist in the liquid-crystalline (l_C) phase (Brown and London, 1998a, b). However, recent studies have shown that heterogeneity among lipids leads to lateral organization and phase separation, forming a liquid-ordered phase (l_O), and causes asymmetric distribution of different lipids in exoplasmic versus cytoplasmic leaflets (Simons and Ikonen, 1997). In polarized epithelial cells, plasma membranes are organized into apical and basolateral domains separated by tight junctions. Both the lipid and the protein compositions of apical and basolateral membranes vary. The apical plasma membrane is enriched in sphingolipids (glycosphingolipids and sphingomyelin), whereas the basolateral membrane predominantly contains glycolipids, phosphatidylcholines. Tight junctions in polarized cells prevent lateral diffusion and commingling of sphingolipids and phosphatidylcholines, both of which are present in the exoplasmic leaflet of the lipid bilayer of apical and basolateral membranes, respectively. Glycosphingolipid clusters are formed as microdomains within the exoplasmic leaflet of the Golgi membrane and lateral interactions among the glycosphingolipids are stabilized by intercalation of cholesterol molecules, which function as space fillers. These closely packed sphingolipid–cholesterol complexes are transported from the TGN to the apical plasma membrane and are present as islands of microheterogeneity on that membrane. Cholesterol is not limited to the exoplasmic leaflet, but is also present in the cytoplasmic leaflet and functions as a spacer, filling the voids created by interdigitating fatty acid chains. Although specific lipid elements such as cholesterol and sphingomyelins are critical for the formation and stability of the liquid-ordered phase in lipid rafts, the transmembrane peptides may also play an important role in maintaining the ordered lipid phase. Synthetic transmembrane domain (TMD) peptide of influenza virus hemagglutinin (HA) was found to increase the order of acyl chains of the lipid bilayer. Ordered lipids may attract the HA TMD peptides, which in turn may further order the lipids surrounding them (Tatulian and Tamm, 2000). However, whether the ability of the TMD peptides to order the lipids surrounding them into the liquid-ordered phase is also universal for other apical raft-associated peptides or unique to HA TMD remains to be seen.

These sphingolipid–cholesterol clusters are usually small, vary in size (2.5–10 nm), and are dispersed in the plasma membrane. They are dynamic in nature, however, and can coalesce or dissociate, with individual lipids moving in and out of the complex (Kusumi and Sako, 1996). Large oligomeric proteins like caveolin can stabilize rafts in the liquid-ordered phase, suggesting dependence of raft size on protein oligomers. Cholesterol–sphingolipid microdomains are partially insoluble in neutral detergents such as Triton X-100 (TX-100) at low temperature (4°C) and can be isolated following TX-100 extraction as detergent-resistant membranes (DRMs), also called detergent-insoluble, glycolipid-enriched complexes (DIGs) (Brown and London, 1998a). Because of their high lipid content and low density, DIGs float to the top in sucrose flotation gradients and therefore can be separated from TX-100-soluble lipids and proteins as well as from other, TX-100-insoluble complexes such as cytoskeletal elements. In addition, DIGs are soluble in mild ionic detergents such as octylglucoside (OG), which also serves to distinguish the DIGs from the detergent-insoluble cytoskeletal complexes following TX-100 extraction. These TX-100-insoluble sphingolipid–cholesterol-enriched DIGs are also called "lipid rafts" (Simons and Ikonen, 1997). These lipid rafts are found in plasma membranes, Golgi complexes, and transport vesicles. The lipid composition of MDCK cells and TX-100-insoluble vesicles has been determined. The relative percentages of different lipids in these vesicles show that they are distinctly different and do not represent the average composition of total cellular lipids. They are enriched in sphingomyelin, cholesterol, and cerebrosides, but contain less phospholipids in general (Table I). Caveolae, 50- to 70-nm plasma membrane invaginations, contain a cholesterol-binding protein, caveolin (22-kDa protein), and exhibit similar biochemical properties to lipid rafts. However, caveolae are predominantly present on the basolateral membrane, whereas lipid rafts are predominantly present in apical membranes of polarized epithelial cells such as MDCK cells (Simons

TABLE I
LIPID COMPOSITION OF CELLULAR MEMBRANES AND INFLUENZA VIRUS PARTICLES

	Mole Percentage Lipid Composition		
Lipid	Whole Cells	Vesicles	Influenza Viruses
Phospholipids			
Phosphatidylethanolamine	78.84	17.55	15.85
Phosphatidylcholine	22.85	12.44	6.71
Phosphatidylserine	2.17	2.22	11.65
Phosphatidylinositol	4.59	2.67	0.67
Cardiolipin	1.77	ND	ND
Sphingomyelin	1.09	11.11	9.11
Cholesterol	11.63	32.67	43.97
Triglyceride	23.43	ND	2.49
Neutral glycolipids			
Cerebrosides	1.57	7.55	9.55
Forssman antigen	1.11	6.67	ND
Lactosyl ceramide	0.21	2.44	ND
Acidic glycolipids			
Sulfatides	0.48	2.89	ND
Gangliosides	0.25	1.78	ND

[a] Adapted from Brown and Rose (1992) and Zhang *et al.* (2000).

and Ikonen, 1997). Lipid rafts are also present in nonpolarized cells such as fibroblasts. In this review, terms such as DRMs, DIGs, TX-100-insoluble lipids, and lipid rafts are used interchangeably.

III. Functions of Lipid Rafts

A. Protein Transport, Protein Sorting, and Protein Targeting

Lipid rafts have been shown to be associated with multiple cellular functions (Simons and Ikonen, 1997). These include protein transport, protein sorting, protein targeting, membrane trafficking, membrane signaling, and clathrin-independent endocytosis, as well as in the assembly and budding of some enveloped viruses. In polarized epithelial cells such as MDCK cells, integral membrane proteins are targeted and delivered to either the apical or the basolateral surface. Basolateral sorting signals have been characterized in greater detail and include tyrosine- or dileucine-based motifs present in the cytoplasmic tail of basolateral transmembrane proteins. These motifs are similar, but not identical to those implicated in clathrin-dependent endocytosis, suggesting that these basolateral proteins use similar mechanisms for targeting and recruiting vesicular carriers, which are destined from the TGN to the cell surface. Basolateral carriers, after being released from the TGN, move along microtubules to the basolateral surface (Lafont *et al.,* 1994), where they may dock and fuse with the SNARE (soluble *N*-ethyl maleimide-sensitive factor [NSF] attachment protein [SNAP] receptor)–Rab machinery (Rothman, 1994; Ikonen *et al.,* 1995). Apical carriers also move along microtubules to the apical surface and require microfilament disassembly to dock and fuse with apical membranes (Muallem *et al.,* 1995). The nature of the apical vesicles and the process of docking and fusion of apical vesicles have not been fully characterized. Because of the divergent nature of the apical determinants of the cargo proteins, the carrier vesicles are likely be divergent and vary both in lipid and protein contents. The majority of these apical vesicles contain lipid rafts as evidenced by their lipid composition and TX-100 insolubility. However, some apical proteins, such as bovine enteropeptidase (Zheng *et al.,* 1999), intestinal maltase–glucoamylase, lactase–phlorizin hydrolase (Danielsen, 1995), as well as influenza C virus HEF and influenza A virus M2 protein (Hughey *et al.,* 1992), both of which are acylated (Veit *et al.,* 1990, 1991), do not use lipid rafts (Zhang *et al.,* 2000) for apical delivery.

Influenza virus HA, a model transmembrane apical protein, has been used as a marker to identify a number of proteins present in apical carrier vesicles which are involved in the transport of HA to both the cell surface in nonpolarized cells such as fibroblasts as well as the apical surface in polarized epithelial cells such as MDCK cells. These proteins include the following.

1. Caveolin-1 / VIP21

Caveolae are present predominantly on the basolateral surface and contain caveolin-1 and caveolin-2. Whereas caveolin-1 is present on both apical and basolateral membranes, caveolin-2 is predominantly present on the basolateral surface. Caveolin-1/VIP21 (but not caveolin-2) binds to cholesterol (Murata *et al.,* 1995), exhibits TX-100 insolubility (Kurzchalia *et al.,* 1992), and is targeted to the apical surface of polarized epithelial cells as larger homooligomer complexes (Scheiffele *et al.,* 1998). Antibody to caveolin-1 inhibits the transport of influenza virus HA to the cell surface and to the apical membrane without affecting the basolateral transport of vesicular stomatitus virus (VSV) G protein, implying the presence of caveolin-1 in the apical vesicle and its functional role in apical transport (Scheiffele *et al.,* 1998). However, it should be noted that caveolin-1 is not an essential element for apical transport of HA. In Fisher rat thyroid (FRT) cells, which do not express caveolin-1/VP21 (Zurzolo *et al.,* 1994), HA as well as glycosylphosphatidylinositol (GPI)-anchored placental alkaline phosphatase (PLAP) are efficiently transported to the apical plasma membrane (Martin-Belmonte *et al.,* 2000).

2. Annexin XIIIb

Annexin XIIIb, a member of the annexin XIII subfamily, in its myristoylated form is present in sphingolipid–cholesterol rafts and is distributed along the apical route of the exocytic pathway as well as on the apical membrane. Furthermore, annexin XIIIb appears to function in apical transport, since it colocalizes with influenza virus HA and stimulates its apical transport (Lafont *et al.,* 1998). Annexin XIIIb also facilitates the association of Nedd4 with the lipid raft and thereby targets Nedd4 to the apical membrane (Plant *et al.,* 2000).

3. MAL / VIP17

The tetraspanning membrane proteolipid MAL/VIP17 is a nonglycosylated integral membrane protein which is present in polarized epithelial cells, myelin-forming cells, and T lymphocytes. MAL is present in lipid rafts and also functions as an element in the apical transport pathway. Depletion of MAL inhibits both the cell surface transport and apical transport of influenza virus HA and in addition causes partial missorting of HA to the basolateral surface (Puertollano *et al.,* 1999).

4. Phospholipase D2

Phospholipase D2 (PLD2) is present in TX-100-insoluble lipid rafts and is stimulated by phosphatidylinositol-4,5-biphosphate (PIP_2) and inhibited by neomycin. It is upregulated by caveolin-1 and therefore may function as a component of apical vesicles (Czarny *et al.,* 1999). In addition, Thy-1 CD59 as well as ganglioside GM1 are preferentially partitioned into lipid rafts, whereas CD45, a 200-kDa transmembrane phosphatase protein, is excluded from lipid rafts. GM1 is a ganglioside-specific marker for lipid rafts. However, the function of CD59 protein or GM1 ganglioside in the apical transport pathway is not well defined.

In addition to these protein components of apical transport vesicles, different classes of proteins can bind to lipid rafts either as cargo or passenger molecules involved in different functions. Among these are GPI-anchored proteins transmembrane proteins, and doubly acylated tyrosine kinase of the Src family. Lipid anchors such as saturated fatty acyl chains of GPI anchors and Src kinases are responsible for raft association. For transmembrane apical proteins, the membrane-spanning domains (TMDs) confer the affinity for lipid raft association (Kundu *et al.,* 1996; Scheiffele *et al.,* 1997). Mutational analysis shows that specific amino acids within the TMDs of both type I (influenza virus HA) and type II (influenza virus neuraminidase, NA) proteins provide the determinants for raft association. However, the common structural or functional properties of these TMD amino acids, the peptide sequences, and the structural bases of peptides that lead to raft association are yet to be determined. Furthermore, raft association and apical transport determinants do not involve the same sequence(s), although both are present in the TMD of HA or NA and may overlap each other (Lin *et al.,* 1998; Barman and Nayak, 2000). In addition, apical sorting is not the exclusive property of TMDs. For example, some proteins may use N-glycans for binding to raft-associated lectins for apical sorting (Scheiffele *et al.,* 1995; Benting *et al.,* 1999; Lipardi *et al.,* 2000), whereas others may use TMDs to directly partition into lipid rafts (Kundu *et al.,* 1996; Lin *et al.,* 1998). Furthermore, some TMDs may possess apical detriments, but may not become raft-associated (Dunbar *et al.,* 2000).

Lipid rafts are involved not only in protein sorting and protein targeting to apical membranes in polarized cells, but also in post-Golgi transport of apical proteins to the cell surface in nonpolarized cells. A number of studies have shown that removal or depletion of cholesterol, which leads to disorganization of lipid rafts, causes missorting of apical proteins like HA, and also reduces the overall transport of HA to the cell surface in both polarized and nonpolarized cells (Keller and Simons, 1998).

B. Membrane Signaling and Endocytosis

In addition to involvement of lipid rafts in protein transport and protein sorting from the TGN to the cell surface, lipid rafts are also involved in endocytosis of proteins from the apical surface via endocytic pathways independent of clathrin-coated vesicles or caveolae (Danielsen and van Deurs, 1995) as well as in endocytosis and transcytosis of proteins from the basolateral surface via caveolae (Schnitzer *et al.,* 1994). Lipid rafts are also involved in membrane signaling. Several signaling molecules, such as trimeric G proteins and Ras, are also associated with lipid rafts (Li *et al.,* 1995; Song *et al.,* 1996). The Src-family kinases Lck ($p56^{lck}$) and Fyn ($p59^{fyn}$), implicated in T cell activation, are also associated with DIGs. However, how lipid raft association aids in signal transduction is unclear. It is likely that raft association may concentrate several members of proteins involved in signal transduction and thus facilitate cross-talk among the protein molecules involved.

C. Virus Assembly and Budding

Lipid rafts play an important role in assembly and budding of several viruses in at least three ways: (1) sorting, targeting, and delivering the viral components to the assembly site via exocytic pathways, (2) constituting the budding site on the plasma membrane, and (3) facilitating the budding process and release of virus particles.

1. Role of Lipid Rafts in Sorting, Targeting, and Delivering Viral Components to the Assembly Site

In the majority of enveloped viruses which bud from the apical surface in polarized epithelial cells, lipid rafts are involved in transporting, sorting, and delivering the envelope proteins from the TGN to the budding site on the plasma membrane. In addition, lipid rafts are also involved in transporting internal core viral components, including matrix proteins and nucleocapsids, possibly indirectly as they associate with envelope proteins during exocytic transport (Ali *et al.,* 2000). These viral envelope proteins are transported to the apical surface of polarized epithelial cells and use lipid rafts as platforms for exocytic transport. In addition, these viruses are also likely to bud from lipid rafts present on the plasma membrane in both polarized and nonpolarized cells and therefore will possess an envelope mimicking the lipid raft. Furthermore, some basolateral viruses such as human immunodeficiency virus (HIV) and simian immunodeficiency virus (SIV) and viral proteins, although basolateral, undergo fatty acid modifications, become lipid raft associated and selectively use raft-associated lipids in the liquid-ordered phase in their envelope during budding (Aloia *et al.,* 1993; Nguyen and Hildreth, 2000).

Influenza viruses and influenza viral proteins have been studied extensively as models for raft association in protein sorting, protein transport, virus assembly, and budding. Influenza A viruses possess three transmembrane envelope proteins, HA, NA, and M2, all of which possess apical signals and are capable of targeting to the apical plasma membrane when expressed independently. Among these, both HA, a type I transmembrane protein, and NA, a type II transmembrane protein, use lipid rafts as platforms for apical transport; M2, although an apical protein, does not use lipid rafts for apical transport (Zhang *et al.,* 2000). However, M2 is a minor component of the viral envelope and does not affect the ordered lipid phase of the viral envelope. Influenza virus HA was shown to associate with lipid rafts and acquire TX-100 insolubility late in the exocytic transport pathway after addition of the complex sugars in the Golgi complex. TX-100-insoluble HA floated to the top of the gradient and became soluble after octylglucoside (OG) treatment, demonstrating the association of HA with lipid rafts. Association of HA with lipid rafts was not dependent on oligosaccharide modification, its association in other viral proteins, or its assembly into virus particles, since in cDNA-transfected cells, HA also became associated with DIG. However, some linkage of HA with NA has been suggested with respect to raft association in virus-infected cells (Zhang *et al.,* 2000). Association of HA with lipid rafts was not dependent on the polarity of cells, as HA became TX-100-insoluble in nonpolar MDCK cells in suspension as well as in nonpolar chick embryo fibroblasts (Skibbens *et al.,* 1989). Cholesterol was critically involved in detergent insolubility and interaction of HA with lipid rafts. Removal of cholesterol with lovastatin or methyl-β-cyclodextrin caused increased solubility of HA after TX-100 treatment, accompanied by the reduction in cell surface transport of HA and the increased missorting of HA to the basolateral surface (Keller and Simons, 1998). The TMD of HA contained the critical determinants for its interaction with lipid rafts, since chimeric HGH or HCH proteins containing the TMD of VSV G protein (G) or herpes simplex virus (HSV) C protein (C), respectively, and the ectodomain and the cytoplasmic tail of HA protein (H) did not interact with TX-100-resistant lipid rafts.

Mutational analysis of the amino acids in the HA TMD showed that the hydrophobic residues in the exoplasmic half of HA TMD were critical for interaction and stable association with lipid rafts (Scheiffele *et al.*, 1997). Further analysis of these HA TMD mutants in polarized MDCK cells demonstrated that although the HA TMD contained the apical determinants, association with lipid rafts did not completely correlate with apical transport of HA. Some HA mutants exhibiting wild-type lipid raft association were randomly sorted, and a mutant (2A517) exhibiting a predominantly apical distribution ($\sim$61%) did not associate with TX-100-insoluble membrane. However, among the mutants which were missorted basolaterally, none was TX-100-insoluble, indicating that they were not associated with lipid rafts (Lin *et al.*, 1998).

It was also found that the HA TMD peptide was predominantly in α-helical conformation in detergent micelles and in phospholipid bilayers (Tatulian and Tamm, 2000). Helicity of the HA peptide was also increased in lipid bilayers composed of pure phosphatidylcholine. Furthermore, the HA peptide increased the acyl chain order in lipid bilayers, which may be related to the preferential association of HA with lipid rafts. In addition to TMD, the cytoplasmic tail of HA also may have an effect on its association with lipid rafts, HA lacking the cytoplasmic tail (HAt−) exhibited markedly reduced association with DIG in both HA cDNA-transfected cells and virus-infected cells, although deletion of the cytoplasmic tail did not affect the apical transport of HA (Zhang *et al.*, 2000). These results also show that the determinants of apical targeting and raft association were not identical, although both resided in the HA TMD. In addition to the HA TMD peptide, three palmitoylated cysteine residues in HA were also important in raft association. Furthermore, in HA/NAt− virus-infected cells, the presence of NAt− (tail minus NA) significantly reduced the raft association of HA, suggesting a possible linkage of HA with NA for interaction with lipid rafts (Zhang *et al.*, 2000).

Influenza virus NA, a type II transmembrane protein, was shown to possess two apical determinants, one in the ectodomain, which is likely to be N-glycan, and the other in the NA TMD, which is 29 amino acids long and is flanked by a highly conserved 6-amino-acid cytoplasmic tail at the NH_2 terminus and the ectodomain at the COOH terminus (Kundu *et al.*, 1996). Chimeric analysis using human transferrin receptor (TR), a type II basolateral protein, demonstrated that the NA TMD possessed apical determinants and was able to direct a reporter protein (TR ectodomain) directly to the apical surface of MDCK cells. Furthermore, it was also shown that the NA TMD interacted with TX-100-resistant membranes, suggesting that the lipid raft may function as a platform for interaction with the NA TMD and thereby deliver the protein to the apical surface. In addition, since NA, unlike HA, does not undergo fatty acid modification in TMD or cytoplasmic tail, it interacts with directly lipid rafts via lipid-protein interaction. NA TMD sequences were further analyzed by making chimeric TMDs between the NA TMD and TR TMD sequences and by mutating the NA TMD sequences (Barman and Nayak, 2000). These analyses showed that, like HA TMD, the COOH-terminal half of the NA TMD was significantly involved in interacting with the exoplasmic lipid leaflet containing cholesterol and glycosphingolipids for raft association. However, in addition, highly conserved NH_2-terminal NA TMD sequences also played a significant role in raft association. Removal of the conserved cytoplasmic tail of NA also reduced the stable raft association of NA TMD, supporting the notion that the cytoplasmic tail and adjacent amino acids in the TMD may aid in stabilizing the lipid/protein interaction between the TMD sequences and the lipid rafts in both HA and NA (Barman and Nayak, 2000; Zhang *et al.,* 2000).

Signals for apical transport in NA TMD were not short peptide sequences, but consisted of multiple regions extended over the TMD. However, the entire TMD was not required for apical transport, since 19 amino acids (aa 9–27) of NA TMD were sufficient for apical transport. Also, as in HA, determinants for both apical transport and lipid raft association resided in the NA TMD and the apical signal overlapped with those for lipid raft association, but these signals were not identical and varied independently. Similarly, the degree of apical transport did not correlate with the lipid raft association. Some NA TMD mutants exhibited wild-type apical transport, but only intermediate TX-100 insolubility (~30% compared to 60% in wild-type NA TMD) and some had even very low TX-100 insolubility (~22%) (Barman and Nayak, 2000). However, all NA TMD mutants exhibiting random or basolateral sorting also exhibited poor raft association as determined by TX-100 insolubility. Taken together, these results show that although raft association facilitates apical transport of the NA TMD, raft association is not obligatory for apical transport, and the NA TMD (and possibly the HA TMD) possesses apical determinants which can interact with apical sorting machineries outside the lipid raft.

In summary, analysis of both HA and NA TMDs shows that both TMDs possess signals for apical transport and raft association. Although these signals overlap, they are not identical and vary independently. However, none of the randomly sorted or basolaterally sorted mutant proteins exhibited a high degree of TX-100 resistance, showing that all missorted proteins were excluded from lipid rafts. In addition, some apical proteins were highly TX-100-soluble, indicating that apical sorting signals in the NA TMDs and possibly in HA TMDs can be recognized by the apical sorting machinery outside the lipid raft and that a high degree of lipid raft association was not obligatory for apical sorting. Furthermore, influenza type A virus M2, a type III protein, is an apical protein, but is not lipid raft-associated (Hughey *et al.*, 1992; Zhang *et al.*, 2000). Since M2 protein does not possess glycan in its ectodomain, it is unclear how M2 is targeted apically.

Like orthomyxoviruses, members of the paramyxoviruses are apical in budding. The glycoproteins of these viruses are also likely to be raft-associated in the TGN during exocytic transport and in the plasma membrane. Sendai virus (SV) F and HN proteins associate with lipid rafts, although the degree of TX-100 resistance of membranes containing F and HN is less than that of HA and NA (Ali and Nayak, 2000). The TMD of HN protein of SV5 has been shown to possess an apical determinant (Huang *et al.*, 1997). However, detailed analyses of raft association of SV5 or other, related paramyxoviral glycoproteins and preferential budding of these viruses from lipid rafts as well as incorporation of lipid rafts in the viral envelope have not been done. Similarly, respiratory syncytial virus is also known to bud from apical surfaces in polarized Vero C1008 and MDCK cells (A. Roberts *et al.*, 1995). However, lipid raft association of its glycoproteins and the presence of ordered lipids in the viral envelope have not been determined. In measles virus, which also buds apically, the roles of lipid rafts in transporting H and F glycoproteins and in the budding process are complex (see below).

2. *Role of Lipid Rafts at the Budding Site*

The lipid composition of influenza virions from either nonpolarized MDCK cells or the apical surface of polarized MDCK cells is similar, but is different from that of total cellular lipids. Similarly, VSV budding from either nonpolarized cells or the basolateral surface of polarized cells has a similar lipid composition, which is different from the total cellular lipid composition (van Meer and Simons, 1982). In addition, some apical viruses, such as measles viruses, have varying requirements for lipid raft association for the apical transport of glycoproteins (Naim *et al.*, 2000), but use lipid rafts, though not exclusively, as the budding sites (Manié *et al.*, 2000). Some basolateral viruses, such as HIV and SIV, have also been shown to bud from cholesterol-enriched membranes containing lipid rafts (Aloia *et al.*, 1993; Nguyen and Hildreth, 2000). These results clearly demonstrate that viruses do not bud randomly from either the apical or the basolateral membrane, but bud from selective domains within the apical or the basolateral plasma membrane. Analysis of virion envelope lipids shows that influenza viruses contain lipids in a state similar to the liquid-ordered

(l_O) phase (Scheiffele *et al.,* 1999). Cholesterol, sphingomyelin (SM), and phosphatidylcholine (PC) present in the virion envelope are TX-100-detergent insoluble. Furthermore, the presence of cholesterol is required for the detergent insolubility of these lipids in the virion membrane. The cholesterol dependence of the insolubility of SM and PC and of influenza virus HA in the virion envelope suggests the requirement of cholesterol in the formation of the ordered lipid domain of the membrane. A high protein/lipid ratio in the viral membrane also suggests that transmembrane viral proteins such as HA and possibly NA are not just passive passengers in the ordered phase of the membrane lipids, but may actively participate in the membrane structure. In addition, the specific amino acid sequence and structure of the TMD peptide of HA and NA, palmitoylation of cystine residues, and the cytoplasmic tail of HA may affect the ordered phase of lipids. However, incorporation of HA alone is not sufficient to organize the ordered lipid environment, since HA incorporated in the VSV envelope was TX-100-soluble (Scheiffele *et al.,* 1999). On the other hand, lipid rafts are not exclusively present in the apical membrane, since a minor fraction of GPI-anchored PLAP (an apical protein) present in the basolateral membrane of polarized MDCK cells is TX-100-resistant and therefore raft-associated (Arreaza and Brown, 1995).

Both the biochemical properties and the lipid composition show that the influenza virus envelope possesses a similar lipid composition to that of the TX-100-insoluble membranes and TX-100-insoluble intracellular transport vesicles, particularly with respect to phosphatidylethanolamine, cholesterol, and sphingomyelin, and is distinctly different from the whole-cell lipid composition (Table I), suggesting that these viruses incorporate and maintain the ordered lipid phase of lipid rafts in the virion envelope.

As mentioned above, one possible reason that the lipid composition of viruses does not match with that of the host cell is that viruses do not bud randomly, but bud from specific organelles or plasma membranes. Furthermore, virions do not incorporate lipids randomly, but instead incorporate specific lipids selectively in their envelope. Consequently, the virion lipid composition is different from that of the host membrane. For example, when viruses bud from the apical or the basolateral membranes of polarized epithelial cells, their lipid composition varies and resembles more closely that of the budding membrane, but even so, does not completely match the total lipid composition of the budding membrane. This would suggest that virus budding occurs from selective domains within the membrane. For example, when influenza viruses bud from nonpolarized cells, for example, MDCK cells in suspension, the lipid composition of both transporting vesicles and the viral envelope shows not only that influenza viral components are transported to the plasma membrane by lipid raft-associated vesicles, but also that influenza viruses are also assembled from the lipid raft domains on the plasma membrane. In other words, after reaching the plasma membrane, viral components do not dissociate and diffuse out of the lipid raft, but instead remain associated with lipid rafts. Coalescence of lipid rafts containing influenza viral components in the liquid-ordered phase may facilitate virus budding from the plasma membrane and may form the platform for the budding site by concentrating the viral glycoproteins on the outer side of the plasma membrane and the interacting proteins such as matrix proteins and nucleocapsid on the inner side of the plasma membrane. Such interaction of viral components will crowd out the majority of host membrane proteins from the budding site. Furthermore, increased viscosity and reduced fluidity of the liquid-ordered phase in lipid rafts on the plasma membrane may induce curvature outward and thereby facilitate budding. However, recent studies have shown that M1 protein of influenza virus alone is sufficient for generating budding particles (Gomez-Puertas *et al.,* 2000), suggesting that interaction with other host factors such as cytoskeletal elements with viral components may provide the pushing force and thereby aid in the budding process.

In addition to influenza virus, a number of other viruses have been shown to bud from the apical surface. Measles viruses (MV) bud from the apical surface in polarized epithelial cells (Blau and Compans, 1995). However, both raft association and apical budding of MV appear to result from complex interactions of viral proteins during intracellular transport, sorting, and plasma membrane budding. In MV-infected cells, both glycoproteins (F and H proteins) became raft-associated during Golgi maturation and in addition, M and N proteins also became lipid raft-associated. However, when expressed alone, measles virus F protein, but not H protein, became raft-associated, suggesting that H protein required other viral proteins to become raft-associated in virus-infected cells. Upon coexpression of H protein with F protein, H protein also became lipid raft-associated, supporting the idea that F protein facilitated raft association of H protein (Vincent *et al.,* 2000). Furthermore, it appears that M and N proteins did not depend on H and F proteins for lipid raft association (Manié *et al.,* 2000). However, how M and N proteins, which are not transported by exocytic pathways, became raft-associated remains unclear. Furthermore, relative ratios of raft-associated cell-surface MV glycoproteins such as H protein were lower than those of intracellular proteins, suggesting that H protein became dissociated from the lipid raft after it reached the cell surface. Analysis of the virus envelope from released virus particles showed that about 30% of H and 10% of F glycoproteins were raft-associated within the viral envelope. Since the lipid composition of the virion envelope is likely to represent the membrane of the budding site, this would suggest that MV budding did not occur exclusively from lipid rafts on the plasma membrane or that H and F proteins became dissociated from lipid rafts during or after budding and release of MV particles. Since specific domains of glycoproteins are involved in raft association, it is expected that both the TMD of F and as well as palmitoylation of the membrane-proximal cysteine residue of F protein are likely to be involved in lipid raft association.

The mechanism of apical transport of F and H glycoproteins in MV-infected cells appears to be complex. When H and F proteins were expressed in the absence of M protein they were either basolateral or nonpolar in cell surface transport, but in the presence of M protein, MV particles predominantly budded from the apical surface, suggesting that M protein was targeted to the apical domain and was responsible for specifically targeting F and H proteins to the apical surface, the budding site of measles virus (Naim *et al.,* 2000; Maisner and Reidl, 1998; Reidl and Maisner, 2000).

Rubella virus, a member of the Togaviridae family (genus *Rubivirus*), buds from the Golgi membrane, but is released from the apical side of cultured polarized epithelial cells (Garbutt *et al.,* 1999). Both viral proteins E1 and E2 are targeted to the Golgi compartment, where virus budding occurs. Little is known about the interaction of E1 and E2 proteins with the lipid raft or the role of the lipid raft in the transport of envelope proteins in virus release. Two other togaviruses, Sindbis virus and Semliki forest virus (SFV), bud apically in FRT cells, but basolaterally from CaCo-2 cells, and the distribution of p62/E2 envelope proteins is also apical in FRT cells, but basolateral in CaCo-2 cells (Zurzolo *et al.,* 1992). Again, the roles of the lipid raft in the transport of p62/E2 protein and the budding of these viruses are not known.

Hepatitis B virus surface antigen (HBsAg) is an apical protein, but its transport via the apical pathway is not dependent on N-glycan signals, since tunicamycin did not affect its apical transport (Marzolo *et al.,* 1997), nor does it follow the pathway of GPI-anchored proteins, since in FRT cells, HBsAg was transported to the apical surface in the presence of mannosamine, an inhibitor of the GPI anchor (Lisanti *et al.,* 1991). However, whether HbsAg interacts with the lipid raft in a GPI-independent manner remains to be determined.

Both HIV and SIV present an interesting problem in protein transport and in virion assembly and budding. In polarized epithelial cells, these viruses bud from the basolateral surface. Activated T cells during HIV infection become polarized and viruses do not bud randomly, but bud from specific sites in these cells. Analysis of the lipid composition and the fluidity of the HIV envelope showed that the HIV envelope is enriched in cholesterol and sphingomyelin relative to their level in the host cell membrane. Further analysis of HIV assembly in Jurkat cells showed that the HIV envelope incorporated GPI-linked, lipid raft-associated proteins such as Thy-1 and CD59 and ganglioliside GM1, a marker of lipid rafts. Furthermore, the lipid raft partitioning lipid analogue $DilC_{16}$ colocalizes with HIV proteins on the uropods, the assembly site of HIV in infected Jurkat cells (Nguyen and Hildreth, 2000). In addition, HIV proteins MA/p17 and gp 41 are both present in detergent-insoluble membranes and therefore associate with lipid rafts. Myristoylated Gag proteins also predominantly associate with lipid rafts. Similarly, myristoylated HIV and SIV Nef proteins along with Lck are known to be raft-associated (Flaherty *et al.,* 1998). Although the gp41 cytoplasmic tail contains a basolateral signal, palmitoylation of two cysteine residues in the cytoplasmic tail of gp41 may target gp41 to the lipid raft. Therefore it has been proposed that HIV and possibly SIV and other retroviruses bud from specific cellular microdomains containing lipid rafts. During infection, T cells become activated and polarized, capping GPI-anchored proteins. Proteins such as the CD59 as well as myristoylated proteins of viral origin such as gp166/gp41, Gag/p17, and Nef (Wang *et al.,* 2000) become associated with lipid rafts enriched in cholesterol and sphingomyelin. Interaction among viral proteins will initiate budding with the inclusion of specific cellular proteins such as Thy-1 and CD59 and the exclusion of proteins such as CD45. Gag/p17, Nef proteins, and Env proteins are likely to be transported through different pathways. The Env protein gp160/gp41 determines the basolateral budding site in polarized MDCK cells (Owens *et al.,* 1991). Basolateral proteins are usually excluded from lipid rafts. However, though gp160/gp41 proteins are basolateral, they may become lipid raft-associated because of myristoylation. It has been shown that a small fraction of PLAP is basolateral, but the basolateral PLAP, like its apical counterpart, is still raft-associated (Arreaza and Brown, 1995).

Vesicular stomatitis virus (VSV), a member of rhabdovirus family, contains G protein (a transmembrane type 1 protein) and M protein (membrane-associated, matrix protein) in its envelope. VSV buds from the basolateral surface in polarized MDCK cells and VSV G protein is basolateral and TX-100-soluble (i.e., non-raft-associated). However, although the VSV virus envelope does not contain lipids in the ordered phase as determined by TX-100 insolubility, it is enriched in phosphatidylserine (PS) and sphingomyelin (Sph), phosphatidic acid (PA), and cholesterol, but is low in phosphatidylcholine (PC) compared to the host plasma membrane (Luan *et al.,* 1995). Its specific lipid composition suggests that VSV also buds from microdomains in the plasma membrane enriched in specific lipids. Using *in vitro* vesicles containing different lipids and fluorescence digital microscopy, it has been shown that the presence of both VSV G and M is required for the formation of lipid domains enriched in phospholipid and cholesterol composition similar to that found in the viral envelope, suggesting that such lipid domains represent areas in the plasma membrane from which VSV buds (Luan *et al.,* 1995). However, since these domains and the VSV envelope are not TX-100-insoluble, they do not represent typical lipid rafts.

3. Role of Lipid Rafts in the Budding Process

Lipid rafts are not only involved in transporting the viral components to the budding site and function as the budding platform, but, in addition, may facilitate the budding process. The budding process requires the formation of outward curvature of the budding membrane due to pushing and pulling forces at the assembly site and eventual pinching-off of the virion, the last step in the release of virus particles. This last step requires the fusion of apposing viral envelope and cellular envelope, causing the separation of virions from the host cell membrane. These processes of bud formation and membrane fusion leading to pinching-off of virus particles are among the least understood processes in virus biology. These processes have similarity to receptor-directed endocytosis in the reverse direction, that is, outward instead of inward budding in endocytosis, and may require viral as well as host components. Among the viral components, the pulling force of the viral glycoproteins and pushing force of the matrix proteins are important. However, the most critical factor appears to be the matrix protein, which alone can bud from cells expressing the matrix proteins of a number of viruses (Lenard, 1996), whereas glycoproteins expressed alone or viruses that are deficient or defective in matrix proteins are extremely inefficient in budding. Matrix protein in the absence of glycoproteins can induce efficient budding in the case of rabies virus (Mebatsion *et al.,* 1996, 1999), VSV (Justice *et al.,* 1995; Roberts *et al.,* 1999), HIV (Gheysen *et al.,* 1989), influenza virus (Gomez-Puertas *et al.,* 2000), SIV (Delchambre *et al.,* 1989), and parainfluenza virus (Coronel *et al.,* 1999). However, viral glycoproteins have been shown to increase the efficiency of budding and it is generally believed that the site of virus budding and assembly on the plasma membrane as well as the lipid composition of the viral envelope are determined by the envelope proteins of the virus. This view, however, is being questioned for a number of viruses, including measles virus (Naim *et al.,* 2000), influenza virus (Mora *et al.,* 2002, Adhikary *et al.,* 2001), VSV (Zimmer *et al.,* 2002). In measles virus infected cells, the matrix (M) protein has been shown to determine the site of budding by modifying transport and targeting of glycoproteins (Naim *et al.,* 2000). Consequently, matrix proteins in measles virus and in HIV appear to interact independently of lipid rafts and can be directed to the assembly site. Therefore both viral proteins as well as host components including lipid rafts may actively participate in the budding process of virions. Furthermore, the nature of the membrane of budding particles containing the matrix protein alone may not represent that of infectious virions.

Among the host factors involved in virus budding from plasma membranes, three components appear to play major role in the budding process:

1. *Cytoskeletal components,* both microtubules and actin, are known to be involved in protein transport. In addition, actin may also provide the pushing force in the budding process. Indeed actin filaments have been found in buds of measles virus (Bohn *et al.,* 1986; Tyrrel and Ehrnst, 1979) and actin is present in released particles for a number of infectious viruses (Loza-Tulimowska *et al.,* 1981; Ott *et al.,* 1996). However, if actin is involved in the budding process, fusion of the membranes and release of virion particles will require disassembly of actin filaments at the last stage of the budding process. In influenza virus-infected HeLa 229 cells, which are defective in the budding process, cytochalasin B, which depolymerizes actin filaments, was shown to aid in releasing virus particles (Gujuluva *et al.,* 1994). Disruption of microfilaments by cytochalasin D also increased the ratio of spherical to filamentous particles from influenza and parainfluenza virus in polarized epithelial cells (P. C. Roberts and Compans, 1998; Yao and Compans, 2000).

2. *Ubiquitin,* which interacts with the Gag protein of retroviruses, has been shown to participate in the budding and release process (Strack *et al.,* 2000; Patnaik *et al.,* 2000), since drugs which blocked proteasome function interfered with the budding and release of virus particles. In addition, the PPxY motif present in many viruses, including VSV (Harty *et al.,* 1999) and Ebola virus (Harty *et al.,* 2000), interacts with WW domains of host proteins like ubiquitin ligase Nedd4. However, how the ubquitination of Gag protein aids in the release of virus particles remains unclear.

3. Tsg 101, Vps4, members of the vacuolar protein sorting pathway, which are involved in the formation of multi-vesicular bodies in the endosome have been shown to play a critical role in the fusion

of budding membranes and release of HIV (Garrus *et al.,* 2001, Martin-Serrano *et al.,* 2001). However, the role of these proteins in budding of other enveloped viruses remains to be determined.

4. *Lipid rafts,* as mentioned earlier, are known to be preferred sites for some apical and basolateral virus budding. Increased efficiency of virus budding in the presence of glycoproteins and the similar composition of virion lipids to that of the lipid rafts suggest a role of the lipid rafts in the budding process. However, how the lipid rafts facilitate budding is unclear; as indicated earlier, the process may be the reverse of the clathrin-independent endocytosis process, where the membranes are enriched in lipid rafts. Indeed coalescence of lipid rafts containing influenza viral components in the liquid-ordered phase may facilitate budding from the plasma membrane. The fluidity of the influenza virus envelope was significantly less than that of VSV or SFV (Scheiffele *et al.,* 1999) . The liquid-ordered phase of lipids in lipid rafts may induce curvature either inward or outward depending on the nature of the membrane proteins in the lipid rafts. The viscosity and fluidity of the budding membrane appear to affect the budding rate of influenza viruses, and ATP depletion, which affects the physical state of the membrane, also decreases influenza virus budding (Hui and Nayak, 2001).

IV. Assembly of Influenza Virus

Influenza viruses and viral proteins have been studied as a model for protein transport and virus assembly. Therefore we will discuss some of the steps involved in the assembly and budding process of influenza viruses. For influenza virus assembly and budding to occur, all viral components must interact with each other at the assembly site. Therefore viral glycoproteins must interact with matrix protein (M1) and M1 with vRNP complex (RNA, NP, and polymerase proteins). Although the M1–vRNP complex has been demonstrated for influenza and Sendai viruses both in virus particles as well as in virus-infected cells, the interaction of M1 with glycoproteins has been difficult to demonstrate because both M1 and glycoproteins associate with membrane and M1–glycoprotein complexes are labile for coimmunoprecipitation by monospecific antibodies. Recently, however, TX-100 detergent has been used to distinguish the membrane-bound M1 alone from the membrane-bound M1–glycoprotein complex (Ali *et al.,* 2000). Since M1 binds to membranes randomly, but mature HA and NA specifically associate in the TGN with lipid rafts enriched in glycosphingolipids and cholesterol, it is expected that M1 interacting with mature HA and NA will become more resistant to TX-100 due to either direct or indirect association of M1 with lipid rafts than the membrane-bound M1 alone. The membrane-bound M1 from cells expressing M1 alone or M1 coexpressing with heterologous protein such as Sendai virus F was completely TX-100-soluble. However, in influenza virus-infected cells as well as cells coexpressing M1 with HA, NA, or both, a major fraction of membrane-bound M1 became detergent-resistant because of its interaction with HA and NA present in the lipid raft. These results demonstrated that both HA and NA either independently or together can interact with M1.

Further attempts were made to define the domains of glycoproteins which interact with M1 (Ali *et al.,* 2000). When chimeric proteins—FFH, containing the cytoplasmic tail of HA (H) and the ectodomain and transmembrane domain of Sendai virus F protein, and HHF, containing the ectodomain and transmembrane domain of HA and the cytoplasmic tail of F protein—were coexpressed with M1, a significant fraction of the membrane-bound M1 became detergent-resistant. FHH, containing both the TMD and the cytoplasmic tail of HA, behaved like HA, whereas Sendai virus F protein (FFF) did not render M1 TX-100-resistant. These results demonstrated that both the cytoplasmic tail (FFH) and the TMD (HHF) of HA played an important role in rendering the membrane-bound M1 TX-100-resistant, supporting the interaction of M1 with both the cytoplasmic tail and the TMD of HA. Analysis by confocal microscopy also showed that in virus-infected MDBK cells in the absence of monensin, a fraction of M1 and HA was colocalized throughout the cell cytoplasm and the cell periphery. In the presence of monensin, HA was present predominantly in the perinuclear Golgi region and absent in the plasma membrane due to transport block of the exocytic pathway by monensin. However, M1 was also more concentrated in the perinuclear region and less on the cell periphery, supporting the colocalization of M1 and HA in the Golgi region in influenza virus-infected cells. However, colocalization of M1 and HA was not complete in these cells because of the different intracellular localization and function of these proteins and varying rate of synthesis in different phases of the infectious cycle. Recent studies have shown that Sendai virus M also interacts with the transmembrane domain and cytoplasmic tail of the F protein (Ali and Nayak, 2000).

Based on these biochemical and morphological analyses, we propose a model showing the interaction of M1 with lipid membranes, HA, NA, and vRNP in the influenza virus-infected cells (Fig. 1). In the absence of HA or NA, M1 is predominantly present in membranes randomly and is therefore not TX-100-resistant. In the presence of mature HA and NA, M1 and M1 bound to vRNP interact with HA and NA present

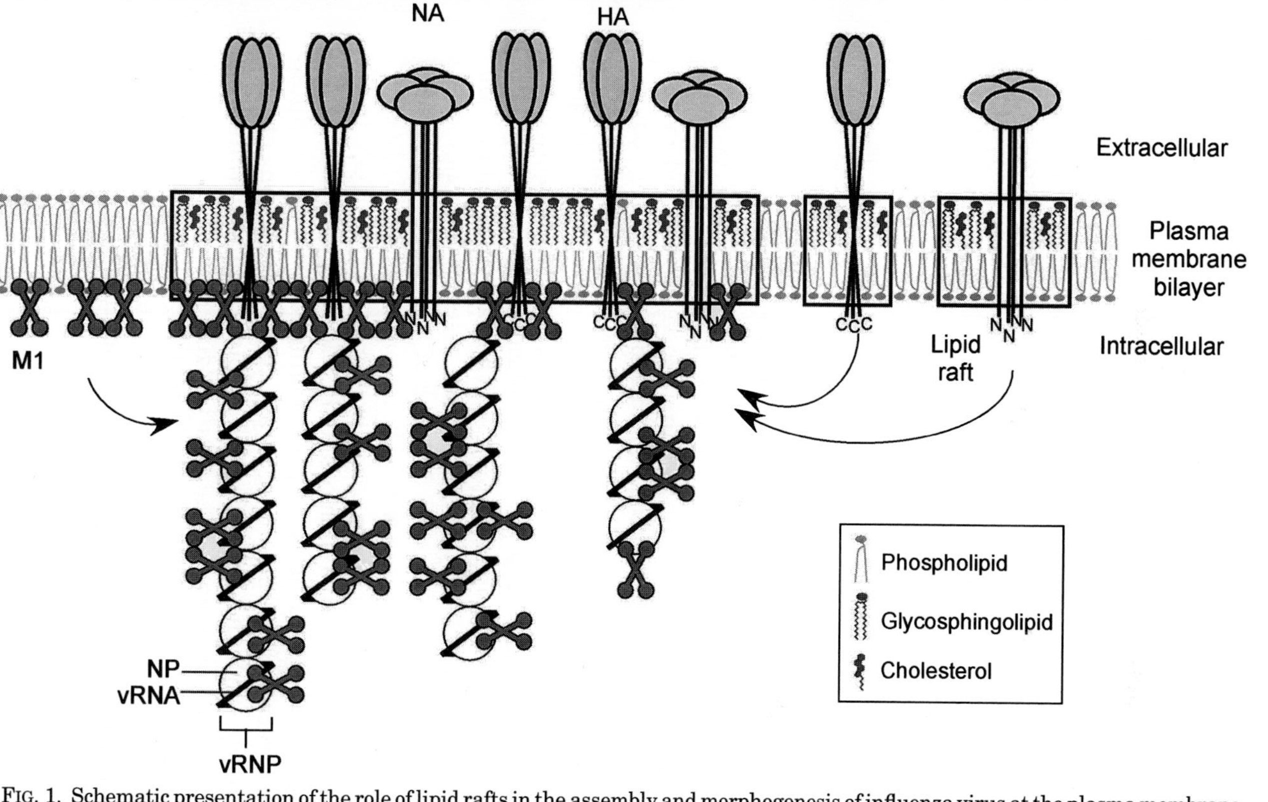

FIG. 1. Schematic presentation of the role of lipid rafts in the assembly and morphogenesis of influenza virus at the plasma membrane of virus-infected cells. The plasma membrane contains islands of lipid rafts enriched in influenza virus HA and NA. The M1 protein alone interacts randomly with lipid membranes and is TX-100-soluble. Mature HA and NA bind to lipid rafts enriched in cholesterol and glycosphingolipids in the TGN during exocytic transport and are transported to the plasma membrane. However, when M1 binds to the cytoplasmic tail and the TMD of HA and NA, it becomes associated with lipid rafts and becomes TX-100-resistant. Furthermore, interaction among M1 proteins also brings vRNPs to the assembly site and the lipid raft serves as a platform for concentrating viral components and thereby aids in initiating the budding process.

in the lipid raft enriched in cholesterol, and M1 becomes associated with the lipid raft. Low TX-100 resistance of M1 expressed alone and increased detergent resistance of M1 in the presence of mature HA and NA both in virus-infected and in coexpressing cells as discussed in this review would support this model. Although these interactions are shown at the plasma membrane of the virus-infected cells, intracellular colocalization of HA and M1 by confocal microscopy in the presence of monensin also demonstrate that such interactions can take place during the exocytic transport of glycoproteins, which can further direct M1 and vRNP to the assembly and budding sites (Nayak and Hui, 2002).

V. Conclusion

Lipid rafts containing sphingolipid–cholesterol clusters in the liquid-ordered phase are TX-100-insoluble at low temperature. They are dynamic in nature, vary in size (2.5–10 nm), and are dispersed in the plasma membrane and *trans*-Golgi network. Lipid rafts are involved in various cell functions including clathrin-independent endocytosis, transcytosis, membrane trafficking, and membrane signaling required for many biological functions. In addition, these lipid rafts along with other host components play an important role and actively participate in the assembly and budding of many enveloped viruses, particularly for viruses budding from the plasma membrane. Lipid rafts provide multiple functions in the budding process. They are involved in transporting and targeting viral components to the assembly site. Lipid rafts interact with the transmembrane domains of both type I and type II apical glycoproteins, and along with cellular proteins like MAL/VIP17 actively participate in targeting these proteins from the TGN to the apical domain of the plasma membrane. In addition, lipid rafts aid in concentrating viral components at the budding site, thereby facilitating their interaction and functioning as a platform for budding of these viruses. Finally, the liquid-ordered phase of lipids in the lipid rafts along with viral components may facilitate the budding process itself, including the release of virus particles. Clearly, further work in defining the role of lipid rafts and other host components, including cytoskeletal elements, in membrane signaling will aid in elucidating the steps and the processes involved in virus budding.

Acknowledgments

This work was supported by U.S. PHS grants AI 16348, AI 41681, and AI 35570. We thank Eric K.-W. Hui for the illustration and Eleanor Berlin for typing the manuscript.

References

Ali, A., and Nayak, D. P. (2000). Assembly of Sendai virus: M protein interacts with F and HN proteins and with the cytoplasmic tail and transmembrane domain of F protein. *Virology* **276,** 289–303.

Ali, A., Avalos, R. T., Ponimaskin, E., and Nayak, D. P. (2000). Influenza virus assembly: Effect of influenza virus glycoproteins on the membrane association of M1 protein. *J. Virol.* **74,** 8709–8719.

Aloia, R. C., Tian, H., and Jensen, F. C. (1993). Lipid composition and fluidity of the human immunodeficiency virus envelope and host cell plasma membranes. *Proc. Natl. Acad. Sci. USA* **90,** 5181–5185.

Arreaza, G., and Brown, D. A. (1995). Sorting and intracellular trafficking of a glycosylphosphatidylinositol-anchored protein and two hybrid transmembrane proteins with the same ectodomain in Madin–Darby canine kidney epithelial cells. *J. Biol. Chem.* **270,** 23641–23647.

Barman, S., and Nayak, D. P. (2000). Analysis of the transmembrane domain of influenza virus neuraminidase, a type II transmembrane glycoprotein, for apical sorting and raft association. *J. Virol.* **74,** 6538–6545.

Benting, J. H., Rietveld, A. G., and Simons, K. (1999). N-Glycans mediate the apical sorting of a GPI-anchored, raft-associated protein in Madin–Darby canine kidney cells. *J. Cell Biol.* **146,** 313–320.

Blau, D. M., and Compans, R. W. (1995). Entry and release of measles virus are polarized in epithelial cells. *Virology* **210,** 91–99.

Bohn, W., Rutter, G., Hohenberg, H., Mannweiler, K., and Nobis, P. (1986). Involvement of actin filaments in budding of measles virus: Studies on cytoskeletons of infected cells. *Virology* **149,** 91–106.

Brown, D. A., and London, E. (1998a). Functions of lipid rafts in biological membranes. *Annu. Rev. Cell Dev.Biol.* **14,** 111–136.

Brown, D. A., and London, E. (1998b). Structure and origin of ordered lipid domains in biological membranes. *J. Membrane Biol.* **164,** 103–114.

Brown, D. A., and Rose, J. K. (1992). Sorting of GPI-anchore proteins to glycolipid-enriched membrane subdomains during transport to the apical cell surface. *Cell.* **68,** 533–544.

Compans, R. W. (1995). Virus entry and release in polarized epithelial cells. *Curr. Top. Microbiol. Immunol.* **202,** 209–219.

Coronel, E. C., Murti, K. G., Takimoto, T., and Portner, A. (1999). Human parainfluenza virus type 1 matrix and nucleoprotein genes transiently expressed in mammalian cells induce the release of virus-like particles containing nucleocapsid-like structures. *J. Virol.* **73,** 7035–7038.

Czarny, M., Lavie, Y., Fiucci, G., and Liscovitch, M. (1999). Localization of phospholipase D in detergent-insoluble, caveolin-rich membrane domains. Modulation by caveolin-1 expression and caveolin-182-101. *J. Biol. Chem.* **274,** 2717–2724.

Danielsen, E. M. (1995). Involvement of detergent-insoluble complexes in the intracellular transport of intestinal brush border enzymes. *Biochemistry* **34,** 1596–1605.

Danielsen, E. M., and van Deurs, B. (1995). A transferrin-like GPI-linked iron-binding protein in detergent-insoluble noncaveolar microdomains at the apical surface of fetal intestinal epithelial cells. *J. Cell Biol.* **131,** 939–950.

Delchambre, M., Gheysen, D., Thines, D., Thiriart, C., Jacobs, E., Verdin, E., Horth, M., Burny, A., and Bex, F. (1989). The GAG precursor of simian immunodeficiency virus assembles into virus-like particles. *EMBO J.* **8,** 2653–2660.

Dunbar, L. A., Aronson, P., and Caplan, M. J. (2000). A transmembrane segment determines the steady-state localization of an ion-transporting adenosine triphosphatase. *J. Cell Biol.* **148,** 769–778.

Flaherty, M. T., Barber, S. A., and Clements, J. E. (1998). Neurovirulent simian immunodeficiency virus incorporates a Nef-associated kinase activity into virions. *AIDS Res. Hum. Retrovir.* **14,** 63–70.

Garbutt, M., Chan, H., and Hobman, T. C. (1999). Secretion of rubella virions and virus-like particles in cultured epithelial cells. *Virology* **261,** 340–346.

Garoff, H., Hewson, R., and Opstelten, D. E. (1998). Virus maturation by budding. *Microbiol. Mol. Biol. Rev.* **62,** 1171–1190.

Garrus, J. E., von Schwedler, U. K., Pornillos, O. W., Morham, S. G., Zavitz, K. H., Wang, H. E., Wettstein, D. A., Stray, K. M., Cote, M., Rich, R. L., Myszka, D. G., and Sundquist, W. I. (2001). Tsg 101 and the vacuolar protein sorting pathway are essential for HIV-1 budding. *Cell.* **107,** 55–65.

Gheysen, D., Jacobs, E., de Foresta, F., Thiriart, C., Francotte, M., Thines, D., and De Wilde, M. (1989). Assembly and release of HIV-1 precursor Pr55gag virus-like particles from recombinant baculovirus-infected insect cells. *Cell* **59,** 103–112.

Gomez-Puertas, P., Albo, C., Perez-Pastrana, E., Vivo, A., and Portela, A. (2000). Influenza virus matrix protein is the major driving force in virus budding. *J. Virol.* **74,** 11538–11547.

Gujuluva, C. N., Kundu, A., Murti, K. G., and Nayak, D. P. (1994). Abortive replication of influenza virus A/WSN/33 in HeLa229 cells: Defective viral entry and budding processes. *Virology* **204,** 491–505.

Haffar, O., Garrigues, J., Travis, B., Moran, P., Zarling, J., and Hu, S. L. (1990). Human immunodeficiency virus-like, nonreplicating, Gag–Env particles assemble in a recombinant vaccinia virus expression system. *J. Virol.* **64,** 2653–2659.

Harty, R. N., Paragas, J., Sudol, M., and Palese, P. (1999). A proline-rich motif within the matrix protein of vesicular stomatitis virus and rabies virus interacts with WW domains of cellular proteins: Implications for viral budding. *J. Virol.* **73,** 2921–2929.

Harty, R. N., Brown, M. E., Wang, G., Huibregtse, J., and Hayes, F. P. (2000). A PPxY motif within the VP40 protein of Ebola virus interacts physically and functionally with a ubiquitin ligase: Implications for filovirus budding. *Proc. Natl. Acad. Sci. USA* **97,** 13871–13876.

Huang, X. F., Compans, R. W., Chen, S., Lamb, R. A., and Arvan, P. (1997). Polarized apical targeting directed by the signal/anchor region of simian virus 5 hemagglutinin–neuraminidase. *J. Biol. Chem.* **272,** 27598–27604.

Hughey, P. G., Compans, R. W., Zebedee, S. L., and Lamb, R. A. (1992). Expression of the influenza A virus M2 protein is restricted to apical surfaces of polarized epithelial cells. *J. Virol.* **66,** 5542–5552.

Hui, E. K.-W., and Nayak, D. P. (2001). Role of ATP in influenza budding. *Virology.* **290,** 329–34.

Ikonen, E., Tagaya, M., Ullrich, O., Montecucco, C., and Simons, K. (1995). Different requirements for NSF, SNAP, and Rab proteins in apical and basolateral transport in MDCK cells. *Cell* **81,** 571–580.

Justice, P. A., Sun, W., Li, Y., Ye, Z., Grigera, P. R., and Wagner, R. R. (1995). Membrane vesiculation function and exocytosis of wild-type and mutant matrix proteins of vesicular stomatitis virus. *J. Virol.* **69,** 3156–3160.

Keller, P., and Simons, K. (1998). Cholesterol is required for surface transport of influenza virus hemagglutinin. *J. Cell Biol.* **140,** 1357–1367.

Kundu, A., Avalos, R. T., Sanderson, C. M., and Nayak, D. P. (1996). Transmembrane domain of influenza virus neuraminidase, a type II protein, possesses an apical sorting signal in polarized MDCK cells. *J. Virol.* **70,** 6508–6515.

Kurzchalia, T. V., Dupree, P., Parton, R. G., Kellner, R., Virta, H., Lehnert, M., and Simons, K. (1992). VIP21, a 21 kD membrane protein, is an integral component of trans-Golgi-network-derived transport vesicles. *J. Cell Biol.* **118,** 1003–1014.

Kusumi, A., and Sako, Y. (1996). Cell surface organization by the membrane skeleton. *Curr. Opin. Cell Biol.* **8,** 566–574.

Lafont, F., Burkhardt, J. K., and Simons, K. (1994). Involvement of microtubule motors in basolateral and apical transport in kidney cells. *Nature* **372,** 801–803.

Lafont, F., Lecat, S., Verkade, P., and Simons, K. (1998). Annexin XIIIb associates with lipid microdomains to function in apical delivery. *J. Cell Biol.* **142,** 1413–1427.

Lenard, J. (1996). Negative-strand virus M and retrovirus MA proteins: All in a family. *Virology* **216,** 289–298.

Li, S., Okamoto, T., Chun, M., Sargiacomo, M., Casanova, J. E., Hansen, S. H., Nishimoto I., and Lisanti, M. P. (1995). Evidence for a regulated interaction between heterotrimeric G proteins and caveolin. *J. Biol. Chem.* **270,** 15693–15701.

Lin, S., Naim, H. Y., Rodriguez, A. C., and Roth, M. G. (1998). Mutations in the middle of the transmembrane domain reverse the polarity of transport of the influenza virus hemagglutinin in MDCK epithelial cells. *J. Cell Biol.* **142,** 51–57.

Lipardi, C., Nitsch, L., and Zurzolo, C. (2000). Detergent-insoluble GPI-anchored proteins are apically sorted in Fischer rat thyroid cells, but interference with cholesterol or sphingolipids differentially affects detergent insolubility and apical sorting. *Mol. Biol. Cell* **11,** 531–542.

Lisanti, M. P., Field, M. C., Caras, I. W., Menon, A. K., and Rodriguez-Boulan, E. (1991). Mannosamine, a novel inhibitor of glycosylphosphatidylinositol incorporation into proteins. *EMBO J.* **10,** 1969–1977.

Loza-Tulimowska, M., Michalak, T., and Semkow, R. (1981). Attempts at detection of actomyosin associated with influenza virus. *Acta Virol.* **25,** 251–253.

Luan, P., Yang, L., and Glaser, M. (1995). Formation of membrane domains created during the budding of vesicular stomatitis virus. A model for selective lipid and protein sorting in biological membranes. *Biochemistry* **34,** 9874–9883.

Maisner, A., and Riedl, P. (1998). Studies on measles virus matrix protein in polarized epithelial cells. *J. Virol.* **71,** 5276–5278.

Maisner, A., Klenk, H., and Herrler, G. (1998). Polarized budding of measles virus is not determined by viral surface glycoproteins. *J Virol.* **72,** 5276–5278.

Manié, S. N., Debreyne, S., Vincent, S., and Gerlier, D. (2000). Measles virus structural components are enriched into lipid raft microdomains: A potential cellular location for virus assembly. *J. Virol.* **74,** 305–311.

Martin-Belmonte, F., Puertollano, R., Millan, J., and Alonso, M. A. (2000). The MAL proteolipid is necessary for the overall apical delivery of membrane proteins in the polarized epithelial Madin–Darby canine kidney and Fischer rat thyroid cell lines. *Mol. Biol. Cell* **11,** 2033–2045.

Martin-Serrano, J., Zang, T., and Bieniasz, P. D. (2001). HIV-1 and Ebola virus encode small peptide motifs that recruit Tsg 101 to sites of particle assembly to facilitate egress. *Nat. Med.* **7,** 1313–1319.

Marzolo, M. P., Bull, P., and González, A. (1997). Apical sorting of hepatitis B surface antigen (HBsAg) is independent of N-glycosylation and glycosylphosphatidylinositol-anchored protein segregation. *Proc. Natl. Acad. Sci. USA* **94,** 1834–1839.

Mebatsion, T., Konig, M., and Conzelmann, K.-K. (1996). Budding of rabies virus particles in the absence of the spike glycoprotein. *Cell* **84,** 941–951.

Mebatsion, T., Weiland, F., and Conzelmann, K.-K. (1999). Matrix protein of rabies virus is responsible for the assembly and budding of bullet-shaped particles and interacts with the transmembrane spike glycoprotein G. *J. Virol.* **73,** 242–250.

Mora, R., Rodriguez-Boulan, E., Palese, P., and Garcia-Sastre, A. (2002). Apical budding of a recombinant influenza A virus expressing a hemagglutinin protein with a basolateral localization signal. *J. Virol.* **76,** 3544–3553.

Muallem, S., Kwiatkowska, K., Xu, X., and Yin, H. L. (1995). Actin filament disassembly is a sufficient final trigger for exocytosis in nonexcitable cells. *J. Cell Biol.* **128,** 589–598.

Murata, M., Peranen, J., Schreiner, R., Wieland, F., Kurzchalia, T. V., and Simons, K. (1995). VIP21/caveolin is a cholesterol-binding protein. *Proc. Natl. Acad. Sci. USA* **92,** 10339–10343.

Naim, H. Y., Ehler, E., and Billeter, M. A. (2000). Measles virus matrix protein specifies apical virus release and glycoprotein sorting in epithelial cells. *EMBO J.* **19,** 3576–3585.

Nayak, D. P. (2000). Virus morphology, replication, and assembly. *In* "Viral Ecology" (C. Hurst, Ed.), pp. 63–124. Academic Press, New York.

Nayak, D. P., and Hui, E. K.-W. (2002). Assembly and morphogenesis of influenza viruses, in press. Recent research developments in Virology Transworld Research Network, India.

Nguyen, D. H., and Hildreth, J. E. K. (2000). Evidence for budding of human immunodeficiency virus type 1 selectively from glycolipid-enriched membrane lipid rafts. *J. Virol.* **74,** 3264–3272.

Ott, D. E., Coren, L. V., Kane, B. P., Busch, L. K., Johnson, D. G., Sowder, II, R. C., Chertova, E. N., Arthur, L. O., and Henderson, L. E. (1996). Cytoskeletal proteins inside human immunodeficiency virus type 1 virions. *J. Virol.* **70,** 7734–7743.

Owens, R. J., Dubay, J. W., Hunter, E., and Compans, R. W. (1991). Human immunodeficiency virus envelope protein determines the site of virus release in polarized epithelial cells. *Proc. Natl. Acad. Sci. USA* **88,** 3987–3991.

Patnaik, A., Chau, V., and Wills, J. W. (2000). Ubiquitin is part of the retrovirus budding machinery. *Proc. Natl. Acad. Sci. USA* **97,** 13069–13074.

Plant, P. J., Lafont, F., Lecat, S., Verkade, P., Simons, K., and Rotin, D. (2000). Apical membrane targeting of Nedd4 is mediated by an association of its C2 domain with annexin XIIIb. *J. Cell Biol.* **149,** 1473–1484.

Puertollano, R., Martín-Belmonte, F., Millán, J., de Marco, M. C., Albar, J. P., Kremer, L., and Alonso, M. A. (1999). The MAL proteolipid is necessary for normal apical transport and accurate sorting of the influenza virus hemagglutinin in Madin–Darby canine kidney cells. *J. Cell Biol.* **145,** 141–151.

Reidl, P., and Maisner, A. (2000). *Studies on measles virus matrix protein in polarized epithelial cells.* Presented at the 11th Conference on Negative Strand Virus, Qubec City, Canada [Abstract #21].

Rietveld, A., and Simons, K. (1998). The differential miscibility of lipids as the basis for the formation of functional membrane rafts. *Biochim. Biophys. Acta* **1376,** 467–479.

Roberts, A., Buonocore, L., Price, R., Forman, J., and Rose, J. K. (1999). Attenuated vesicular stomatitis viruses as vaccine vectors. *J. Virol.* **73,** 3723–3732.

Roberts, P. C., and Compans, R. W. (1998). Host cell dependence of viral morphology. *Proc. Natl. Acad. Sci. USA* **95,** 5746–5751.

Rothman, J. E. (1994). Mechanisms of intracellular protein transport. *Nature* **372,** 55–63.

Scheiffele, P., Peranen, J., and Simons, K. (1995). N-Glycans as apical sorting signals in epithelial cells. *Nature* **378,** 96–98.

Scheiffele, P., Roth, M. G., and Simons, K. (1997). Interaction of influenza virus haemagglutinin with sphingolipid–cholesterol membrane domains via its transmembrane domain. *EMBO J.* **16,** 5501–5508.

Scheiffele, P., Verkade, P., Fra, A. M., Virta, H., Simons, K., and Ikonen, E. (1998). Caveolin-1 and -2 in the exocytic pathway of MDCK cells. *J. Cell Biol.* **140,** 795–806.

Scheiffele, P., Rietveld, A., Wilk, T., and Simons, K. (1999). Influenza viruses select ordered lipid domains during budding from the plasma membrane. *J. Biol. Chem.* **274,** 2038–2044.

Schnitzer, J. E., Oh, P., Pinney, E., and Allard, J. (1994). Filipin-sensitive caveolae-mediated transport in endothelium: Reduced transcytosis, scavenger endocytosis, and capillary permeability of select macromolecules. *J. Cell Biol.* **127,** 1217–1232.

Simons, K., and Ikonen, E. (1997). Functional rafts in cell membranes. *Nature* **387,** 569–572.

Skibbens, J. E., Roth, M. G., and Matlin, K. S. (1989). Differential extractability of influenza virus hemagglutinin during intracellular transport in polarized epithelial cells and nonpolar fibroblasts. *J. Cell Biol.* **108,** 821–832.

Song, K. S., Shengwen, L., Okamoto, T., Quilliam, L. A., Sargiacomo, M., and Lisanti, M. P. (1996). Co-purification and direct interaction of Ras with caveolin, an integral membrane protein of caveolae microdomains. Detergent-free purification of caveolae microdomains. *J. Biol. Chem.* **271,** 9690–9697.

Strack, B., Calistri, A., Accola, M. A., Palu, G., and Gottlinger, H. G. (2000). A role for ubiquitin ligase recruitment in retrovirus release. *Proc. Natl. Acad. Sci. USA* **97,** 13063–13068.

Tatulian, S. A., and Tamm, L. K. (2000). Secondary structure, orientation, oligomerization, and lipid interactions of the transmembrane domain of influenza hemagglutinin. *Biochemistry* **39,** 496–507.

Tyrrel, D. L., and Ehrnst, A. (1979). Transmembrane communication in cells chronically infected with measles virus. *J. Cell Biol.* **81,** 396–402.

van Meer, G., and Simons, K. (1982). Viruses budding from either the apical or the basolateral plasma membrane domain of MDCK cells have unique phospholipid compositions. *EMBO J.* **1,** 847–52.

Veit, M., Herrler, G., Schmidt, M. F., Rott, R., and Klenk, H. D. (1990). The hemaglutinating glycoproteins of influenza B and C viruses are acylated with different fatty acids. *Virology* **177,** 807–811.

Veit, M., Klenk, H. D., Kendal, A., and Rott, R. (1991). The M2 protein of influenza A virus is acylated. *J. Gen Virol.* **72,** 1461–1465.

Verkade, P., Harder, T., Lafond, F., and Simons, K. (2000). Induction of caveolae in the apical plasma membrane of Madin–Darby canine kidney cells. *J. Cell Biol.* **148,** 727–739.

Vincent, S., Gerlier, D., and Manié, S. N. (2000). Measles virus assembly within membrane rafts. *J. Virol.* **74,** 9911–9915.

Wang, J. K., Kiyokawa, E., Verdin, E., and Trono, D. (2000). The Nef protein of HIV-1 associates with rafts and primes T cells for activation. *Proc. Natl. Acad. Sci. USA* **97,** 394–399.

Yao, Q., and Compans, R. W. (2000). Filamentous particle formation by human parainfluenza virus type 2. *J. Gen Virol.* **81,** 1305–1312.

Zhang, J., Pekosz, A., and Lamb, R. A. (2000). Influenza virus assembly and lipid raft microdomains: A role for the cytoplasmic tails of the spike glycoproteins. *J. Virol.* **74,** 4634–4644.

Zheng, X., Lu, D., and Sadler, J. E. (1999). Apical sorting of bovine enteropeptidase does not involve detergent-resistant association with sphingolipid–cholesterol rafts. *J. Biol. Chem.* **274,** 1596–1605.
Zimmer, G., Zimmer, K. P., Trotz, I., and Herrler, G. (2002). Vesicular Stomatitis virus glycoprotein does not determine the site of virus release in polarized epithelial cells. *J. Virol.* **76,** 4103–4107.
Zurzolo, C., Polistina, C., Saini, M., Gentile, R., Aloz, L., Migliaccio, G., Bonatti, S., and Nitsch, L. (1992). Opposite polarity of virus budding and of viral envelope glycoprotein distribution in epithelial cells from different tissues. *J. Cell Biol.* **117,** 551–564.
Zurzolo, C., van't Hof, W., van Meer, G., and Rodriguez-Boulan, E. (1994). VIP21/caveolin, glycosphingolipid clusters and the sorting of glycosylphosphatidylinositol-anchored proteins in epithelial cells. *EMBO J.* **13,** 42–53.

ADVANCES IN VIRUS RESEARCH, VOL. 58

NOVEL VACCINE STRATEGIES

Lorne A. Babiuk, Shawn L. Babiuk,
and Maria E. Baca-Estrada

Veterinary Infectious Disease Organization
Saskatoon, Saskatchewan S7N 5E3 Canada

I. Introduction
II. The Ideal Vaccine
III. Principles of Vaccination
IV. Conventional Vaccines
A. Live Attenuated Vaccines
B. Killed Vaccines
V. Genetically Engineered Vaccines
A. Subunit Vaccines
B. Live Vaccines
C. Polynucleotide Vaccines
VI. Marker Vaccines
VII. Vaccine Formulation and Delivery
A. Needle-Free Delivery Devices
B. Vaccine Formulation
VIII. Epilogue
References

I. INTRODUCTION

Infectious diseases continue to cost society billions of dollars annually in lost productivity due to morbidity and mortality. Although it is at least partially possible to reduce these economic losses by using antimicrobial agents (antiviral, antiparasitic, and antibacterial), the problems with this approach are that for some pathogens no effective therapeutic agents are available and for others the pathogens are managing to rapidly acquire resistance to the agent. The best examples are the pathogen responsible for chloroquine-resistant malaria and multidrug-resistant bacteria. Indeed, in the case of bacteria, resistance is becoming a major concern to health authorities (Barry, 1999; D. K. Chen *et al.*, 1999). This has resulted in the use of the most effective antibiotics only as a last resort of treatment. In addition to the rapid increase in drug resistance, our modern lifestyles with their increased urbanization and farming practices are resulting in the crowding of people and

0065-3527/02 $35.00

animals, thereby ensuring rapid spread of infection not only between individuals of the same species, but between species. For example, the newly emerged Nipah virus is a result of the encroachment of civilization on the environment of fruit bats and the subsequent transmission of the virus to humans and pigs (Centers for Disease Control [CDC], 1999; Parashar *et al.,* 2000). Furthermore, we are now finding that diseases that were not previously considered to be caused by infectious agents have a potential microbial origin. For example, *Helicobacter pylori* is clearly a cause of gastrointestinal infections, ulcers, and cancers (Marshall, 1983; Munos, 1997; Blaser, 1999). Similarly, human papillomaviruses 16 and 18 are clearly associated with cervical cancers (zur Hausen, 1994). Indeed, it is estimated that 16% of all cancers can be attributed to infectious agents (Pisani *et al.,* 1997; Ames *et al.,* 1995). The observation of chlamydial antigens in atherosclerotic plaques has made us rethink its role in cardiovascular disease (Kuo *et al.,* 1993). With better diagnostics, we will possibly identify other diseases that are at least exacerbated or initiated by infectious agents.

Based on these factors, it is clear that better methods of disease control are required. It is our contention that vaccination, which has been the most effective method of disease control to date, will become even more important in the future. The present review briefly summarizes the principles of vaccination and the vaccines being employed today and describes the novel approaches being developed to make better and safer vaccines in the future.

II. The Ideal Vaccine

The primary concern to regulators, biopharmaceutical companies, and consumers is that vaccines be safe and effective. Indeed, safety has become a major lightning rod around which antivaccination groups have rallied (Poland and Jacobson, 2001). Since it is very difficult to achieve 100% safety as well as 100% efficacy in all vaccines, one must look at the risk–benefit ratio for vaccinations. In most instances, antivaccine groups do not do this. For example, smallpox vaccination results in a postvaccination reaction at a rate of 1 per 250,000–300,000 vaccinated individuals. At the time that 2 million people were dying of smallpox annually, this risk was clearly worth taking. However, today, when there is no smallpox disease in the world, the risk is unacceptable and therefore society no longer immunizes individuals against smallpox. Similarly, with polio virus vaccination, individuals develop varying degrees of paralysis at a rate of 1 per 1 million immunized individuals. Indeed, the vaccine may even spread to in-contact individuals (Greensfelder, 2000). Thus, as we control polio and approach eradication, oral polio vaccination is being replaced by inactivated polio vaccines. Thus, although vaccine safety is a major concern, it should not be considered as an absolute, and the benefits of a vaccine also need to be considered. Second, safety must be considered in light of the various types of individuals who might be exposed to the vaccine. The vaccine may be perfectly safe in individuals with a normal immune system. However, individuals who are immunosuppressed for whatever reason, such as through concurrent infections or stress, may be at a higher risk of adverse consequences due to vaccination. Pregnancy may also increase the risk either to the mother or to the fetus. Thus, some vaccines may be contraindicated for use in pregnant women.

Regardless of whether vaccines are used for animals or humans, injection site reactions are of major concern. These reactions can be due to the vaccine itself, to contaminating components such as lipopolysaccharides (LPS), or to the adjuvants required to improve the efficacy of the vaccine, as well as the method of delivery. Ideally, vaccines should not be given by the intramuscular or the subcutaneous routes. This is both expensive and traumatic for some individuals. For example, in humans, the fear of needles may dramatically influence compliance. In livestock, where restraining large animals is required, intramuscular injection leads to significant tissue damage due to adjuvants as well as broken needles. There are also reports that vaccines containing aluminum hydroxide may later lead to fibrosarcomas, at least in cats (Lester *et al.,* 1996; Burton and Mason, 1997). Whether similar issues will arise in other species and their impact on humans await further investigation. Thus, it is imperative that better vaccines, adjuvants, and delivery systems be developed to reduce these unwanted side effects and improve the safety of vaccination.

Regarding safety, it is also important that the correct immune responses are generated following vaccination. Skewing the immune response to either a Th1- or Th2-type response may lead to enhanced disease because of either enhanced pathogenesis or inability to clear the pathogen, with subsequent immune complex disease (McGuire *et al.,* 1986; Graham *et al.,* 1993; Hussell *et al.,* 1996; Tripp *et al.,* 1999).

One may create a perfectly safe vaccine, but if it does not meet the efficacy criterion of inducing rapid immunity after a single administration, it may not be used. Although very few vaccines meet the criterion of 90–100% efficacy after a single dose, that is what we strive to achieve. However, whether this will ever be achieved is doubtful, and we generally expect protection to be achieved after two injections. This immunity should be of long duration, so that multiple boosters are not required. The more boosters that are required, the lower is the expected level of compliance. Following a booster immunization, the individual should be protected from all variants or serotypes of that specific pathogen that are in circulation. Rarely will these vaccines provide sterile immunity, but they should curtail the spread of the pathogen within the individual (systemic immunity) as well as reduce the quantity of pathogen shed into the environment. This reduction in shedding, combined with as it is believed, the fact that vaccinated animals require higher levels of exposure to the pathogen to become diseased, results in what is called "herd immunity" (Wittman *et al.,* 1982; Diekmann *et al.,* 1990). Reduced shedding and fewer individuals becoming infected should eventually lead to elimination of the pathogen from the specific environments or niches, thereby breaking the cycle of infection.

Since most pathogens enter via mucosal surfaces, mucosal immunity reduces the probability of infection, increases the dose of pathogen required for infection, and, finally, reduces the shedding of the pathogen into the environment (Mestecky *et al.,* 1994). Therefore, vaccines that induce mucosal immunity are more efficacious than those that only induce systemic immunity, although systemic immunity should also be induced in order to reduce systemic spread should the pathogen have been able to initiate infection.

Neonates are less responsive to some vaccines, especially those containing carbohydrate antigens, as a result of the presence of maternal antibodies, which interfere with the development of active immunity (Murphy *et al.,* 1986). Thus, the ideal vaccine should be effective in neonates both in the presence and absence of maternal antibodies. Similarly, as individuals age their response to a vaccination decreases, so vaccines should be able to provide protection throughout a person's life regardless of age or immune status. Clearly, this is a Herculean task, especially in individuals who are immunosuppressed for various reasons either due to therapy or concurrent infection.

III. Principles of Vaccination

The observation that recovery from an infection results in resistance to subsequent infection and clinical disease is the basis for all vaccines used today. This observation led to the early development of live and killed vaccines to many agents even before we were aware of T cells, B cells, and all the complex interactions among cells of the immune system and the cytokine cascades involved in immune induction, or before knowledge of the specific antigens that were important in inducing specific immune responses was available. Although this seems like a simple concept, a thorough understanding of the host responses and the pathogen is needed if the appropriate immune response is to stimulated. For example, administration of vaccines via the intramuscular or subcutaneous route (the routes most often used today) may induce good systemic immunity, but will not induce mucosal immunity, and mucosal surfaces are the sites of entry of many pathogens (Israel *et al.,* 1992; McGee *et al.,* 1992). Therefore, implementation of effective immunization regimes requires information regarding both the specific immune responses involved in inducing protection and the specific antigens that elicit the appropriate immune response.

Recent advances in our understanding of immune responses clearly demonstrate that both specific (humoral and cellular) and nonspecific responses act in concert to help clear an infection and ensure increased resistance to reinfection (Babiuk *et al.,* 1996). Indeed, there is significant redundancy in the immune system. For example, both cellular and humoral components of the immune system can lyse virus-infected cells, and individuals with natural deficiencies in one arm of the immune system or the other can still cope with many infections.

Following a primary infection, there is a rapid expansion of both B and T cells that recognize antigenic components (epitopes) of the invading pathogen. The magnitude and kinetics of this response and the speed with which the pathogen replicates and spreads in the host will influence the final outcome. If the pathogen is extremely virulent, it may outstrip the host's capacity to cope with the pathogen, leading to death either as a result of massive tissue damage by the pathogen or its components or by the host's immune system itself. Indeed, the major cause of the damage induced by many pathogens is the host's immune response (immunopathology). This is critical since the aim of the immune system is to eliminate the pathogen before damage occurs at such a level that the host suffers unduly. It is this battle between the host's immune system and the pathogen that leads in most cases to clinical signs and symptoms. Thus, vaccination does not need to induce sterile immunity; it just needs to limit the degree of replication of the pathogen below the threshold which induces disease.

As a result of infection or exposure to specific foreign antigens of a pathogen, the number of specific cells capable of recognizing the same or related antigens is dramatically expanded. These "memory" cells can then respond extremely quickly upon reexposure of the host to the same or related pathogens. Therefore, a classical "memory" response (both humoral and cellular) occurs not only more rapidly, but also at a much higher level than following a primary infection (Kinman *et al.,* 1989; Bradley *et al.,* 1993). Thus, the immune system can reduce the amount of damage to the host caused by the excessive replication of the pathogen. Following exposure a second time to a pathogen, the immune response is increased not only in magnitude, but in duration as well. Whether periodic stimulation with specific antigens, cross-reactive antigens, is sufficient, or whether cytokines induced by other, unrelated antigens are required for maintenance of long-term memory remains controversial (Maruyama *et al.,* 2000; Sprent and Surh, 2001). However, the existence of long-term memory is clearly present following initial exposure to an antigen even in the absence of specific antigen restimulation. Although CD4+ and CD8+ T cells play an important role in resolution of infection, especially those caused by viruses, the major mediators of resistance to reinfection are antibodies. Thus, vaccination should maintain sufficient levels of antibodies and cytotoxic T cells in circulation to either prevent infection or rapidly limit the degree of replication.

As a consequence of a secondary immune response, the host marshals an armamentarium of defenses much more rapidly to clear the infection before the pathogen can infect and cause disease. For example, in the case of viruses and bacteria, if there is antibody already present in circulation and especially at mucosal surfaces (the site of entry of most pathogens), the agent will be neutralized before it can infect the host. Antibodies can also aggregate the pathogen, thereby decreasing the effective dose and also making it easier for phagocytic cells to clear the organism. Thus, a much higher infective dose will be required to even initiate the infection.

Since most pathogens must attach to host cells to replicate, blocking attachment will dramatically reduce infections. For example, monoclonal antibodies to K99 fed to calves prior to infection can totally prevent disease (Sherman *et al.,* 1983). Similarly, antibodies to EspA and Tir can prevent the interaction of *Escherichia coli* O157: H7 with intestinal epithelial cells (DeGrado *et al.,* 1999; Celli *et al.,* 2000). Similarly, all viruses must enter cells to replicate. Antibodies neutralize virus by either blocking initial attachment, preventing the release of the nucleic acid into the cytoplasm that initiates viral replication, or inactivating the virus directly, either in the presence or in the absence of complement (Babiuk *et al.,* 1996). Once the virus enters the cells and initiates its replication cycle, antibodies can also play a role by lysing virus-infected cells in concert with complement. Since antigenic components are expressed on the surface of virus-infected cells well before viral assembly is complete, lysis of the infected cells will often decrease the amount of infected virus released, thereby reducing overall infectivity. Other antibodies can actually prevent the release of virus from cells (Yokomori *et al.,* 1992) or the assembly of virus, especially in epithelial cells where immunoglobulin A (IgA) is present (Mazanec *et al.,* 1992). Those viruses that are released can interact with antibody and are more rapidly cleared from the circulation or infectious particles can be converted to noninfectious empty capsids (Delaet and Boeye, 1993). Finally, antibodies can act in concert with Fc receptor-bearing cells to lyse cells by antibody-dependent cytotoxicity (Rouse *et al.,* 1976).

NK cells do not prevent initiation of infection as do antibodies, but are involved in clearing an established infection. Although NK cells are involved in clearing infection, this type of immunity is not specific and does not exhibit immunological memory; therefore, this response is not enhanced directly by immunization. However, indirect enhancement of cytokine responses as a result of immunization can result in enhancement of NK and LAK cell activity and resolution of an infection.

In addition to enhancing the speed with which antibody responses occur following exposure to the same or related pathogens, the number of epitopes recognized and the affinity of the antibody to a specific antigen are increased following secondary exposure to the pathogen, thus providing a rationale for multiple immunization regimes in order to get an optimal, broad immune response.

The redundancy of the immune response is clearly demonstrated in experiments of nature or in knock-out animals where a specific arm of the immune system is deficient, yet individuals generally recover from most infections. However, there are instances such as in agammaglobinemia where individuals exhibit increased susceptibility to bacterial infections, but not to viral infections. In contrast, T cell-deficient individuals are more prone to fatal viral infections such as measles (Nahmias *et al.,* 1967). The best defense is to have all components of the immune system intact.

Thus, although viral, bacterial, and parasitic agents require slightly different constellations of immune components to provide complete protection, vaccination can and does stimulate all of these components. Thus, the principle of immunization against all infectious agents is the same: to increase the number of cells that can produce antibody and those that can respond in a cell-mediated immune response to the pathogen as early as possible to limit the amount of tissue damage and immunopathology produced by the interaction of the host and pathogen.

IV. Conventional Vaccines

The majority of vaccines used today are produced by conventional methods using principles initially developed by Jenner and Pasteur just over 200 and 100 years ago, respectively (Tizard, 1999). Both of these types of vaccines (live attenuated and killed) have proven to be effective in at least partially reducing the clinical manifestations following exposure to virulent field strains of the pathogen. Furthermore, they have helped curtail the spread of the pathogens to other individuals by reducing the quantity of pathogen shed into the environment.

A. Live Attenuated Vaccines

The principle of live attenuated vaccination is that the pathogen is crippled sufficiently such that it cannot cause disease. The greatest concerns in developing conventional attenuated vaccines are the level of attenuation and the potential for reversion to virulence. Using conventional technology, attenuation is achieved by passaging the agent *in vitro* either in the presence of mutagenizing agents or under various *in vitro* culturing conditions. Following a number of different passages, the variants are selected for reduced virulence. Since the agent must replicate *in vivo* in order to induce an effective immune response, overattenuation will limit replication, and the magnitude and quality of the immune response would not be adequate to provide protection against virulent challenge. In contrast, underattenuation will result in clinical disease. Thus, there is a fine balance between attenuation of sufficient magnitude to reduce clinical signs and overattenuation, which would limit the efficacy of the vaccine. Unfortunately, this approach is purely empirical and depends on serendipity with no true knowledge of which genes are altered as a result of the mutation. This leads to two problems. First, each mutant must be tested *in vivo* to ensure that its level of attenuation is sufficient not to cause disease, but is not such that it cannot stimulate immunity and memory. This is a very expensive process. Second, since the attenuation occurs at random and is uncharacterized, the possibility exists that the agent can backmutate and revert to virulence following *in vivo* replication (Appel, 1978; Minor *et al.,* 1986). Thus, although these vaccines are generally very effective, there has been concern about their potential safety. This is especially the case in some individuals who may be partially immunosuppressed due to stress or other factors which may make them susceptible even to attenuated vaccines. Some attenuated vaccines cannot be used in pregnant animals since they may cause abortion (Straub, 1990). It is for this reason that some companies, producers, and countries do not favor live attenuated vaccines.

If genetic stability for live vaccines could be well controlled, they would be considered to be better than killed vaccines, since they induce a broader immune response (cellular and humoral), similar to that induced by natural infection. Furthermore, they can be delivered via the natural route of infections, such as mucosal surfaces, thereby providing a broader range of immune responses and consequently better protection, especially at the site of entry of the pathogen. This is critical since greater than 90% of all pathogens enter via mucosal surfaces. Even if a vaccine does not induce sterile immunity, immunity at mucosal surfaces should reduce shedding of the pathogen into the environment and may provide greater herd immunity than killed vaccines, which do not generally induce mucosal immunity. Another advantage of live vaccines is that they replicate in the host and generally induce a longer duration of immunity. There are also additional disadvantages, such as the presence of contaminating extraneous viruses in vaccines grown in tissue culture. One of the best examples of this problem is the introduction of bovine viral diarrhea (BVD) into vaccines grown in bovine cells (King and Harkness, 1975; Thornton, 1986; Wensvoort and Terpstra, 1988). The fact that the attenuated vaccines are alive requires their appropriate storage between manufacturing and administration. Thus, maintenance of a cold chain is a critical factor in ensuring that the vaccines maintain their viability and efficacy. This may be a real impediment in remote areas.

Since the vaccines must replicate in order to induce an immune response, the presence of passive antibody in the animal will limit the replication and the development of immunity to the vaccine. Unfortunately, in many instances, the level of passive antibody that is sufficient to interfere with vaccination is below that which will prevent infection with wild-type field strains of the pathogen (Kinman *et al.,* 1989). Thus, one needs to wait until passive antibody decreases to a specific level before vaccination. This leaves the young vulnerable to disease for extended periods of time. Another problem is that many individuals, especially in veterinary medicine, become susceptible or exposed to multiple diseases simultaneously either due to management conditions or decay of antibody. To reduce the chances of infection by these varied agents, individuals are vaccinated with multiple vaccines at the same time. Although this is a common practice, there is a potential for interference among the vaccines present in the cocktail (Harland *et al.,* 1992; Roth, 1999).

B. Killed Vaccines

An approach used to overcome the problem of safety and reversion to virulence is the use of killed vaccines. Killed vaccines are produced by inactivating the infectious agent so that it cannot replicate in the host without altering the immunogenicity of the protective proteins. However, recent studies have shown that most inactivating agents do have an impact on the immunogenicity of the protective proteins, and a balance is therefore sought between inactivation and reduction in immunogenicity (Duque *et al.,* 1989; Ferguson *et al.,* 1993). In some instances, field outbreaks of disease have occurred as a consequence of incomplete inactivation. This problem is particularly critical with pathogens that aggregate, thereby preventing the penetration of the inactivating agent into the center of the aggregates (Brown, 1993). Advances are being made at employing better inactivating agents and testing the level of inactivation during the quality assurance phase of product development. However, even testing would not have prevented the introduction of scrapie into sheep following immunization with an inactivated looping ill vaccine (Gordon, 1946). Similar concerns are being expressed with regard to transmissible spongiform encephalitis as a result of bovine spongiform encephalopathy in Britain. Thus, unless one knows exactly what extraneous agents might be present in the vaccine, they may never be tested for.

A second major problem with inactivated vaccines, especially in bacteria, is that some of the protective antigens are not produced *in vitro* or require special cultural conditions for their expression. For example, some iron-regulated outer membrane proteins shown to be protective *in vivo* are only produced under restricted iron growth conditions (Deneer and Potter, 1989a, b). Similarly, bacterin preparations often lack secreted proteins. The RTX toxins and Tir are considered to be crucial virulence factors in some bacterial infections, but are absent in bacterin preparations because they are secreted into the growth media (Shewan and Wilkie, 1982; Celli *et al.,* 2000). To overcome this deficiency, some new products are being enriched with these secreted toxins.

A major disadvantage of killed vaccines is that they are not very immunogenic; therefore, they need to be combined with strong adjuvants to improve their efficacy (Audibert and Lise, 1993). Although there is a continuing search for new adjuvants, only a few have been shown to be effective and most of them are expensive and often lead to adverse side reactions (tissue damage or residues; see Section VII.B.2). For example, mineral oil-based adjuvants are often considered to be extremely effective, but unfortunately they are not metabolized and remain at the injection site. Similarly, aluminum hydroxide adjuvants have been associated with fibrosarcomas in cats (Lester *et al.,* 1996; Burton and Mason, 1997). Finally, even if these adjuvants induce good immunity, the spectrum of immunity is generally narrow, with good systemic humoral immunity being present, but limited, if any, cell-mediated immunity or mucosal immunity. These latter types of immune responses are considered to be critical for protecting animals and aiding in recovery from many infectious agents. In addition to causing adverse reactions due to the presence of adjuvants at the injection site, there is a concern of broken needles being left in carcasses and the spread of infection from animal to animal by repeated use of the same needle for vaccination.

V. Genetically Engineered Vaccines

Advances in molecular biology, antigen identification, and expression have ushered in a new era in vaccinology. These advances have the potential to not only make vaccines that are more effective in inducing the correct type of immune response, but also make the vaccine safer. This section will describe some of these new vaccines including (1) subunit vaccines made from single proteins, peptides, or virus-like particles, (2) live, genetically modified vaccines, including gene-deleted replication-competent and replication-incompetent vaccines as well as vectored vaccines, and (3) polynucleotide vaccines. Examples of each type of vaccine for treating infectious diseases and cancer and for regulating growth and reproduction in livestock will be highlighted.

A. Subunit Vaccines

Subunit vaccines are defined as those containing one or more pure or semipurified antigens. The foundation for the development of subunit vaccines is the understanding of the pathogenesis of the pathogen and the proteins, glycoproteins, or carbohydrates involved in inducing protective immunity. Thus, it is critical to identify the individual components of a pathogen that are involved in inducing protective immunity.

Recent advances in genomics and the availability of complete sequences of many pathogens have allowed the identification of new surface or secreted proteins that might be useful for future vaccines. Furthermore, the expression of these novel genes *in vivo* is allowing a more systematic and directed approach to identifying putative protective proteins for use in vaccines. This is critical since parasites and bacteria possess thousands of genes, many of which are irrelevant for induction of protective immunity. In addition, putative antigenic peptides which bind to major histocompatibility complex (MHC) class I and II molecules may be determined using computational algorithms of protein sequences. An even greater possibility is the use of functional genomics and computer algorithms to determine the host T cell repertoire. Thus, knowledge of the host immune system, combined with genomics of the pathogen, should allow for more rapid identification of vaccine antigens. Genomics combined with recombinant DNA technology to produce proteins from individual genes and monoclonal antibodies to test the potential of each protein as a vaccine or for purification of antigen is making subunit vaccines a reality.

Subunit vaccines can be produced by employing recombinant DNA technology or by purifying a specific component from a conventionally produced vaccine. For example, extracts of bacteria or concentrates of excreted proteins can be used as vaccines or to "spike" conventional vaccines to increase the concentration of a particular protective component in a vaccine. Many bacteria produce an extracellular toxin, which is involved in pathogenesis and tissue damage (Shewan and Wilkie *et al.*, 1982). Furthermore, it has been shown that immunity to toxins is extremely critical for inducing protection against a number of bacterial diseases. Although it is possible to isolate and purify these extracellular toxins from conventionally grown bacteria, it is more economical to use recombinant DNA technology for production of these products.

Before recombinant DNA technology can be used for vaccine production, it is critical to be able to produce these proteins in a commercial setting. This has generally been easy for bacterial proteins, where large quantities can be made in prokaryotic expression systems grown to high density in fermenters. Unfortunately, even though viral or parasitic antigens can be produced in high concentrations in prokaryotic systems, the need for posttranslational modification of viral and parasitic proteins makes prokaryotic systems less than ideal for vaccine production of these specific proteins. For example, foot-and-mouth disease virus VP1 can be expressed in *E. coli* to levels where up to 17% of the protein produced in the bacteria is VP1, however, the immunogenicity of this protein was less effective than native protein (Kleid *et al.*, 1981). Similar reports have been made with numerous viral antigens where yield and antigenicity are excellent, but immunogenicity and protective immunity were either poor or nonexistent (van Drunen Little-van den Hurk *et al.*, 1993). It is for these reasons that most viral proteins are produced in eukaryotic systems.

The most appropriate eukaryotic expression system for ensuring that all the posttranslational steps are optimized is the mammalian cell. Unfortunately, this is also the most expensive and difficult to scale up for commercial production. However, even with these production issues a number of proteins/glycoproteins have been produced in mammalian cells (Laskey *et al.*, 1984; Kowalski *et al.*, 1993). However, the level of expression is not the only factor that needs to be considered. Protein must be purified from the host's cell component. Many viral proteins involved in inducing protective immunity, especially from enveloped viruses, are glycoproteins that are inserted into the host's cellular membrane. Thus, isolation of these glycoproteins requires lysis of the cells to release the glycoproteins followed by purification of the individual component. This is a very expensive process since it requires a large quantity of cells to be cultured, and once they reach the required cell density and they are lysed, one needs to start over and grow up new batches of cells for the next vaccine production run.

To overcome this inefficient production approach, it is possible to use recombinant DNA technology to remove the transmembrane anchors, if this does not alter the conformation of the glycoprotein. The glycoprotein is then secreted into the cell culture medium (Kowalski *et al.*, 1993). This not only allows the production of proteins/glycoproteins that may be toxic to cells if produced at high concentration, but also facilitates very efficient purification of the proteins (vaccine) from the host cells. More importantly, the cells can be grown in continuous culture to repeatedly harvest the subunit protein from the culture. This reduces the expense of growing large quantities of cells for vaccine production.

In addition to expression of subunit vaccines in mammalian cells, expression in yeast, insect, and plant cells has been pursued (Valenzuela *et al.*, 1982; Ding *et al.*, 1987; Maeda, 1989; Marciani *et al.*, 1991; Arntzen, 1995; Berman *et al.*, 1998). Indeed, the world's first genetically engineered vaccine, hepatitis B surface antigen (HBsAg), was produced in yeast. This vaccine should be safer than the HBsAg purified from plasma since there is no possibility of contamination with other human pathogens. Furthermore, the characteristics and immunogenicity of yeast-produced HBsAg are indistinguishable from those of the plasma-produced product.

HbsAg assembled as a virus-like particle (VLP) can also be used as a carrier for epitopes from other viruses (Netter *et al.*, 2001; Kunkel *et al.*, 2001). Similar VLPs have been produced from a variety of infectious agents (Jiang *et al.*, 1992; Kirnbauer *et al.*, 1992; Konishi *et al.*, 1992; Bansal *et al.*, 1993). The particulate nature of VLPs appears to induce a more effective immune response than soluble proteins, making this an attractive approach to vaccine production and delivery. The reasons for this enhanced immunogenicity are many, including the particulate nature of the immunogen and, more importantly, the fact that VLPs often present confirmation epitopes to the immune system in a manner that mimics the natural agent, thereby stimulating the appropriate immune responses. More importantly, it is possible to introduce specific

B and T cell epitopes into the VLPs to further enhance the immune response. Another advantage of VLPs is that no inactivating agents are required which can alter the immunogenicity of the protein. One possible limitation to such an approach is the limitation of the size of the epitope that can be inserted into assembled VLPs.

Recently, using hepatitis B core VLPs, it was possible to insert a large protein, such as green fluorescent protein (GFP), and a malarial antigen into the VLP without the disruption of the assembly process. A variation on the chimeric VLPs uses the ability to attach various epitopes and proteins to them to provide a binding peptide or chemical coupling (Redmond *et al.,* 1993). This demonstrates the versatility of such particles as vaccines and vaccine carriers.

Expression of viral antigens using the baculovirus *Autographica california* nuclear polyhedrosis virus has been used extensively for research purposes and for some vaccine trials. However, to date there are no licensed vaccines employing the baculovirus expression system. Two potential problems with the baculovirus expression of protein are the fact that insect cells do not posttranslationally modify proteins in exactly the same manner as mammalian cells, which is a real problem in numerous vaccine studies, and that there is less experience with growing large quantities of insect cells in fermenters than with mammalian cells. Thus, production remains a challenge.

Similar to the case of insect cells, production of proteins in plants does not result in identical posttranslational modification of viral proteins. However, the economics of production of vaccines in plants is extremely attractive for a number of reasons. First, they can be produced in large quantities and, more importantly, at least in some instances, they can induce immunity following oral delivery (Koprowski and Yusibov, 2001; Streatfield *et al.,* 2001; Lauterslager *et al.,* 2001). Since oral feeding of vaccines would dramatically reduce the cost of delivery and vaccine formulation and, it is hoped, improve compliance, this is proving to be a relatively attractive approach to vaccine production. Initially, oral delivery was shown with a particulate antigen, HBsAg; more recently, it has also been shown with soluble glycoproteins and with bacterial toxins (Arakawa *et al.,* 1998). Since plant pathogens generally do not infect humans or animals, the concern of spreading adventitious agents by vaccination with plant-derived vaccines is very low. This is currently a concern with vaccines grown in media containing bovine serum components, which may be contaminated with transmissible spongiform encephalopathy.

Techniques are also being developed to target recombinant vaccine antigens to oil bodies in plants. Since techniques are available for rapid purification of oil bodies in plants, this makes purification of the vaccine for parenteral delivery also extremely easy (van Rooijen and Moloney, 1995). Second, the association of the vaccine with oil bodies provides for a built-in adjuvant. Thus, in one single step the vaccine is purified and formulated in a metabolizable oil adjuvant. Possibly the greatest advantage of such an approach is that the vaccine can be produced and stored indefinitely and only purified from the seed oil bodies as required. Thus, the shelf life could theoretically be indefinite with no need for a cold chain during storage before formulation.

The advantages of using subunits as vaccines are increased safety, lowered antigenic competition between irrelevant proteins (since only a few components are included in the vaccine), ability to target the vaccine to a site where immunity is required, and ability to differentiate vaccinated animals from infected animals (marker vaccines; see Section VI). Most subunit vaccines being developed are targeted against viral and bacterial pathogens. However, in addition to identifying potential targets of specific pathogens, it is also possible to develop subunit vaccines to parasites or use vaccines to reduce the transmission of some pathogens (Opdebeeck *et al.,* 1988; Willadsen and Kemp, 1988). For example, many infectious agents are transmitted by blood-feeding vectors. Studies have shown that immunization of animals against antigens present in the gut of these vectors results in the disruption of intestinal cell function of the vector after exposure to blood containing antibodies directed against these intestinal antigens. Thus, although the vaccine is not directly focused against a specific pathogen, the vaccines are effective in reducing the transmission of certain pathogens. The best example of such an approach is immunizing animals against a 86K glycoprotein present in the mid-gut of ticks. When the tick takes a blood meal from an immunized animal, the antibody to the 86K glycoprotein interacts with the protein in the tick's mid-gut and interferes with the reproductive capacity as well as the viability of the tick. Thus, the environmental reduction of tick infestation reduces the spread of disease and therefore reduces the incidence of disease in the host population. Without the use of subunit vaccines and the knowledge of these "concealed" antigens, it would not have been possible to control tick infestation by conventional vaccination (Willadsen and Kemp, 1988). It is hoped that similar approaches could be used in vectors involved in transmitting malaria as well as other pathogens.

In addition to producing vaccines containing whole proteins or VLPs, it is also possible to produce vaccines containing only a single epitope of interest. Since many proteins possess immunodominant epitopes, it is possible to identify these epitopes and synthesize them chemically

(Muller *et al.,* 1982; Neurath *et al.,* 1989; Bittle *et al.,* 1982). Although in theory this appears to be an attractive approach and neutralizing antibodies have been induced with synthetic peptides and, more importantly, protection has been demonstrated in animals immunized with these peptides, no synthetic peptides are currently licensed. One of the impediments to such an approach is the poor immunogenicity of peptides. This is partially being overcome by adding both B and T cell epitopes to the peptide or presenting the peptide in the context of a VLP (Redmond *et al.,* 1993) or in a replicating virus or bacterial vector. A second problem with peptide vaccines is the elasticity of most pathogens, where quasi-species or mutants can rapidly develop that circumvent the immune response directed at a single epitope. Finally, due to the outbred nature of human and animal populations, the specificity of peptide binding to specific MHC molecules would preclude mass immunization. These last two problems are partially being overcome by developing a "string-of-beads" vaccine containing multiple epitopes (Whitton *et al.,* 1993). However, even with this approach, no synthetic peptide vaccines are yet in use and it is our contention that only in a few isolated cases will such vaccines prove effective.

In addition to developing subunit vaccines for infectious diseases, recent success in identifying unique tumor-specific or tumor-associated antigens (TAAs) is ushering in the possibility of also developing immunization strategies against various tumors (Rosenberg, 1999; Offringa *et al.,* 2000). These strategies include passive immunizations with monoclonal antibodies targeting specific antigens. Examples include antibody (17-1A) against metastatic colorectal cancer (anti-CD20) for B cell lymphomas and anti-Her 2 neu for metastatic breast cancer (Scott and Welt, 1997). Active immunizations are also possible.

In the case of active immunization, the main objective is to induce a strong cell-mediated immune response whereby cytotoxic T cells kill the cancer cells. Preliminary results demonstrating the "proof of principle" in mice have been sufficiently encouraging that human trials are being prepared (Rosenberg, 1999; Greten and Jaffee, 1999). However, it must be emphasized that although mice may be interesting models, they are even less reliable for human cancer studies than they have been for the study of infectious diseases.

In contrast to infectious disease vaccines, where the objective is to prevent infection (prophylaxis), cancer vaccines are primarily designed as therapeutic vaccines. As such, the tumor burden will probably need to be relatively low for effective control. Even if such vaccines do not totally eliminate the tumor, slowing down the progression of disease and thereby prolonging life and more importantly the quality of life will be critical.

As with infectious diseases, the key is to identify a tissue-specific antigen to which cytotoxic lymphocytes can be targeted. This is being achieved by both immunological (van den Eynde and Brichard, 1995; Rosenberg, 1999; Sahin *et al.*, 1997; Old and Chen, 1998) and biochemical approaches by purifying and characterizing peptides associated with MHC class I molecules. The advent of genomics and microarrays is allowing us to analyze the specific genes expressed in tumors versus normal tissue to further identify targets for vaccination.

Originally it was thought that each tumor may have a specific TAA or mutation. This would have made vaccination a real challenge. However, the recognition that there are various tumors that share antigens is encouraging with respect to prophylactic vaccination on a mass scale. For example, MAGE (van der Bruggen *et al.*, 1991) and CEA (Gold and Freeman, 1965) antigens are present on a variety of different tumors. Other antigens, such as Melan A/Mart 1 and tyrosinase TRPI, are overexpressed on melanomas as compared to normal parental cells. Cancer vaccines can also be directed at carbohydrate components such as ganglioside GM2 or CD2 coupled to strong carriers. These types of vaccines are being used against specific tumors such as melanomas or prostate cancers.

Once a specific TAA is identified, it can be produced as a subunit vaccine or delivered by viral vectors (Tsang *et al.*, 1995; Zhai *et al.*, 1996; Rosenberg *et al.*, 1998; Conry *et al.*, 1999; Bonnet *et al.*, 2000). These vectors have also been used to deliver cytokines (Soiffer *et al.*, 1998; Simons and Mikhak, 1998; Simons *et al.*, 1999) directly into the tumor mass.

Tumors where antigenic alterations are specific, for example, mutated Ras or Her 2 neu, will require individualized vaccine development. For these types of tumors it will be more difficult to design vaccines for mass use.

Some cancers are hormone-dependent; thus, if it were possible to develop vaccines against specific hormones, this would also possibly reduce tumor growth. Vaccines against hormones are being developed to modulate growth and reproduction in animals. Thus, such a possibility does not appear to be out of the question for treating cancers.

Currently, castration is the main approach used in livestock management to enhance production and modify the behavior of male animals. However, castration is considered to be painful and not humane by many in our society. Since it is now possible to immunize against the sex hormones, a number of companies are developing vaccines to replace castration. This has been made possible by our understanding of the physiology of growth and reproduction combined with our knowledge of vaccine delivery/carriers and adjuvants. Most advanced vaccines for altering reproduction are designed to neutralize hormones such as gonadotrophin-releasing hormone (GnRH) (Robertson *et al.*, 1982). Male animals immunized against GnRH showed testicular involution, suppression of testosterone secretion, reduced aggressive behavior, improved meat quality, and significantly faster growth than surgically castrated animals.

Another area where immunization against hormones is being pursued is to enhance the production of eggs in turkeys. When female turkeys begin laying, they have a tendency to follow their natural instinct to incubate the eggs (Sharp, 1997). This results in fewer eggs being laid by the turkeys. By immunizing them against vasoactive intestinal peptide, the broodiness is reduced and the number of eggs is significantly increased (Caldwell *et al.*, 1999).

These are just some examples of how vaccines are being used to alter reproduction, growth, and productivity. Others include immunization against zona pellucida to reduce fertility (Millar *et al.*, 1989) or against inhibin, a hormone which inhibits secretion of follicle-stimulating hormone; neutralization of inhibin increases follicle-stimulating hormone production and ovulation rates (Forage *et al.*, 1987).

B. Live Vaccines

Since the majority of infectious agents enter via mucosal surfaces, immunity at these surfaces is most effective in restricting the rate of infection by increasing the infectious dose required to initiate infection and reducing the level of replication of the pathogen at the site of entry. Although there are various delivery systems that can be used to induce mucosal immunity, live vaccines appear to be the most effective at present. In addition to being able to induce immunity at the portal of entry, thereby providing protection from infectious diseases, mucosal immunization with live vectors has other advantages. These include lower reactogenicity and absence of injection site reactions. As a result, these vaccines should be more acceptable to the public and increase vaccine compliance. Increasing vaccine coverage is clearly a goal in order to reduce the spread of infection through "herd immunity."

Although current techniques of attenuation of microorganisms using repeated pathogen *in vitro* under various growth conditions (low temperature or in the presence of mutagens) will continue, newer approaches directed at modifying specific genes are gaining popularity for generating live attenuated vaccines. This has become possible as a direct result of our increased knowledge of the virulent genes of many pathogens. This knowledge is allowing us to introduce multiple directed mutations or deletions in specific virulence genes in order to reduce virulence. For example, mutations in viral surface proteins, capsid proteins, or nucleic acid metabolism genes have all been shown to alter virulence (Coulon *et al.,* 1983; Spriggs *et al.,* 1983; Macadam *et al.,* 1991; Liang *et al.,* 1997). Thus, attenuating genes may be structural, nonstructural, or even in the noncoding regions of the virus (Macadam *et al.,* 1992; Kuhn *et al.,* 1992). Finally, attenuation can be also achieved by making chimeric viruses with complementary, but not fully compatible genes from related viruses (Kuhn *et al.,* 1991).

In addition to mutating or exchanging specific genes, in some instances the entire gene may be deleted. These approaches allow different levels of attenuation to be achieved by a more directed approach. Using such approaches, it should be possible to develop more stable mutations which cannot revert to virulence (Kit *et al.,* 1987), a potential problem with conventional attenuation. Such a scenario could not occur with modern directed approaches to attenuation since it would be very difficult to reacquire an entire gene or portions of a gene. As a result, genetically engineered vaccines should prove to be safer than conventionally developed vaccines.

The most frequently used strategy is to make deletions of nonessential genes. In this case, the pathogen could be grown *in vitro* and used as a live vaccine which replicates *in vivo* to induce the appropriate immune responses. However, due to the deletion, it will not be able to replicate sufficiently to cause disease. Genetic introduction of specific temperature-sensitive (ts) mutations into respiratory pathogens would restrict the pathogen's replication to the upper respiratory tract with limited or no replication to the lower lungs. Since replication to the lower lungs is where significant damage occurs, leading to pneumonia, such a vaccine would be safer. By introducing a number of site-specific ts mutations, one would have vaccines that are much more stable than those obtained using conventional attenuation (Subbarao *et al.,* 1993).

Another approach is to delete genes involved in establishing persistent infection. For example, in herpesvirus infections deletion of genes involved in replication of neurons should reduce the chances of latency (Whitley *et al.,* 1993). Similarly, mutagenizing or deleting the integrase gene from retroviruses should prevent the establishment of persistent infection (Vogel *et al.,* 1993).

A further degree of attenuation can be achieved by deleting genes considered to be essential for *in vivo* replication. This type of mutation results in a replication-incompetent organism. Such an approach has been used for viral vaccines (Jacobs *et al.,* 1992; Nguyen *et al.,* 1992;

Zakhartchouk *et al.,* 1999; Reddy *et al.,* 1999). In this case, the virus can be propagated *in vitro* in a cell line transfected with an essential gene. Under these conditions, the transfected cell line provides essential functions for *in vitro* replication of the virus and production of the vaccine. However, upon introduction of the vaccine into an individual, the virus can only undergo an abortive infection since the host does not possess the complementary gene to allow completion of the replication cycle (Forrester *et al.,* 1992; Farrell *et al.,* 1994). These defective infectious single-cycle (DISC) viruses could be produced against a number of different viruses.

A variation of this approach can also be achieved by developing suicide vectors where the replication of the pathogen is dependent on a specific metabolite or compound. This approach is being pursued in bacterial vaccines, where the vaccine can be grown *in vitro* in the presence of the required compound (Donnenberg and Kaper, 1991). To allow the vaccine to replicate *in vivo,* compounds are introduced with the vaccine or individuals are fed the specific compound for a period of time to allow the bacteria to replicate *in vivo*. Upon removal of the required compound, the bacteria die. These types of vaccines should remove the concern with respect to their shedding into the environment since they would rapidly die because of lack of complementary growth components in the environment.

Another advantage of live, genetically engineered vaccines is the fact that these vaccines can be used as vectors to carry protective proteins from other pathogens (Yilma *et al.,* 1988). Thus, in addition to immunizing the individual against a single pathogen, protection could be induced against a variety of pathogens. Thus, a single immunization could protect individuals from a variety of diseases. This should make vaccine production and delivery much more economical since a single vector carrying multiple vaccines can be manufactured in a single step. Furthermore, there should be no interference between the antigens as is often seen with multiple agents delivered simultaneously.

Currently, a variety of viral and bacterial vectors carrying multiple genes are being tested. These include herpesviruses, adenoviruses, poxviruses, and bacteria (*Yersinia enterocolytica, Listeria monocytogenes, Lactobacillus* spp., *Streptococcus* spp., *Mycobacteria* spp., etc.). Depending on the vector and species of interest, these vectors lend themselves to mass administration in aerosols, feed, or water. This is especially attractive in poultry, where management conditions are conducive to this approach. This approach overcomes the need to handle individual animals and also overcomes the disadvantage of needle delivery. In addition, these vectors can be used to induce immunity *in ovo*

(Gagic *et al.,* 1999; Sharma, 1999). These vectors have been shown to induce mucosal immunity even in the presence of high levels of systemic immunity to the vector (Zakhartchouk *et al.,* 1999). This makes these vectors extremely attractive.

The efficacy of such vaccine vectors can be further increased by incorporation of cytokines or immunomodulatory genes into the vector. These immunomodulatory genes may have two functions: Their insertion can attenuate the pathogen (Sambhi *et al.,* 1991; Kurilla *et al.,* 1993; Giavendoni *et al.,* 1992, 1997) as well as enhance immune responses to the pathogen, making these vectors extremely useful even in immunocompromised individuals.

One of the best examples of a vectored vaccine is vaccinia virus carrying the rabies virus glycoprotein gene (Brochier *et al.,* 1994). The thermal stability of vaccinia virus and its ability to infect foxes by oral ingestion provided the basis of a vaccine campaign to reduce rabies infection in foxes in Western Europe. Baits containing the vaccinia virus carrying the rabies glycoprotein gene were distributed from airplanes. Foxes consuming the bait were immunized against rabies virus. As a result of this approach, a number of Western European countries have dramatically reduced wildlife rabies and, concomitantly, reduced the transmission to domestic animals and humans. Without the combination of a thorough understanding of vaccinia virus biology and its genetics, combined with rabies virus epidemiology in wild foxes, such an approach would not have been possible.

A variation on the vaccinia virus theme is the use of related viruses that are replication-defective in mammals. Such vectors could be safe even in immunocompromised individuals. The best example of such an approach is the use of avian poxviruses in mammals (Somogyi *et al.,* 1993). These viruses (canary pox and fowl pox) can be engineered to carry any variety of genes from foreign pathogens and grow to high titers *in vitro* in avian cells. These avian poxviruses can infect mammalian cells and express the transgene, leading to humoral and cellular immune responses to the transgene, but since they undergo an abortive infection no infectious virus particles are produced. These make these vectors extremely safe.

Although these vectored vaccines have many advantages, consideration must be given to the antigen being delivered. For example, one would not chose a viral vector if one needed to develop an immune response to carbohydrate or lipopolysaccharide antigens. In contrast, a bacterial vector would be ideal for such a vaccine. Similarly, one would not choose a bacterial vector to deliver a viral protein whose conformation is very sensitive to posttranslational modifications.

C. Polynucleotide Vaccines

The most recent development in vaccinology is the delivery of genes encoding putative protective antigens in the form of naked DNA (plasmids) or as replicons coated with alphavirus envelopes. Although the first report of transfection of cells with DNA *in vitro* occurred over 40 years ago, it was not until the 1990s that the concept of using plasmids as vaccines gained a significant following. This interest was fueled by the report of Wolff *et al.* (1990) that a reporter gene could be expressed sufficiently following *in vivo* delivery. Based on this observation, the era of polynucleotide vaccination was launched with hundreds of reports of plasmid immunization against numerous pathogens (viral, bacterial, parasitic) in various species (Donnelly *et al.,* 1993; Lewis and Babiuk, 1999; Babiuk *et al.,* 2000). Since any cell can theoretically take up DNA, various delivery systems and routes of delivery have been used for DNA immunization. These include intramuscular and intradermal as well as mucosal routes (respiratory, ocular, and vaginal).

The attractiveness of DNA immunization is its simplicity. Expression of the gene of interest is driven by a strong promoter capable of expressing the gene in mammalian cells. Another requirement is the presence of CpG sequences in the plasmid backbone. These sequences are critical in not only creating the cytokine microenvironment required for induction of immunity, but also providing the Th1–Th2 balance and long-term memory observed with DNA vaccines (Davis *et al.,* 1998; Krieg *et al.,* 1998; Jones *et al.,* 1999). Improved expression levels and immune responses have been achieved by introducing introns into the plasmid, resynthesizing genes to remove cryptic splice sites, and improving codon biases (Vinner *et al.,* 1999). The use of introns and removal of cryptic splice sites are often critical for genes from RNA viruses and bacteria, which are not, under normal circumstances, expressed in the nucleus.

One of the most interesting features of DNA immunization is that only very small quantities of protein are needed for effective priming of the immune response. The reason for this effectiveness is not fully understood, but it could be related to the continuous production of antigen, which can drive the expansion of the immune response. Unfortunately, it is not possible to quantitate the level of antigen production *in vivo,* but at least in some cases antigen is produced for several months after DNA immunization (Yankauckas *et al.,* 1993).

The attractiveness of DNA immunization, especially for viral diseases, lies in the fact that the antigens are produced endogenously and therefore all the posttranslational modifications are similar to those occurring during a natural viral infection. As a result, the antigens are authentic with all the conformational epitopes required for protection being expressed. Second, since the viral antigens are presented efficiently in the context of MHC class I antigens, they induce a broad range of immune responses, including cytoxic T lymphocytes (CTLs) and antibodies (Ulmer *et al.,* 1996). Prevention of infection and recovery from many diseases require both cellular and humoral immunity, and DNA vaccines meet the criterion of inducing a broad immune response. The observation that various manipulations of the genes and coadministration of plasmids encoding cytokine genes with the vaccines either as separate plasmids or in the same plasmid shift the response to the desired Th1 or Th2 type suggests that this technology is amenable to tailoring the immune response to ensure that the most effective protective responses are generated. By delivering the plasmids to mucosal surfaces, it is also possible to induce mucosal immune responses with DNA vaccines (Fynan *et al.,* 1993; Kuklin *et al.,* 1997).

Another advantage is that the individual acts as a bioreactor, and thus there is no need for downstream processing or formulation of the vaccine. Since adjuvants are not needed, there are none of the adjuvant-associated problems or injection site reactions with DNA vaccines.

A real advantage of DNA vaccines is that they can induce immune responses in neonates even in the presence of passive antibody (Lewis *et al.,* 1999; van Drunen Littel-van den Hurk *et al.,* 1999). This is a true advantage over conventional or genetically engineered vaccines. Although the mechanism of overcoming passive antibodies is not known, the phenomenon has been reported in various species against different antigens. This observation is extremely encouraging since it will allow us to vaccinate individuals at a very early age and ensure that they are fully protected at the time passive antibodies decay. Indeed, it is even possible to vaccinate individuals *in utero,* resulting in the birth of individuals who are immune to certain diseases (Gerdts *et al.,* 2000). This is extremely attractive for diseases such as herpesvirus simplex, chlamydia, hepatitis B, and those due to cytomegalovirus, group B *Streptococcus, Hemophilus,* and human immunodeficiency virus (HIV), where infection occurs in the birth canal or shortly after birth. The World Health Organization estimates that over 5 million infants get infected during their first week of life. Since many diseases are vertically transmitted, active immunization of the fetus may provide an effective approach to reducing the high risk of neonatal infection.

Duration of immunity continues to be a concern with all vaccines. Currently, DNA vaccines appear to induce long-term immunity, possibly due to the extended period of protein expression. Whether this will translate into life-long immunity is unknown. However, even if boosting is required, it could be done with conventional or genetically engineered vaccines, which would rapidly boost the level of immunity. Indeed, the practice of priming with DNA followed by boosting (prime–boost strategy) with protein is being practiced more and more. Indeed, it is possible to induce higher levels of protection with a prime–boost strategy than with either DNA immunization or protein immunization individually (Amara *et al.,* 2001).

In addition to regulatory hurdles associated with DNA vaccination, possibly the greatest hurdle is that of vaccine delivery. Currently, the majority of the plasmid is degraded before it enters the nucleus and results in transcription and translation of the gene of interest. Until this impediment is overcome, the cost of DNA vaccination may be prohibitive. Various delivery systems are being explored, including (1) mechanical devices, including injection by needles or microneedles, pressure injection, or particle bombardment, (2) electrical means, including electroporation or iontophoresis combined with mechanical delivery, and (3) chemical approaches, using liposomes, dendrimers, or encapsulation of DNA in various polymers. Unfortunately, no ideal delivery system has been developed.

Recent developments in replicon technology based on using the alphavirus coat/envelope to deliver nucleic acid directly to cells offer a possible way of overcoming some of the impediments of delivery (Grieder *et al.,* 1995; Pushko *et al.,* 1997; Schultz-Cherry *et al.,* 2000). Using alphavirus-based replicons, not only are the genes protected by the artificial viral coat/envelope, but the envelope can be engineered in such a way that it can target the antigen-presenting cells (dendritic cells). This latter development is a possible method for improving vaccination and delivery of nucleic acid-based vaccines.

VI. Marker Vaccines

Genetically engineered (subunit, live, or polynucleotide) vaccines lend themselves to codevelopment of diagnostic tests that can differentiate vaccinated individuals from those who are naturally infected and may be carriers of the disease. These vaccines have been called "marker vaccines" or more recently "DIVA vaccines" (Differentiate Infected from Vaccinated Animals). These vaccines/diagnostic tests are especially valuable in the livestock industry and where countries want to eradicate specific diseases. The basis for DIVA vaccines is that animals that are immunized with a specific antigen develop immune responses to that antigen. If one develops a subunit or a gene-deleted vaccine, it is possible to develop a parallel diagnostic test which tests for the presence or the absence of antibodies in animals. For example, animals that are infected naturally with pseudo-rabies virus or bovine herpesvirus will develop antibodies to all of the proteins of the virus. However, if an animal is vaccinated with a single glycoprotein (gD) or a gE gene-deleted live virus, it will not develop antibodies to gE. Thus, if an animal demonstrated that it had antibodies to gE, it would be considered a latent carrier of the virus (Kaashoek and van Oirschot, 1996). Such vaccines are being used to eradicate pseudo-rabies and bovine herpesviruses from various countries. However, the concept of DIVA vaccination can be applied to any disease situation (Anderson *et al.,* 1993; Bergmann *et al.,* 1993; Kaashoek and van Oirschot, 1996; Birch-Marchin *et al.,* 1997; Smedegaard Madsen *et al.,* 1997).

Another use of such vaccination/diagnostic tests is in countries where a specific disease does not normally occur, but the country wants to protect itself from accidental introduction of the disease and maintain its disease-free status. An excellent example is the recent outbreak of foot-and-mouth disease virus in the United Kingdom and parts of Western Europe. Since Western Europe, North America, and Australia do not have foot-and-mouth disease virus, they do not vaccinate, and control of the introduction of the disease is done by limiting the importation of potentially infected products from countries where foot-and-mouth disease is endemic. However, accidental introduction of the disease under the current management systems can result in the rapid spread of infection. If animals were vaccinated with marker vaccines, the spread might be curtailed and the disease not become established. However, to maintain a disease-free status in the presence of vaccination would require continuous monitoring of the national herd to see if the agent was actually introduced. This would be extremely expensive and would probably not be implemented. However, the vaccine could be used as a complement to quarantine and slaughter policies in countries where an exotic disease was accidentally introduced. Thus, animals on the periphery of the infected area could be vaccinated to reduce the spread and help contain the disease. As the disease is contained and eliminated, the differential diagnostic test could be used to aid in culling the potential infected carrier animals and return the country to disease-free status quickly. Society might find such an approach more appealing than watching the slaughter and destruction of large numbers of healthy animals. Important criteria for choosing a marker for such a vaccine are that it must be one that induces a rapid and long-lived antibody response in infected animals and not be present in vaccinated animals even after repeated immunizations with it.

VII. Vaccine Formulation and Delivery

Regardless of the quality of the vaccine developed, if it is not formulated and delivered properly, the full benefit of the vaccine may not be achieved. Currently, the majority of vaccines are delivered parenterally. As stated previously, this mode of delivery does not induce mucosal immunity, and therefore one important component of the immune system is not primed. Other concerns regarding parenteral delivery of vaccines with needles center around the safety issue. These include broken needles in animal carcasses as well as the tissue damage from the needle. Other safety concerns are accidental needle sticks, spread of disease in developing countries from reusing needles, and disposal concerns. Finally, because of the pain associated with needles, compliance may be reduced.

In addition to safety concerns, there is also a need to improve the efficacy of immunity following vaccination, which is especially true with DNA vaccines (see above). Therefore, the goal is to improve the efficacy of vaccines by noninvasive methods as well as to improve compliance. It is for these reasons that a number of alternative delivery methods have been or are being explored. The real challenge is to enhance immunity in the least invasive way possible.

A. Needle-Free Delivery Devices

Since the skin is designed to protect the body from external insults, it is not surprising that uptake of vaccines transdermally is not very efficient. To overcome this impediment, a number of devices to break the skin barrier with minimal pain are being developed. One of the first such devices uses jet propulsion. Vaccines delivered by jet propulsions are formulated in liquid excipients, which are delivered under pressure through the stratum corneum. Such devices can be designed to deliver multiple or single doses. Obviously, multiple-dose delivery would be the preferred approach for mass immunization. Depending on the size of the orifice of the injector and the velocity of the jet, the vaccine can be delivered intradermally, subcutaneously, or intramuscularly. The real advantages of needle-free devices are that they eliminate the possibility of unintentional needle stick accidents, reduce the cost of injection compared to needle injection systems, eliminate needle phobia, especially in children, eliminate the medical waste problem associated with needles and syringes, and allow for high-speed delivery systems to save time and labor. This is especially important where large vaccine campaigns are being conducted, allowing countries to vaccinate their populations extremely quickly.

Although these devices have been around for a number of decades, "splash back" and the potential cross-infection of individuals remains a concern and therefore these devices are not used. Indeed, both the WHO and the U.S. Department of Defense have rejected the use of multiple-use injectors due to the concern of potential transmission of blood-borne diseases due to "splash back" (Canter *et al.,* 1990). As a result, companies are developing novel devices that will prevent cross-contamination.

The problem of disease transmission with single-use jet injectors does not exist, but the current cost makes them unsuitable for mass administration, especially in developing countries or in the face of an epidemic where hundreds of thousands of individuals must be immunized in a very short time.

A variation on the jet injector is the gene gun. The gene gun was originally designed to transfer genetic material into plant cells using plasmids coated on gold beads. With the advent of DNA vaccination, the gene gun became a useful tool for delivery of plasmid-based vaccines into individuals (Pertmer *et al.,* 1995). Gene-gun-delivered plasmids are much more efficient at inducing immune responses than an equivalent amount of plasmid delivered by a needle. The principle upon which gene gun delivery is based is that gold beads coated with DNA are "fired" through the skin and directly into cells, where the DNA is released from the gold particles. Since the DNA is introduced directly into cells as well as the nucleus, this method of plasmid delivery is relatively efficient. However, the gene gun also has some disadvantages. The first disadvantage relates to the quantity of DNA that can be loaded onto gold beads. Excessive quantities of DNA cause clumping of beads and reduce the efficacy of the plasmid delivered. The need to limit the quantity results in the need for multiple shots, which is inconvenient. A second disadvantage is the extensive preparation of vaccine required for keeping the gold particles dehydrated after coating. Recently, particle bombardment by gene gun has been adopted for the delivery of protein antigens as vaccines (D. Chen *et al.,* 2001). However, the same technical difficulties that arise with particle bombardment of DNA apply to proteins, along with the additional protein formulation problems. Finally, the gene guns themselves are not simple to use. These technical limitations will have to be overcome before particle bombardment becomes extensively used outside the research environment.

The ideal delivery system would be to deliver vaccines transdermally by topical delivery. Unfortunately, the stratum corneum generally prevents such an approach. However, there has been some good success in humans with both cholera toxin and *E. coli* heat-labile toxin applied topically (Glenn *et al.,* 1999, 2000). Similarly, immune responses have been elicited with topically applied plasmids (Fan *et al.,* 1999; Shi *et al.,* 1999). These early successes, even though they showed weak immune responses, demonstrate the possibility of topical administration of vaccines, and studies are continuing in this direction. Topical administration may be enhanced by formulation with different lipid-based delivery systems containing permeation enhancers such as cadherin agonists to assist in disrupting host cell tight junctions and the stratum corneum. In addition, topical administration is enhanced with microneedles and electroporation devices designed to disrupt the stratum corneum. Recently, microneedle arrays, with each microneedle 150 μm in length, were shown to increase the permeability of human skin (Henry *et al.,* 1998). These microneedle arrays have the advantage of reducing trauma, being practically pain-free, and allowing precise control over the depth of penetration by varying the needle length. Since the most effective antigen-presenting cells (dendritic cells) are present in the epidermis/dermal layers, any antigen introduced into the skin is rapidly sampled and presented with the host's immune system. It is for this reason that topical administration of vaccines is an extremely attractive immunization strategy.

Electroporation has been used successfully for years as a means of improving transfection *in vitro*. Recently, this technique has been used successfully to increase gene expression *in vivo* following delivery of plasmids into the skin, liver, and muscle (Titomirov *et al.,* 1991; Heller *et al.,* 1996; Widera *et al.,* 2000; Bachy *et al.,* 2001). The basis of electroporation is that the cells are pulsed with an electrical field and the nucleus and plasmid membranes are temporarily made permeable, allowing plasmids and possibly proteins to enter the cell.

Figure 1 illustrates the correlation between invasiveness and efficacy for current delivery systems. The goal is to develop delivery systems that reduce the degree of invasion, but increase the level of efficacy. Whether this will be possible remains to be determined.

B. Vaccine Formulation

The type of vaccine being administered will greatly influence the method of delivery as well as whether special formulations are required for insuring efficacy. For example, live vaccines will require minimal formulation and can be delivered orally, intranasally, or parenterally by various delivery devices. However, for conventional, killed, or subunit vaccines, special formulations are required. Thus, even though advances in molecular biology and immunology have contributed greatly to the identification and characterization of protective antigens from

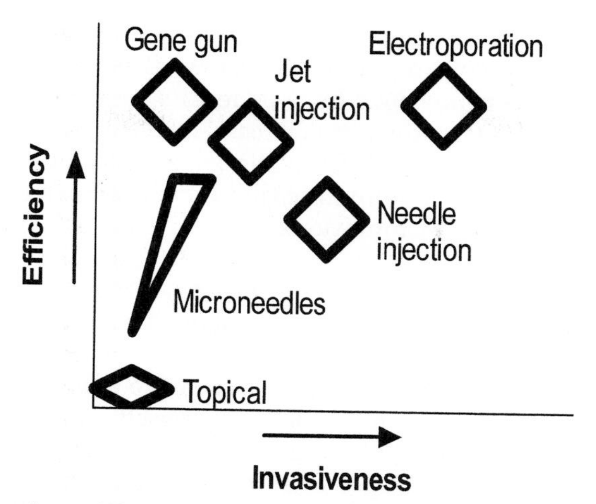

FIG. 1. Comparison of the efficiency and invasiveness of different delivery methods used with DNA vaccines. Gene expression efficiency increases vertically and the level of invasiveness increases to the right.

pathogenic microorganisms, they have not eliminated the need for formulation. Recombinant subunit vaccines are even less immunogenic than conventional vaccines and therefore generally require formulation with better (stronger) adjuvant or delivery systems to stimulate appropriate immune responses. The type and the magnitude of the immune response induced by a particular antigen are the result of multiple factors such as the biochemical properties of the antigen, the type of adjuvant, and the method of delivery. It is for this reason that a single adjuvant or delivery system may not be universally available. Thus, a wide range of adjuvants and delivery systems have been designed and tested experimentally. However, only few adjuvants are licensed for human and veterinary use because most adjuvants induce adverse reactions and are associated with local inflammatory responses and strong tissue damage. This is an important feature when vaccines are intended to protect healthy individuals. Therefore, it is critical to balance the safety and efficacy of various adjuvants with their ability to stimulate the immune response.

Although the exact mechanism by which adjuvants exert their effect is largely unknown, they can be classified according to their mode of action (Cox and Coulter, 1997; Singh and O'Hagan, 1999). In general, adjuvants exert their activity by one or more of the following mechanisms: (1) immunomodulation, (2) targeting antigen to antigen-presenting cells (APCs), and, (3) enhancing antigen presentation.

Unfortunately, many adjuvants do not possess all the features required for optimal induction of immunity. Therefore, it is not surprising that the most effective formulations have generally resulted from the combination of two or more adjuvants (Baca-Estrada *et al.,* 1997; Thoelen *et al.,* 1998; Weeratna *et al.,* 2000). The following sections briefly describe the various types of adjuvants that have been used either experimentally or in licensed formulations.

1. Mineral Salts

Aluminum hydroxide and aluminum phosphate are particulate adjuvants widely used in human and veterinary vaccines; they have a relatively good safety record and are inexpensive and easy to manufacture. However, recent concern has been expressed regarding the role of these adjuvants in inducing fibrosarcomas in cats (Burton and Mason, 1997). Since their efficacy depends on the molecular characteristics of the antigen, which facilitate the adsorption to the precipitated salt, it is not surprising that aluminum salts are not effective with every antigen (Claesson *et al.,* 1988). The mechanism of action of aluminum salts appears to be related to their ability to form a depot where soluble antigen is concentrated and presented in a particulate form to antigen-presenting cells. In addition, aluminum salts promote the stimulation of strong Th2 responses, an important feature when stimulation of antibodies is desirable, for example, in the neutralization of toxins, but are less effective in vaccines where induction of cell-mediated immunity is needed, for example, in combatting viruses. In addition, aluminum salts promote the stimulation of IgE antibodies, potentially inducing hypersensitivity reactions (Mancino and Ovary, 1980; Mark *et al.,* 1997). Therefore, despite the success of mineral salts as vaccine adjuvants and their extensive use in licensed vaccines used in humans and animals, there is a need to develop new adjuvants that promote a wide range of immune responses with minimal side effects.

2. Synthetic Adjuvants

Recently, a number of chemicals have been shown to stimulate or activate cells of the immune system, thereby suggesting their potential as adjuvants in vaccine formulations. For example, avridine, a lipid amine, can induce the synthesis of interleukin 1 (IL-1) and interferon α (IFN-α) *in vitro. In vivo,* avridine has been shown to enhance both cellular and humoral immunity to vaccines, especially if incorporated in oil emulsions (Hughes *et al.,* 1991). Similarly, DDA, another lipid amine, can enhance antibody responses and these responses can be augmented if DDA is coadministered with trehalose dimycolate (Dzata *et al.,* 1991).

Since these compounds are generally less toxic than many bacterial products, they may be safer or it may be possible to combine them with lower concentrations of bacterial products to reduce the toxicity, but maximize the benefits.

3. Oil Emulsions

Emulsions were first used as vaccine adjuvants as early as 1916 (reviewed in Gupta *et al.,* 1993); later, more stable formulations were developed, including Freund's adjuvant (Freund *et al.,* 1937). Addition of mycobacteria to the emulsion resulted in a more potent adjuvant (complete Freund's adjuvant); however, both complete and incomplete Freund's adjuvants have been restricted to use in experimental animals due to their reactogenicity. Indeed, there is increased pressure to ban or restrict the use of complete Freund's adjuvant for laboratory animals since less toxic and equally effective alternative formulations are now commercially available. These adjuvants are based on water-in-oil (w/o) or oil-in-water (o/w) emulsions and contain immunomodulators; examples of these are TiterMax, Ribi, and Lipovant. MF59A, a w/o emulsion, has been developed by Chiron Vaccines for human applications. Clinical trials have proven that MF59 is effective in a large number of human vaccines (Ott *et al.,* 1995) and it is already incorporated in a commercial vaccine.

The addition of surfactants to o/w emulsions provides stability and, depending on the surfactant used, different formulations can be obtained (Aucouturier *et al.,* 2001). Mineral oil and emulsions have also been used widely as adjuvants for veterinary vaccines; however, there is a concern regarding the cost of injection site reactions (van Donkersgoed *et al.,* 1999). Second, since mineral oils are not metabolized, there is a concern regarding the residues induced by this type of adjuvant. Since plant-derived oils are metabolized, these are being investigated as replacements for mineral oils. A further refinement on plant oils as adjuvants is the production of vaccines in plants and association of recombinant proteins with plant oil bodies (van Rooijen and Moloney, 1995). Since these oil bodies can be easily purified, it should be possible to economically produce vaccines having both the adjuvant and the vaccine in one simple step.

4. Plant Products

In addition to providing oil components for vaccine adjuvants, some plant components also have immunomodulatory properties through their ability to stimulate cytokine release (Yamaguchi, 1992) and macrophage activation. The best-characterized plant derivative is probably saponin. Isolation and use as an adjuvant of saponin (Quil A) from *Quillaha saponaria* has demonstrated its adjuvant inactivity against both T-dependent and T-independent antigens (Kensil *et al.,* 1991). Although crude Quil A still has toxic activity, different fractions have been shown to have reduced toxicity, but still retain adjuvant activity (White *et al.,* 1991). One such fraction, QS-21, is clearly superior to Quil A and is used in various vaccine formulations. The effects can be further enhanced by combining QS-21 with other adjuvants (Wu *et al.,* 1992).

5. Lipid Vehicles and Carriers

Vehicles such as those composed of particulate structures (immune-stimulating complexes, liposomes, virosomes, etc.) provide large surface areas on which antigens can be retained in a two-dimensional matrix, from which they can readily be transferred to antigen-presenting cells. These types of formulations are generally quite safe and induce excellent immune responses. Furthermore, since these components can be made from nonanimal products, the concern of introducing prions or other agents into individuals by vaccination is reduced with these lipid-based carriers/adjuvants.

Immune-stimulating complexes (ISCOMs) are vaccine delivery systems containing a lipid structure that entraps the antigen, and an immunomodulator (Quil-A) (Morein *et al.,* 1984; Cox *et al.,* 1998). Thus, they act as delivery vehicles and also enhance the level of immunity to the antigen incorporated into the ISCOM proteins. Progress using ISCOMs as delivery systems has primarily focused on using glycoproteins from enveloped viruses as the immunogen (Morein *et al.,* 1987). These vaccines have been delivered by parenteral, mucosal, or oral routes (Jones *et al.,* 1988; Mowat *et al.,* 1991). However, to our knowledge, the only licensed ISCOM vaccines are delivered parenterally. Generally, ISCOM-based vaccines induce both good antibody and cellular responses (Hoglund *et al.,* 1989; Takahashi *et al.,* 1990).

During the formation of ISCOMs, proteins that are present in the mixture generally get embedded by their hydrophobic transmembrane anchor membranes into the ISCOM structure. Since most virus envelope proteins have exposed hydrophobic transmembrane domains, this provides an excellent mechanism for incorporation into ISCOMs. A major limitation of this technology is that soluble proteins do not usually possess exposed hydrophobic regions and therefore do not associate well with the ISCOM structure. This is a major impediment to the incorporation of recombinant subunit antigens designed to be secreted into the media. To overcome this difficulty, attempts have been made to expose other hydrophobic regions within the protein by partial denaturation of the protein at low pH (Menza *et al.*, 1991). Unfortunately, this technique has not always worked because it alters the conformationally dependent domains of the protein. Second, the rate of incorporation of these proteins into ISCOMs has been low. Recently, this has been overcome by covalently attaching palmitic acid to facilitate incorporation of soluble proteins into ISCOMs (Reed, 1992). Thus, it now should be possible to alter proteins genetically in such a way as to produce large quantities of soluble proteins, purify them, and then covalently attach palmitic acid to them for incorporation into ISCOMs. An alternative approach would be to incorporate short hydrophobic tails, by genetic engineering methods, that would be of sufficient length to allow association with the ISCOMs, but would not interfere with secretion of the protein into the medium during its synthesis.

Liposomes are composed of phospholipid bilayers that enclose aqueous compartments. Depending on composition and method of preparation, liposomes can be unilamellar or multilamellar, and therefore can influence the method of presentation and the immune response. Indeed, one can incorporate targeting peptides into the liposome to further enhance antigen delivery. Vaccine antigens can be incorporated within the aqueous or lipid bilayers or they can also be associated with the surface of the vesicles. Due to the complexity of the antigen/liposome interactions, the incorporation efficiency will depend on the type of liposome, the structural characteristics of the antigen, and the method of preparation (Lasic, 1998). Liposomes offer the advantages of low toxicity and biodegradability compared to more conventional adjuvants. Improvements in the manufacture of liposomes have resulted in formulations with increased stability and have led to the development of novel parenteral liposome products. Liposomes have been shown to be effective vaccine delivery systems for subcutaneous, mucosal, and transcutaneous delivery of vaccines (Baca-Estrada *et al.*, 1997, 2000a,b). In addition, the adjuvant effect of liposomes can be further enhanced by incorporation of immunomodulatory molecules together with vaccine antigens (Kersten and Crommelin, 1995; Babai *et al.*, 1999; Childers *et al.*, 2000). This approach should not only protect the molecules, but also deliver the two components simultaneously to the same site. In a number of models, it has been shown that liposome vaccine formulations of this nature are more efficient at inducing immunity than simple mixtures of antigen and cytokine alone (Abraham and Shah, 1992; Baca-Estrada *et al.*, 1997).

Liposomes were first used for topical application in the early 1980s by Mezei and coworkers (Mezei and Gulasekharam, 1980). These studies indicated increased bioavailability of liposome-encapsulated drug within the skin and at the same time lower drug levels in the blood, indicating an advantageous localization of the drug at the target site, similar to a depot. The design of liposomes for vaccine application also underwent modifications and improvements to further enhance vaccine delivery. For example, protein cochleate formulations (Mannino and Gould-Fogerite, 1995), virosomes (Ambrosch *et al.*, 1997; Poltl-Frank *et al.*, 1999; Gluck *et al.*, 1999), transfersomes (Paul *et al.*, 1998), and biphasic delivery systems (Baca-Estrada *et al.*, 2000b) have been developed and tested.

All of the lipid-based delivery systems described here have been tested for the induction of mucosal immune responses in many animal models and humans (Gluck *et al.*, 1999; Michalek *et al.*, 1999) and virosome-based vaccines are available commercially. In addition to being useful in delivering proteins/glycoproteins, lipid-based delivery systems are also excellent vehicles for delivery of genes (plasmids) and oligonucleotides into cells. For efficient DNA delivery, liposomes generally incorporate cationic lipids to improve compaction of the DNA and neutralize the negative charge on DNA (Felgner *et al.*, 1987; Legendre and Szoka, 1992; Zelphati and Szoka, 1996).

6. Bacterial Products

The immunostimulatory activity of many bacterial products has led to the development of many experimental adjuvants and more recently to the incorporation of these compounds into human and veterinary vaccines (Krieg, 1999; Thoelen *et al.*, 2001). Although each specific compound may function by a slightly different mechanism, these adjuvants generally function by stimulating the secretion of cytokines by direct or indirect activation of antigen-presenting cells (APCs).

Monophosphoryl lipid A (MPL) is a substructure of bacterial lipopolysaccharide (LPS). Similar to LPS, MPL stimulates APC to secrete IL-1β, IFN-γ, and TNF-α (Kiener *et al.*, 1988; Carozzi *et al.*, 1989), but has been shown to be less toxic than LPS (Ulrich and Myers, 1995). This reduction in toxicity is attributed to its ability to stimulate IL-10 secretion when compared to the effect of LPS (Salkowski *et al.*, 1997). More recently, the synthesis of less toxic analogues of lipid A with adjuvant activity has been reported (Johnson *et al.*, 1999) and clinical trials demonstrate that vaccine formulations of this nature are well tolerated and induce good humoral and cell-mediated immune responses (Thoelen *et al.*, 2001).

The immunostimulatory properties of bacterial DNA were first discovered in studies investigating tumor regression (Tokunaga *et al.*, 1984). The functional sequences of bacterial DNA were identified as

CpG motifs flanked by two 5′ purines and two 3′ pyrimidines. These CpG motifs are found at a much higher frequency in bacterial DNA than mammalian DNA (Krieg, 1999). In addition, these motifs are not methylated in bacterial DNA, whereas they are methylated in mammalian DNA. The difference in methylation is critical for CpG activity since methylation of CpG abolishes their immunostimulatory activity.

CpG motifs have a wide variety of activities, such as triggering B cell proliferation (Krieg *et al.,* 1995), activating macrophages (Stacey *et al.,* 1996), and activating NK cells (Yamamoto *et al.,* 2000). With the activation of these cells a variety of cytokines such as IL-6, IL-12, and IFN-γ are produced (Klinman *et al.,* 1996). In addition, bacterial DNA is a potent stimulator of dendritic cells, promoting Th1 development (Jakob *et al.,* 1998). Although the exact mechanism by which CpG activates cells is unknown, CpG oligodeoxynucleotides (ODNs) must enter cells to elicit its effects. Immobilization of CpG to allow interaction with the cell surfaces, but prevent internalization prevents immune stimulation (Manzel and MacFarlane, 1999). Recently, a Toll-like receptor was found which recognizes bacterial DNA, suggesting that binding to this receptor is essential for CpG immunostimulatory activity (Hemmi *et al.,* 2000).

CpG motifs in the plasmid are required for enhancing the efficacy of DNA vaccines (Sato *et al.,* 1996) and they have been used as adjuvants with subunit (Davis *et al.,* 1998) as well as killed viral vaccines (Moldoveanu *et al.,* 1998). In addition, when CpG ODNs are used as adjuvants they promote Th1-type immune responses. This is extremely valuable since conventional adjuvants such as alum promote Th2-type responses. Indeed, combining CpG ODNs with alum not only results in a synergistic enhancement of immune responses, but shifts the response from a pure Th2-like response induced by alum to a balanced Th1/Th2 response. CpG ODNs have also been shown to act synergistically with other adjuvants (Weeratna *et al.,* 2000).

Muramyl dipeptide (MDP) is a synthetic compound with strong adjuvant activity. It was identified as the minimal structure of the peptidoglycan from mycobacteria that exerted immunostimulatory activity (Ellouz *et al.,* 1974). However, numerous studies have demonstrated that vaccine formulations containing MDP derivatives induced strong adverse reactions (Allison and Byers, 1991; Keitel *et al.,* 1993; Hoffman *et al.,* 1994). MDP and its derivatives are also capable of enhancing mucosal immune responses (Fukushima *et al.,* 1996; Michalek *et al.,* 1992). Due to their side effects, adjuvants containing MDP derivatives may not be suitable for mass immunization programs; however, they may be useful in therapeutic vaccines, for example, cancer vaccines.

Cholera toxin (CT) produced by *Vibrio cholerae* and the *E. coli* labile toxin are essentially identical molecules with potent mucosal adjuvant activity. Similar to many other bacteria-derived immunomodulators, these molecules are also very toxic and a great deal of research has been focused on dissociating their toxic effect from their adjuvant activity. In order to exert its adjuvant activity, CT has to be administered together with the vaccine antigen by the same route (Lycke and Holmgren, 1986). Another disadvantage of these toxins is that they are very immunogenic in their own right, thereby preventing their continued use when multiple administrations are needed.

7. Cytokines

The ability of cytokines to influence not only the magnitude, but also the type of immune response induced when coadministered with a vaccine antigen has been demonstrated in several animal models (reviewed in Heath, 1995). This is especially the case with those cytokines capable of exerting their immunoregulatory activity by direct action upon antigen-presenting cells and/or T cells. It now appears that protection against infectious agents is determined, in part, by the pattern of cytokines secreted by different subsets of T lymphocytes (Mosmann and Sad, 1996). For example, protection against intracellular pathogens is considered to be mediated by cellular immunity provided by Th1 cells, whereas diseases caused by helminths or extracellular bacteria and their products are preferentially regulated by the cytokines secreted by the Th2 subset of lymphocytes. In this context, cytokines capable of influencing the magnitude and phenotype of the immune response offer the potential to selectively tailor vaccines capable of inducing protection against different diseases. More recently, DNA vectors expressing cytokines have been tested for their ability to amplify the immune response to vaccine antigens (Kim *et al.,* 1997; Barouch *et al.,* 1998). However, the potential side effects or persistent cytokine expression need to be addressed before this technology can be employed in humans.

8. Microspheres

Biodegradable microspheres prepared with homo- and copolymers of lactic and glycolic acid (PLA and PLGA) have been tested extensively as vaccine delivery systems in laboratory animal models (Kersten and Gander, 1996). These polymers have been used in resorbable suture materials and are approved for parenteral administration. Due to their biodegradability, they provide sustained release of entrapped antigen, and therefore they have great potential for developing single-dose vaccines (Eldridge *et al.,* 1991). However, many issues remain to be addressed, including establishing manufacturing processes that will preserve antigen structure during encapsulation and developing methods to protect antigen stability during storage and after administration. Nevertheless, encapsulation of biologically active molecules such as cytokines and immunomodulators has been achieved (Egilmez *et al.*, 2000; Puri and Sinko, 2000). Microspheres for induction of mucosal immunity have been especially attractive. Particulate antigen $<10\ \mu m$ in size delivered orally is taken up by specialized cells in the Peyer's patches of the small intestine. These cells, called M cells, are capable of transporting particulate antigen from the lumen across the epithelium into the lymphoid tissue of the Peyer's patches, where humoral and cell-mediated immune responses are initiated. Similarly, intranasal immunization with vaccine antigens incorporated into microparticles results in protection against various respiratory pathogens (Cahill *et al.*, 1995; Eyles *et al.*, 2000). Thus, such delivery systems provide alternatives to parenteral administration and simultaneously induce protection at the portal of entry of many pathogens.

VIII. Epilogue

The major challenges of vaccinology are to not only identify the protective component of various pathogens, but also understand the pathogenesis of the agents involved and devise appropriate prophylactic strategies. This is possible only by elucidating both the specific components of the pathogen that are critical for inducing protective immunity and the components of the immune system that are critical for protection against specific diseases. This can only be achieved by a multidisciplinary approach combining knowledge of molecular biology; immunology, including the antigens involved in inducing protection; chemistry, as it relates to antigen purification; formulation; and, finally, delivery technologies and animal model systems. For example, the recognition that the formulation and mode of delivery of the vaccine and whether the immune response can become polarized to a Th1 or a Th2 response are critical clearly demonstrates the need for a better understanding of the pathogenesis and the impact of vaccine formulation on these responses. Any deviation in immune responses could lead to either a reduced response or a detrimental response with concomitant immunopathology. By characterizing the constellation of cytokines and the kinetics of the immune response produced during an effective protective response, it should be possible to identify vaccine formulation and delivery systems that induce similar responses in immunized individuals. Based on this knowledge, it should be possible to redirect the immune response in such a way as to modify the course of infection to most pathogens. As a result, we are confident that the next decade will witness the continued rapid evolution of the field of vaccinology with better and safer vaccines being developed against human diseases.

References

Abraham, E., and Shah, S. (1992). Intranasal immunization with liposomes containing IL-2 enhances bacterial polysaccharide antigen-specific pulmonary secretory antibody response. *J. Immunol.* **149,** 3719–3726.

Allison, A. C., and Byers, N. E. (1991). Immunological adjuvants: Desirable properties and side-effects. *Mol. Immunol.* **28,** 279–284.

Amara, R. R., Villinger, F., Altman, J. D., Lydy, S. L., O'Neil, S. P., Staprans, S. I., Montefiori, D. C., Xu, U., Herndon, J. G., Wyatt, L. S., Angelito-Candido, M., Kozyr, N. L., Earl, P. L., Smith, J. M., Ma, H. K., Grimm, B. D., Hulsey, M. L., Miller, J., McClure, H. M., McNicholl, J. M., Moss, B., and Robinson, H. L. (2001). Control of a mucosal challenge and prevention of AIDS by a multiprotein DNA/MVA vaccine. *Science* **292,** 69–75.

Ambrosch, F., Wiedermann, G., Jonas, S., Althaus, B., Finkel, B., Gluck, R., and Herzog, C. (1997). Immunogenicity and protectivity of a new liposomal hepatitis A vaccine. *Vaccine* **15,** 1209–1213.

Ames, B. N., Gold, L. S., and Willett, W. C. (1995). The causes and prevention of cancer. *Proc. Natl. Acad. Sci. USA* **92,** 5258–5265.

Anderson, J., Mertens, P. P. C., and Herniman, K. A. J. (1993). A competitive ELISA for the detection of anti-tubule antibodies using a monoclonal antibody against bluetongue virus non-structural protein NS1. *J. Virol. Meth.* **43,** 167–176.

Appel, M. (1978). Reversion to virulence of attenuated distemper virus *in vivo* and *in vitro*. *J. Gen. Virol.* **41,** 385–393.

Arakawa, T., Chung, D., and Langridge, W. (1998). Efficacy of a food plant-based oral cholera toxin B subunit vaccine. *Nature Biotechnol.* **16,** 292–297.

Arntzen, C. J. (1995). Oral immunization with a recombinant bacterial antigen produced in plants. *Science* **268,** 714–716.

Aucouturier, J., Dupuis, L., and Ganne, V. (2001). Adjuvants designed for veterinary and human vaccines. *Vaccine* **19,** 2666–2672.

Audibert, F. M., and Lise, L. D. (1993). Adjuvants: Current status, clinical perspectives and future prospects. *TIPS* **14,** 174–178.

Babai, I., Samira, S., Barenholz, Y., Zakay-Rones, Z., and Kedar, E. (1999). A novel influenza subunit vaccine composed of liposome-encapsulated haemagglutinin/neuraminidase and IL-2 or GM-CSF. I. Vaccine characterization and efficacy studies in mice. *Vaccine* **17,** 1223–1238.

Babiuk, L. A., van Drunen Littel-van den Hurk, S., and Tikoo, S. K. (1996). Immunology of bovine herpesvirus infections. *Vet. Microbiol.* **53,** 31–42.

Babiuk, L. A., Babiuk, S. L., Loehr, B., and van Drunen Little-van den Hurk, S. (2000). Nucleic acid vaccines: Research tools or commercial reality. *Vet. Immunol. Immunopathol.* **76,** 1–23.

Baca-Estrada, M. E., Foldvari, M., Snider, M., van Drunen Littel-van den Hurk, S., and Babiuk, L. A. (1997). Effect of IL-4 and IL-12 liposomal formulations on the induction of immune response to bovine herpesvirus type-1 glycoprotein D. *Vaccine* **15,** 1753–1760.

Baca-Estrada, M., Foldvari, M., Ewen, C., Badea, I., and Babiuk, L. (2000a). Effects of IL-12 on immune responses induced by transcutaneous immunization with antigens formulated in a novel lipid-based biphasic delivery system. *Vaccine* **18,** 1847–1854.

Baca-Estrada, M., Foldvari, M., Snider, M., Harding, K., Kournikakis, B., and Babiuk, L. (2000b). Intranasal immunization with liposome formulated *Yersinia pestis* vaccine enhances mucosal immune responses. *Vaccine* **18,** 2203–2211.

Bachy, M., Boudet, F., Bureau, M., Girerd-Chambaz, Y., Wils, P., Scherman, D., and Meric, C. (2001). Electric pulses increase the immunogenicity of an influenza DNA vaccine injected intramuscularly in the mouse. *Vaccine* **19,** 1688–1693.

Bansal, G. P., Hatfield, J. A., and Dunn, F. E., *et al.* (1993). Candidate recombinant vaccine for human B19 parvovirus. *J. Infect. Dis.* **167,** 1034–1044.

Barouch, D. H., Santra, S., Steenbeke, T. D., Zheng, X. X., Perry, H. C., Davies, M.-E., Freed, D. C., Craiu, A., Strom, T. B., Shiver, J. W., and Letvin, N. L. (1998). Augmentation and suppression of immune responses to an HIV-1 DNA vaccine by plasmid cytokine/Ig administration. *J. Immunol.* **161,** 1875–1882.

Barry, A. L. (1999). Antimicrobial resistance among clinical isolates of *Streptococcus pneumoniae* in North America. *Am. J. Med.* **107,** 28S–33S.

Bergmann, I. E., De Mello, P. A., Neizert, E., Beck, E., and Gomes, I. (1993). Diagnosis of persistent aphtovirus infections and its differentiation from vaccination response in cattle by use of enzyme-linked immuno-electrotransfer blot analysis with bio-engineered non-structural viral antigens. *Am. J. Vet. Res.* **54,** 825–831.

Berman, P. W., Gregory, T., Dowbenko, D., and Lasky, L. (1998). Protection of viral glycoproteins in genetically engineered mammalian cell lines for use as vaccines against immune deficiency retrovirus. *Appl. Virol. Res.* **1,** 17–24.

Birch-Marchin, I., Rowan, A., Pick, J., Mumford, J., and Binns, M. (1997). Expression of the non-structural protein NS1 of equine influenza A virus: Detection of anti-NS1 antibody in post-infection equine sera. *J. Virol. Meth.* **65,** 255–263.

Bittle, J. L., Houghten, R. A., Alexander, H., Shinnick, T. M., Sutcliffe, J. G., Lerner, R. A., Rowlands, D. J., and Brown, F. (1982). Protection against foot-and-mouth disease by immunization with a chemically synthesized peptide predicted from the viral nucleotide sequence. *Nature* **298,** 30–33.

Blaser, M. J. (1999). Hypothesis: The changing relationships of *Helicobacter pylori* and humans: Implications for health and disease. *J. Infect. Dis.* **179,** 1523–1530.

Bonnet, M. C., Tartaglia, J., Verdier, F., Kourisky, P., Lindberg, A., Klein, M., and Moingeon, P. (2000). Recombinant viruses as a tool for therapeutic vaccination against human cancers. *Immunol. Lett.* **74,** 11–25.

Bradley, M. B., Croft, M., and Swain, S. L. (1993). T cell memory: New perspectives. *Immunol. Today* **14,** 197–199.

Brochier, B., Boulanger, D., Costy, F., and Pastoret, P. P. (1994). Towards rabies elimination in Belgium by fox vaccination using a vaccinia rabies glycoprotein recombinant virus. *Vaccine* **12,** 1368–1371.

Brown, F. (1993). Review of accidents caused by incomplete inactivation of viruses. *Dev. Biol. Stand.* **81,** 103–107.

Burton, G., and Mason, K. V. (1997). Do post vaccinal sarcomas occur in Australian cats. *Aust. Vet. J.* **75,** 100–106.

Cahill, E. S., O'Hagan, D. T., Illum, L., Barnard, A., Mills, K. H., and Redhead, K. (1995). Immune responses and protection against *Bordetella pertussis* infection after intranasal immunization of mice with filamentous hemagglutinin in solution or incorporated in biodegradable micro-particles. *Vaccine* **13,** 455–462.

Caldwell, S. R., Johnson, A. F., Yule, T. D., Grimes, J. L., Ficken, M., and Christensen, V. L. (1999). Increased egg production in juvenile turkey hens after active immunization with vasoactive intestinal peptide. *Poult. Sci.* **78**(6), 899–901.

Canter, J., Mackay, K., and Good, L. S., *et al.* (1990). An outbreak of hepatitis B associated with jet injections in a weight reduction clinic. *Arch. Intern. Med.* **150,** 1923–1927.

Carozzi, S., Nasini, M. G., Schelotto C., *et al.* (1989). Effect of monophosphoryl lipid A (MPLA) on peritoneal leukocyte function. *Adv. Pent. Dial.* **5,** 143–150.

Celli, J., Deng, W., and Finlay, B. B. (2000). Enteropathogenic *E. coli* (EPEC): Attachment of epithelial cells. *Cell. Microbiol.* **2,** 10–19.

Centers for Disease Control and Prevention. (1999). CDC Update: Outbreak of Nipah virus: Malaysia and Singapore. *MMWR Morbid. Mortal. Weekly Rep.* **48,** 335–337.

Chen, D., Erickson, C. A., Endres, R. L., Periwal., S. B., Chu, Q., Shu, C., Maa, Y., and Payne, L. G. (2001). Adjuvantation of epidermal powder immunization. *Vaccine* **19,** 2908–2917.

Chen, D. K., McGeer, A., de Avavedo, J. C., and Low, D. E. (1999). Decreased susceptibility of *Streptococcus pneumoniae* to fluoroquinolones in Canada. *N. Engl. J. Med.* **341,** 233–239.

Childers, N. K., Miller, K. L., Tong, G., Llarena, J. C., Greenway, T., Ulrich, J. T., and Michalek, S. M. (2000). Adjuvant activity of monophosphoryl lipid A for nasal and oral immunization with soluble or liposome-associated antigen. *Infect. Immun.* **68,** 5509–5516.

Claesson, B. A., Trollfors, B., Lagergard, T., Taranger, J., Bryla, D., Otterman, G., Cramton, T., Yang, Y., Reimer, C. B., and Robbins, J. B., *et al.* (1988). Clinical and immunologic responses to the capsular polysaccharide of *Haemophilus influenzae* type b alone or conjugated to tetanus toxoid in 18- to 23-month-old children. *J. Pediatr.* **112,** 695–702.

Conry, R. M., Khazaeli, M. B., Saleh, M. N., Allen, K. O., Barlow, D. L., Moore, S. E., Craig, D., Arani, R. B., Schlom, J., and LoBuglio, A. F. (1999). Phase I trial of a recombinant vaccinia virus encoding carcino-embryonic antigen in metastatic adenocarcinoma: Comparison of intradermal vs. subcutaneous administration. *Clin. Cancer Res.* **5,** 2330–2337.

Coulon, P., Rollin, P. E., and Flamand, A. (1983). Molecular basis of rabies virus virulence. II. Identification of a site on the CVS glycoprotein associated with virulence. *J. Gen. Virol.* **64,** 693–696.

Cox, J. C., and Coulter, A. R. (1997). Adjuvants—A classification and review of their modes of action. *Vaccine* **15,** 248–256.

Cox, J. C., Sjolander, A., and Barr, I. G. (1998). ISCOMs and other saponin based adjuvants. *Adv. Drug. Deliv. Rev.* **32,** 247–259.

Davis, H. L., Weeratna, R., Waldschmidt, T. J., Schorr, J., and Krieg, A. M. (1998). CpG DNA is a potent enhancer of specific immunity in mice immunized with recombinant hepatitis B surface antigen. *J. Immunol.* **160,** 870–876.

DeGrado, M., Abe, A., Gauthier, A., Steele-Mortimer, O., and Finlay, B. B. (1999). Identification of the intimin binding domain of Tir of enteropathogenesis *E. coli* (EPEC). *Cell. Microbiol.* **1,** 7–17.

Delaet, I., and Boeye, A. (1993). Monoclonal antibodies that disrupt poliovirus only at fever temperatures. *J. Virol.* **67,** 5299–5302.

Deneer, H. G., and Potter, A. A. (1989a). Iron-repressible outer membrane proteins of *Pasteurella haemolytica*. *J. Gen. Microbiol.* **135,** 435–443.

Deneer, H. G., and Potter, A. A. (1989b). Effect of iron restriction on the outer membrane proteins of *Actinobacillus* (*Haemophilus*) *pleuropneumoniae*. *Infect. Immun.* **57,** 798–804.

Diekmann, O., Heesterbeek, J. A. P., and Metz, J. A. J. (1990). On the definition and computation of the basis reproduction ratio R0 in models for infectious diseases in heterogeneous populations. *J. Math. Biol.* **28,** 365–382.

Ding, M., Wen, D., Schlesinger, M. J., Wertz, G. W., and Ball, A. L. (1987). Expression and glycosylation of the respiratory syncytial virus G protein in *Saccharomyces cerevisiae*. *Virology* **159,** 450–453.

Donnelly, J. J., Ulmer, J. B., and Liu, M. A. (1993). Immunization with polynucleotides: A novel approach to vaccination. *Immunology* **2,** 20–26.

Donnenberg, M. S., and Kaper, J. B. (1991). Construction of an eae deletion mutant of enteropathogenic *Escherichia coli* by using a positive-selection suicide vector. *Infect. Immun.* **59,** 4310–4317.

Duque, H., Marshall, R. L., Israel, B. A., and Letchworth, G. J. (1989). Effects of formalin inactivation on bovine herpesvirus-1 glycoproteins and antibody response elicited by formalin-inactivated vaccines in rabbits. *Vaccine* **7,** 513–520.

Dzata, G. K., Wycroff, J. H., and Confer, A. W. (1991). Immunopotentiation of cattle vaccinated with a soluble *Brucella abortus* antigen with low LPS content: An analysis of cellular and humoral immune responses. *Vet. Microbiol.* **29,** 15–26.

Edelman, R. (1980). Vaccine adjuvants. *Rev. Infect. Dis.* **2,** 370–380.

Egilmez, N. K., Jong, Y. S., Sabel, M. S., Jacob, J. S., Mathiowitz, E., and Bankert, R. B. (2000). *In situ* tumor vaccination with interleukin-12-encapsulated biodegradable microspheres: Induction of tumor regression and potent antitumor immunity. *Cancer Res.* **60,** 3832–3827.

Eldridge, J. H., Staas, J. K., Meulbroek, J. A., McGhee, J. R., Tice, T. R., and Gilley, R. M. (1991). Biodegradable microspheres as a vaccine delivery system. *Mol. Immunol.* **28,** 287–294.

Ellouz, F., Adam, A., Ciorbaru, R., and Lederer, E. (1974). Minimal structural requirements for adjuvant activity of bacterial peptidoglycan derivatives. *Biochem. Biophys. Res. Commun.* **59,** 1317–1325.

Eyles, J. E., Williamson, E. D., Spiers, I. D., and Alpar, H. O. (2000). Protection studies following bronchopulmonary and intramuscular immunization with *Yersinia pestis* F1 and V subunit vaccines co-encapsulated in biodegradable microspheres: A comparison of efficacy. *Vaccine* **18,** 3266–3271.

Fan, H., Lin, Q., Morrissey, G. R., and Khavari, P. A. (1999). Immunization via hair follicles by topical application of naked DNA to normal skin. *Nature Biotechnol.* **17,** 870–874.

Farrell, H. E., McLean, C. S., Harley, C., Efstathiou, S., Inglis, S., and Minson, A. C. (1994). Vaccine potential of a herpes simplex virus type 1 mutant with an essential glycoprotein deleted. *J. Virol.* **68,** 927–932.

Felgner, P. L., Gadek, T. R., Holm, M., Roman, R., Chan, H. W., Wenz, M., Northrop, J. P., Ringold, G. M., and Danielsen, M. (1987). Lipofection: A highly efficient, lipid-mediated DNA-transfection procedure. *Proc. Natl. Acad. Sci. USA* **84,** 7413–7417.

Ferguson, M., Wood, D. J., and Minor, P. D. (1993). Antigenic structure of poliovirus in inactivated vaccines. *J. Gen. Virol.* **74,** 685–690.

Forage, R. G., Brown, R. W., Oliver, K. J, Atrache, B. T., Devine, P. L., Hudson, G. C., Goss, N. H., Bertram, K. C., Tolstochev, P., Robertson, D. M., de Kretser, B. T., Doughton, B., Burger, H. G., and Findlay, J. K. (1987). Immunization against an inhibin subunit produced by recombinant DNA techniques results in increased ovulation rate in sheep. *J. Endocrinol.* **114,** R1–R4.

Forrester, A., Farrell, H., Wilkinson, G., Kaye, J., Davis-Poynter, N., and Minson, T. (1992). Construction and properties of a mutant of herpes simplex virus type 1 with glycoprotein H coding sequences deleted. *J. Virol.* **66,** 341–348.

Freund, J., Casal, J., and Hismer, E. P. (1937). Sensitization and antibody formation after injection of tubercle bacilli and paraffin oil. *Proc. Soc. Exp. Biol. Med.* **37,** 509–515.

Fukushima, A., Yoo, Y. C., Yoshimatsu, K., Matsuzawa, K., Tamura, M., Tono-oka, S., Taniguchi, K., Urasawa, S., Arikawa, J., and Azuma, I. (1996). Effect of MDP-Lys(L18) as a mucosal immunoadjuvant on protection of mucosal infections by Sendai virus and rotavirus. *Vaccine* **14,** 485–491.

Fynan, E. F., Webster, R. G., Fuller, D. H., Haynes, J. R., Santoro, J. S., and Robinson, H. L. (1993). DNA vaccines: Protective immunizations by parenteral, mucosal, and gene gun inoculations. *Proc. Natl. Acad. Sci. USA* **90,** 11478–11482.

Gagic, M., St Hill, C. A., and Sharma, J. M. (1999). *In ovo* vaccination of specific pathogen-free chickens with vaccines containing multiple agents. *Avian Dis.* **43,** 293–301.

Gerdts, V., Babiuk, L. A., van Drunen Little van den Hurk, S., and Griebel, P. (2000). Fetal immunization by a DNA vaccine delivered orally into the amniotic fluid. *Nature Medicine* **6,** 929–932.

Giavedoni, L. D., Jones, L., Gardner, M. B., Gibson, H. L., Ng, C. T., Barr, P. J., and Yilma, T. (1992). Vaccinia virus recombinants expressing chimeric proteins of human immunodeficiency virus and gamma interferon are attenuated for nude mice. *Proc. Nat. Acad. Sci. USA* **89,** 3409–3413.

Giavedoni, L., Ahmad, S., Jones, L., and Yilma, T. (1997). Expression of gamma interferon by simian immunodeficiency virus increases attenuation and reduces post-challenge virus load in vaccinated rhesus macaques. *J. Virol.* **71,** 866–872.

Glenn, G. M., Scharton-Kersten, T., Vassell, R., Matyas, G. R., and Alving, C. R. (1999). Transcutaneous immunization with bacterial ADP-ribosylating extotoxins as antigens and adjuvants. *Infect. Immun.* **67,** 1100–1110.

Glenn, G. M., Taylor, D. N., Li, X., Frankel, S., Montemarano, A., and Alving, C. R. (2000). Transcutaneous immunization: A human vaccine delivery strategy using a patch. *Nature Med.* **6,** 1403–1406.

Gluck, U., Gebbers, J. O., and Gluck, R. (1999). Phase 1 evaluation of intranasal virosomal influenza vaccine with and without *Escherichia coli* heat-labile toxin in adult volunteers. *J. Virol.* **73,** 7780–7786.

Gold, P., and Freeman, S. O. (1965). Specific carcino-embryonic antigens of the human digestive system. *J. Exp. Med.* **122,** 467–468.

Gordon, W. S. (1946). Advances in veterinary research. *Vet. Rec.* **58,** 516–525.

Graham, B. S., Henderson, G. S., Tang, Y. W., Lu., X., Neuzil, K. M., and Colley, D. G. (1993). Priming immunization determines T helper cytokine mRNA expression patterns in lungs of mice challenged with respiratory syncytial virus. *J. Immunol.* **151,** 2032–2040.

Greensfelder, L. (2000). Polio outbreak raises questions about vaccine. *Science* **290,** 1867–1869.

Greten, T. F., and Jaffeee, E. M. (1999). Cancer vaccine. *J. Clin. Oncol.* **17,** 1047–1060.

Grieder, F. B., Davis, N. L., Aronson, J. F., Charles, P. C., Sellon, D. C., Suzuki, K., and Johnston, R. E. (1995). Specific restrictions in the progression of Venezuelan equine encephalitis virus-induced disease resulting from single amino acid changes in the glycoproteins. *Virology* **206,** 994–1006.

Gupta, R. K., Relyveld, E. H., Lindblad, E. B., Bizzini, B., Ben-Efraim, S., and Gupta, C. K. (1993). Adjuvants—a balance between toxicity and adjuvanticity. *Vaccine* **11,** 293–306.

Harland, R. J., Potter, A. A., and van Drunen Little-van den Hurk, S., *et al.* (1992). The effect of subunit or modified live bovine herpesvirus-1 vaccines on the efficacy of a recombinant *Pasteurella haemolytica* vaccine for the prevention of respiratory disease in feedlot calves. *Can. Vet. J.* **33,** 734–741.

Heath, A. (1995). Cytokines as immunological adjuvants. *Pharmaceut. Biotechnol.* **6,** 645–658.

Heller, R., Jaroszeski, M., Atkin, A., Moradpour, D., Gilbert, R., Wands, J., and Nicolau, C. (1996). *In vivo* gene electro-injection and expression in rat liver. *FEBS Lett.* **389,** 225–228.

Hemmi, H., Takeuchi, O., Kawai, T., Kaisho, T., Sato, S., Sanjo, H., Matsumoto, M., Hoshino, K., Wagner, H., Takeda, K., and Akira, S. (2000). A Toll-like receptor recognizes bacterial DNA. *Nature* **408,** 740–745.

Henry, S., McAllister, D. V., Allen, M. R., and Prausnitz, M. R. (1998). Micro-fabricated microneedles: A novel approach to transdermal drug delivery. *J. Pharmaceut. Sci.* **87,** 922–930.

Hoffman, S. L., Edelman, R., Bryan, J. P., Schneider, I., Davis, J., Sedegah, M., Gordon, D., Church, P., Gross, M., and Silverman, C., *et al.* (1994). Safety, immunogenicity, and efficacy of a malaria sporozoite vaccine administered with monophosphoryl lipid A, cell wall skeleton of mycobacteria, and squalane as adjuvant. *Am. J. Trop. Med. Hyg.* **51,** 603–612.

Hoglund, S. K. D., Lorgren, K., Sundquist, B., Osterhaus, A., and Morein, B. (1989). ISCOMs and immunostimulation with viral antigens. *Subcell. Biochem.* **15,** 39–68.

Hughes, H. P. A., Campos, M., Godson, D. L., van Drunen Little-van den Hurk, S., McDougall, L., Rapin, N., Zamb, T., and Babiuk, L. A. (1991). Immunopotentiation of bovine herpesvirus subunit vaccination by interleukin-2. *Immunology* **74,** 461–466.

Hussell, T., Spender, L. C., Georgiou, A., O'Garra, A., and Openshaw, P. J. M. (1996). Th1 and Th2 cytokine induction in pulmonary T-cells during infection with respiratory syncytial virus. *J. Gen. Virol.* **77,** 2447–2455.

Green, S., Fortier, A., Dijkstra, J., Madsen, J., Swartz, G., Einck, L., Gubish, E., and Nacy, C. (1995). Liposomal vaccines. *Adv. Exp. Med. Biol.* **383,** 83–92.

Israel, B. A., Herber, R., Gao, Y., and Letchworth, G. J. (1992). Induction of a mucosal barrier to bovine herpesvirus-1 replication in cattle. *Virology* **188,** 256–264.

Jacobs, S. C., Stephenson, J. R., and Wilkinson, G. W. G. (1992). High-level expression of the tick-borne encephalitis virus NS1 protein by using an adenovirus-based vector: Protection elicited in a murine model. *J. Virol.* **66,** 2086–2095.

Jakob, T., Walker, P. S., Krieg, A. M., Udey, M. C., and Vogel, J. C. (1998). Activation of cutaneous dendritic cells by CpG-containing oligodeoxynucleotides: A role for dendritic cells in the augmentation of Th1 responses by immunostimulatory DNA. *J. Immunol.* **161,** 3042–3049.

Jiang, X., Wang, M., Graham, D. Y., and Estes, M. K. (1992). Expression, self-assembly, and antigenicity of the Norwalk virus capsid protein. *J. Virol.* **66,** 6527–6532.

Johnson, D. A., Keegan, D. S., Sowell, C. G., Livesay, M. T., Johnson, C. L., Taubner, L. M., Harris, A., Myers, K. R., Thompson, J. D., Gustafson, G. L., Rhodes, M. J., Ulrich, J. T., Ward, J. R., Yorgensen, Y. M., Cantrell, J. L., and Brookshire, V. G. (1999). 3-*O*-Desacyl monophosphoryl lipid A derivatives: Synthesis and immunostimulant activities. *J. Med. Chem.* **42,** 4640–4649.

Jones, P. D., Tha-Hla, R., Morein, B., Lovgren, K., and Ada, G. L. (1988). Cellular immune responses in the murine lung to local immunization with influenza A virus glycoproteins in micelles and immunostimulatory complexes (ISCOMs). *Scand. J. Immunol.* **27,** 645–652.

Jones, T. R., Obaldia, N., Gramzinski, R. A., Charoenvit, Y., Kolodny, N., Davis, H. L., Krieg, A. M., and Hoffman, S. L. (1999). Synthetic oligonucleotides containing CpG motifs enhance immunogenicity of a peptide malaria vaccine in Aotus monkeys. *Vaccine* **17,** 3065–3071.

Kaashoek, M. J., and van Oirschot, J. T. (1996). Early immunity by a live gE-negative bovine herpesvirus-1 marker vaccine. *Vet. Microbiol.* **53,** 191–196.

Keitel, W., Couch, R., Bond, N., Adair, S., Van Nest, G., and Dekker, C. (1993). Pilot evaluation of influenza virus vaccine (IVV) combined with adjuvant. *Vaccine* **11,** 909–913.

Kensil, L. R., Patel, U., Lennik, M., and Marciani, D. (1991). Separation and characterization of saponins with adjuvant activity from *Quillaja saponaria molina* cortex. *J. Immunol.* **146,** 431–443.

Kersten, G. F., and Crommelin, D. J. (1995). Liposomes and ISCOMS as vaccine formulations. *Biochim. Biophys. Acta* **1241,** 117–138.

Kersten, G. F., and Gander, B. (1996). Biodegradable microspheres as vehicles for antigens. *In* "Concepts in Vaccine Development" (S. H. Kaufmann, Ed.), p. 265–273. de Gruyter, Berlin.

Kiener, P. A., Marek, F., Rodgers, G., Lin, P. F., Wan, G., and Desiderio, J. (1988). Induction of tumor necrosis factor, IFN-gamma, and acute lethality in mice by toxic and non-toxic forms of lipid A. *J. Immunol.* **141,** 870–874.

Kim, J. J., Ayyavoo, V., Bagarazzi, M. L., Chattergoon, M. A., Dang, K., Wang, B., Boyer, J. D., and Weiner, D. B. (1997). *In vivo* engineering of a cellular immune response by coadministration of IL-12 expression vector with a DNA immunogen. *J. Immunol.* **158,** 816–826.

King, A. A., and Harkness, J. W. (1975). Viral contamination of fetal bovine serum. *Vet. Rec.* **97,** 16.

Kinman, T. G., Westenbrink, F., and Straver, P. J. (1989). Priming for local and systemic antibody memory response to bovine respiratory syncytial virus: Effect of amount of virus, virus replication, route of administration and maternal antibodies. *Vet. Immunol. Immunopathol.* **22,** 1456–160.

Kirnbauer, R., Booy, F., Cheng, N., Lowy, D. R., and Schiller, J. T. (1992). Papillomavirus L1 major capsid protein self-assembles into virus-like particles that are highly immunogenic. *Proc. Natl. Acad. Sci. USA* **89,** 12180–12184.

Kit, S., Sheppard, M., Ichimura, H., and Kit, M. (1987). Second generaton pseudorabies virus vaccine with deletions in thymidine kinase and glycoprotein genes. *Am. J. Vet. Res.* **48,** 780–793.

Kleid, D. G., Yansura, D., and Small, B., *et al.* (1981). Cloned viral protein vaccine for foot-and-mouth disease: Response in cattle and swine. *Science* **214,** 1125–1129.

Klinman, D. M., Yi, A. K., Beaucage, S. L., Conover, J., and Krieg, A. M. (1996). CpG motifs present in bacteria DNA rapidly induce lymphocytes to secrete interleukin 6, interleukin 12, and interferon gamma. *Proc. Natl. Acad. Sci. USA* **93,** 2879–2883.

Konishi, W., Pinxua, A., Paoletti, E., Shope, R. E., Burrage, T., and Mason, P. W. (1992). Mice immunized with a subviral particle containing the Japanese encephalitis virus prM/M and E proteins are protected from lethal JEV infection. *Virology* **188,** 714–720.

Koprowski, H., and Yusibov, V. (2001). The green revolution: Plants as heterologous expression vectors. *Vaccine* **19,** 2735–2741.

Kowalski, J., Gilbert, S. A., van Drunen Littel-van den Hurk, S., van den Hurk, J., Babiuk, L. A., and Zamb, T. (1993). Heat-shock promoter-driven synthesis of secreted bovine herpesvirus glycoproteins in transfected cells. *Vaccine* **11,** 1100–1108.

Krieg, A. M. (1999). Mechanisms and applications of immune stimulatory CpG oligodeoxynucleotides. *Biochim. Biophys. Acta* **1489,** 107–116.

Krieg, A. M., Yi, A. K., Matson, S., Waldschmidt, T. J., Bishop, G. A., Teasdale, R., Koretzky, G. A., and Klinman, D. M. (1995). CpG motifs in bacterial DNA trigger direct B-cell activation. *Nature* **374,** 546–549.

Krieg, A. M., Ae-Kyung, Y., Schorr, J., and Davis, H. L. (1998). The role of CpG dinucleotides in DNA vaccines. *Trends Microbiol.* **6,** 23–26.

Kuhn, R. J., Niesters, H. G. M., Hong, Z., and Strauss, J. H. (1991). Infectious RNA transcripts from Ross River virus cDNA clones and the construction and characterization of defined chimeras with Sindbis virus. *Virology* **182,** 430–441.

Kuhn, R. J., Griffin, D. E., Zhang, H., Niesters, H. G. M., and Strauss, J. H. (1992). Attenuation of Sindbis virus neurovirulence by using defined mutations in non-translated regions of the genome RNA. *J. Virol.* **66,** 7121–7127.

Kuklin, N., Daheshia, M., Karem, K., Manickan, E., and Rouse, B. T. (1997). Induction of mucosal immunity against herpes simplex virus by plasmid DNA immunization. *J. Virol.* **71,** 3138–3145.

Kunkel, M., Lorinczi, M., Rijnbrand, R, Lemon, S. M., and Watowich, S. J. (2001). Self-assembly of nucleocapsid-like particles from recombinant hepatitis C virus core protein. *J. Virol.* **75,** 2119–2129.

Kuo, C. C., Shor, A., Campbell, L. A., Fukushi, H., Patton, D. L., and Grayston, J. T. (1993). Demonstration of *Chlamydia pneumoniae* in atherosclerotic lesions of coronary arteries. *J. Infect. Dis.* **167,** 841–849.

Kurilla, M. G., Swaminathan, S., Welsh, R. M., Kieff, E., and Brutkiewicz, R. R. (1993). Effects of virally expressed interleukin-10 on vaccinia virus infection in mice. *J. Virol.* **67,** 7623–7628.

Lasic, D. D. (1998). Novel applications of liposomes. *Trends Biotechnol.* **16,** 307–321.

Lasky, L. A., Dowbenko, D., Simonsen, C., and Berman, P. W. (1984). Production of a herpes simplex virus subunit vaccine by genetically engineered mammalian cell lines. *In* "Modern Approaches to Vaccines" (R. A. Lerner and R. M. Chanock, Eds.), pp. 189–194. Cold Spring Harbor Laboratory, Cold Spring Harbor Laboratory, New York.

Lauterslager, T. G. M., Florack, D. E. A., van der Wal, T. J., Molthoff, J. W., Langeveld, J. P. M., Bosch, D., Boersma, W. J. A., and Hilgers, L. A. (2001). Oral immunization of naive and primed animals with transgenic potato tubers expressing LT-B. *Vaccine* **19,** 2749–2755.

Legendre, J. Y., and Szoka, Jr., F. C. (1992). Delivery of plasmid DNA into mammalian cell lines using pH-sensitive liposomes: Comparison with cationic liposomes. *Pharmaceut. Res.* **9,** 1235–1342.

Lester, S., Clemett, T., and Burt, H. (1996). Vaccine site-associated sarcomas in cats: Clinical experience and a laboratory review (1982–1993). *J. Am. Anim. Hosp. Assoc.* **32,** 91–95.

Lewis, P. J., and Babiuk, L. A. (1999). DNA vaccines: A review. *Adv. Virus Res.* **94,** 129–188.

Lewis, P. J., van Drunen Littel-van den Hurk, S., and Babiuk, L. A. (1999). Induction of immune responses to bovine herpesvirus type 1 gD in passively immune mice after immunization with a DNA-based vaccine. *J. Gen. Virol.* **80,** 2829–2837.

Liang, X., Chow, B., and Babiuk, L. A. (1997). Study of immunogenicity and virulence of bovine herpesvirus-1 mutants deficient in UL49 homolog, UL49.5 homolog and dUTPase genes in cattle. *Vaccine* **15,** 1057–1064.

Lycke, N., and Holmgren, J. (1986). Strong adjuvant properties of cholera toxin on gut mucosal immune responses to orally presented antigens. *Immunology* **59,** 301–308.

Macadam, A. J., Ferguson, G., Arnold, C., and Minor, P. D. (1991). An assembly defect as a result of an attenuating mutation in the capsid proteins of the poliovirus type 3 vaccine strain. *J. Virol.* **65,** 5225–5231.

Macadam, A. J., Ferguson, G., and Burlinson, J., *et al.* (1992). Correlation of RNA secondary structure and attenuation of Sabin vaccine strains of poliovirus in tissue culture. *Virology* **189,** 415–522.

Maeda, S. (1989). Expression of foreign genes in insect cells using baculovirus vectors. *Ann. Rev. Entomol.* **34,** 351–372.

Mancino, D., and Ovary, Z. (1980). Adjuvant effects of amorphous silica and of aluminium hydroxide on IgE and IgG1 antibody production in different inbred mouse strains. *Int. Arch. Allergy. Appl. Immunol.* **61,** 253–258.

Mannino, R. J., and Gould-Fogerite, S. (1995). Lipid matrix-based vaccines for mucosal and systemic immunization. *Pharmaceut. Biotechnol.* **6,** 363–387.

Manzel, L., and Macfarlane, D. E. (1999). Lack of immune stimulation by immobilized CpG-oligodeoxynucleotide. *Antisense Nucleic Acid Drug Dev.* **9,** 459–464.

Marciani, D. J., Kensil, C. R., Beltz, G. A., Hung, C. H., Cronier, J., and Aubert, A. (1991). Genetically engineered subunit vaccine against feline leukemia virus: Protective immune response in cats. *Vaccine* **9,** 89–96.

Mark, A., Bjorksten, B., and Granstrom, M. (1997). Immunoglobulin E and G antibodies two years after a booster dose of an aluminium-adsorbed or a fluid DT vaccine in relation to atopy. *Pediatr. Allergy Immunol.* **8,** 83–87.

Marshall, B. J. (1983). Unidentified curved bacilli on gastric epithelium in active chronic gastritis. *Lancet* **i,** 1273–1275.

Maruyama, M., Lam, K. P., and Rajewsky, K. (2000). Memory B cell resistance is independent of persisting immunizing antigen. *Nature* **407,** 636–642.

Mazanec, M. B., Kaetzel, C. S., Lamm, M. E., Fletcher, D., and Nedrud, J. G. (1992). Intracellular neutralization of virus by immunoglobulin A antibodies. *Proc. Natl. Acad. Sci. USA* **89,** 6901–6905.

McGee, J. R., Mestecky, J., Dertzbaugh, M. T., Eldridge, J. H., Hirasawa, M., and Kiyono, H. (1992). The mucosal immune system: From fundamental concepts to vaccine development. *Vaccine* **10,** 75–88.

McGuire, T. C., Adams, D. S., Johnson, G. C., Klevjer-Anderson, P., Barbee, D. D., and Gorham, J. R. (1986). Acute arthritis in caprine arthritis–encephalitis virus challenge exposure of vaccinated or persistently infected goats. *Am. J. Vet. Res.* **47,** 537–540.

Menza, M., Sober, J., Sundquist, B., Toots, I., and Morcin, B. (1991). Characterization of purified gp51 from bovine leukemia virus integrated into ISCOM. *Arch. Virol.* **120,** 219–231.

Mestecky, J., Abraham, R., and Ogra, P. L. (1994). Common mucosal immune system and strategies for the development of vaccines at the mucosal surfaces. *In* "Handbook of Mucosal Immunology" (P. L. Ogra, M. E. Lamm, J. R. McGee, J. Mestecky, W. Strober, and J. Bienenstock, Eds.), pp. 357–372. Academic Press, San Diego, California.

Mezei, M., and Gulasekharam, V. (1980). Liposomes—A selective drug delivery system for the topical route of administration. Lotion dosage form. *Life Sci.* **26,** 1473–1477.

Michalek, S. M., Childers, N. K., Katz, J., Dertzbaugh, M., Zhang, S., Russell, M. W., Macrina, F. L., Jackson, S., and Mestecky, J. (1992). Liposomes and conjugate vaccines for antigen delivery and induction of mucosal immune responses. *Adv. Exp. Med. Biol.* **327,** 191–198.

Michalek, S. M., O'Hagan, D., Gould-Fogerite, S., Rimmelzwaan, G. F., and Osterhaus, A. D. M. E. (1999). Antigen delivery systems: Nonliving microparticles, liposomes, cochleates and ISCOMS. *In* "Mucosal Immunology" (P. L. Ogra, J. Mestecky, M. E.

Lamm, W. Strober, J. Bienenstock, J. R. McGhee, Eds.), p. 759–768. Academic Press, San Diego, California.

Millar, S. E., Chamow, S. M., Baur, A. W., Oliver, C., Robey, R., and Dean, J. (1989). Vaccination with a zona pellucida peptide produces long-term contraception in female mice. *Science* **246,** 935–938.

Minor, P. D., John, A., Ferguson, M., and Icenogle, J. P. (1986). Antigenic and molecular evolution of the vaccine strain of type 3 poliovirus during the period of excretion of a primary vaccine. *J. Gen. Virol.* **67,** 693–706.

Moldoveanu, Z., Love-Homan, L., Huang, W. Q., and Krieg, A. M. (1998). CpG DNA, a novel immune enhancer for systemic and mucosal immunization with influenza virus. *Vaccine* **16,** 1216–1224.

Morein, B., Sundquist, B., Hoglund, S., Dalsgaard, K., and Osterhaus, A. (1984). ISCOM, a novel structure for antigenic presentation of membrane proteins from enveloped viruses. *Nature* **308,** 457–460.

Morein, B., Lovgren, K., Hoglund, S., and Sundquist, B. (1987). The ISCOM: An immunostimulating complex. *Immunol. Today* **8,** 333–338.

Mosmann, T. R., and Sad, S. (1996). The expanding universe of T-cell subsets: Th1, Th2 and more. *Immunol. Today* **17,** 138–146.

Mowat, A. M., Donachie, A. M., Reid, G., and Jarrett, D. (1991). Immune stimulating complexes containing Quil A and protein antigen prime class I-MHC restricted T lymphocytes *in vivo* and are immunogenic by oval route. *Immunology* **72,** 317–322.

Muller, G., Shapira, M., and Arnon, R. (1982). Anti-influenza response achieved by immunization with a synthetic conjugate. *Proc. Nat. Acad. Sci. USA* **79,** 569–573.

Munos, N. (1997). Disease-burden related to cancer induced by human viruses and *H. pylori*. *In* "World Health Organization (WHO) Vaccine Research and Development: Report of the Technical Review Group Meeting." WHO, Geneva.

Murphy, B. R., Alling, D. W., and Snyderm, M. H., *et al.* (1986). Effect of age and preexisting antibody on serum antibody response of infants and children to the F and G glycoproteins during respiratory syncytial virus infection. *J. Clin. Microbiol.* **24,** 894–898.

Nahmias, A. J., Griffith, D., Salsbury, C., and Yoshida, K. (1967). Thymic aplasia with lymphopenia, plasma cells, and normal immunoglobulins. Relation to measles virus infection. *JAMA* **201,** 729–734.

Netter, H. J., MacNaughton, T. B., Woo, W. P., Tindle, R., and Gowans, E. J. (2001). Antigencity and immunogenicity of novel chimeric hepatitis B surface antigen particles with exposed hepatitis C virus epitopes. *J. Virol.* **75,** 2130–2141.

Neurath, A. R., Strick, N., and Girard, M. (1989). Hepatitis B virus surface antigen (HbsAg) as a carrier for synthetic peptides having an attached hydrophobic tail. *Mol. Immunol.* **26,** 53–62.

Nguyen, L. H., Knipe, D. M., and Finberg, R. W. (1992). Replication-defective mutants of herpes simplex virus (HSV) induce cellular immunity and protect against lethal HSV infection. *J. Virol.* **66,** 7067–7072.

Offringa, R., van der Burg, S. H., Ossendorp, F., Toes, R. E. M., and Melief, C. J. M. (2000). Design and evaluation of antigen-specific vaccination strategies against cancer. *Curr. Opin. Immunol.* **2,** 576–582.

Old, L. J., and Chen, Y. T. (1998). New paths in human cancer serology. *J. Exp. Med.* **187,** 1163–1167.

Opdebeeck, J. P., Wong, J. Y. M., Jackson, L. A., and Dobson, C. (1988). Hereford cattle immunized and protected against *Boophilus microplus* with soluble and membrane associated antigens form the midgut of ticks. *Parasite Immunol.* **10,** 405–410.

Ott, G., Barchfeld, G. L., Chernoff, D., Radhakrishanan, R., van Hoogevest, P., and Van Nest, G. (1995). Design and evaluation of a safe and potent adjuvant for human vaccines. *In* "Vaccine Design" (M. F. Powell and M. J. Newman, Eds.). Plenum Press, New York, 74–85.

Parashar, U. D., Sunn, L. M., Ong, F., Mounts, A. W., Arif, M. T., Ksiazek, T. G., Kamaluddin, M. A., Mustafa, A. N., Kaur, H., Ding, L. M., Othman, G., Radzi, H. M., Kitsutani, P. T., Stockton, P. C., Arokiasamy, J., Gary, H. E., and Anderson, L. J. (2000). Case–control study of risk factors for human infection with a new zoonotic paramyxovirus, Nipah virus during a 1998–1999 outbreak of severe encephalitis in Malaysia. *J. Infect. Dis.* **181,** 1755–1759.

Paul, A., Cevc, G., and Bachhawat, B. K. (1998). Transdermal immunization with an integral membrane component, gap junction protein, by means of ultradeformable drug carriers, transfersomes. *Vaccine* **16,** 188–195.

Perera, P. Y., Manthey, C. L., Stutz, P. L., Hildebrandt, J., and Vogel, S. N. (1993). Induction of early gene expression in murine macrophages by synthetic lipid A analogs with differing endotoxic potentials. *Infect. Immun.* **61,** 2015–2023.

Pertmer, T. M., Eisenbraun, M. D., McCabe, D., Prayaga, S. K., Fuller, D. H., and Haynes, J. R. (1995). Gene gun-based nucleic acid immunization: Elicitation of humoral and cytotoxic T lymphocyte responses following epidermal delivery of nanogram quantities of DNA. *Vaccine* **13,** 1427–1430.

Pisani, P., Parkin, D. M., Munoz, N., and Ferlay, J. (1997). Cancer and infection: Estimates of the attributable fraction in 1995. *Cancer Epidemiol.* **6,** 387–400.

Poland, G. A., and Jacobson, R. M. (2001). Understanding those who do not understand: A brief review of the anti-vaccine movement. *Vaccine* **19,** 2440–2445.

Poltl-Frank, F., Zurbriggen, R., Helg, A., Stuart, F., Robinson, J., Gluck, R., and Pluschke, G. (1999). Use of reconstituted influenza virus virosomes as an immunopotentiating delivery system for a peptide-based vaccine. *Clin. Exp. Immunol.* **117,** 496–503.

Puri, N., and Sinko, P. J. (2000). Adjuvancy enhancement of muramyl dipeptide by modulating its release from a physicochemically modified matrix of ovalbumin microspheres. II. *In vivo* investigation. *J. Controlled Release* **69,** 69–80.

Pushko, P., Parker, M., Ludwig, G., Davis, N. L., Johnston, R. E., and Smith, J. F. (1997). Replicon-helper systems from attenuated Venezuelan equine encephalitis virus: Expression of heterologous genes *in vitro* and immunization against heterologous pathogens *in vivo*. *Virology* **239,** 389–401.

Reddy, P. S., Idamakanti, N., Chen, Y., Whale, T., Babiuk, L. A., Mehtali, M., and Tikoo, S. K. (1999). Replication-defective bovine adenovirus type-3 as an expression vector. *J. Virol.* **73,** 9137–9144.

Redmond, M. J., Ijaz, M. K., Parker, M. D., Sabara, M. I., Dent, D., Gibbons, E., and Babiuk, L. A. (1993). Assembly of recombinant rotavirus proteins into virus-like particles and assessment of vaccine potential. *Vaccine* **11,** 273–281.

Reed, G. (1992). Soluble proteins incorporated into ISCOMs after covalent attachment of fatty acid. *Vaccine* **9,** 597–602.

Robertson, I. S., Fraser, H. M., Innes, G. M., and Jones, A. S. (1982). Effect of immunocastration on sexual and characteristics in male cattle. *Vet. Res.* **111,** 529–531.

Rosenberg, S. A. (1997). Cancer vaccines based on the identification of genes encoding cancer regression antigens. *Immunol. Today* **18,** 175–182.

Rosenberg, S. A. (1999). A new era for cancer immunotherapy based on the genes that encode cancer antigens. *Immunity* **10,** 281–287.

Rosenberg, S. A., Zhai, Y., Yang, J. C., Schwartzentruber, D. J., Hwu, P., Marincola, F. M., Topalian, S. L., Restifo, N. P., Seipp, C. A., Einhorn, J. H., and Roberts, B., and

White, D. E. (1998). Immunizing patients with metastatic melanoma using recombinant adenoviruses encoding MART-1 or gp100 melanoma antigens. *J. Natl. Cancer Inst.* **90,** 1894–1900.

Roth, J. A. (1999). Mechanistic bases for adverse vaccine reactions and vaccine failures. *Adv. Vet. Med.* **41,** 681–700.

Rouse, B. T., Wardley, R. C., and Babiuk, L. A. (1976). The role of antibody dependent cytotoxicity in recovery from herpesvirus infection. *Cell. Immunol.* **22,** 182–186.

Rudbach, J. A., Cantrell, J. L., Ulrich, J. T., and Mitchell, M. S. (1990). Immunotherapy with bacterial endotoxins. *Adv. Exp. Med. Biol.* **256,** 665–676.

Sahin, U., Tureci, O., and Pfreundschuh, M. (1997). Serological identification of human tumor antigens. *Curr. Opin. Immunol.* **9,** 709–716.

Salkowski, C. A., Detore, G. R., and Vogel, S. N. (1997). Lipopolysaccharide and monophosphoryl lipid A differentially regulate interleukin-12, gamma interferon, and interleukin-10 mRNA production in murine macrophages. *Infect. Immun.* **65,** 3239–3247.

Sambhi, S. K., Kohonen-Corish, M. R. J., and Ramshaw, I. A. (1991). Local production of tumor necrosis factor encoded by recombinant vaccinia virus is effective in controlling viral replication *in vivo*. *Proc. Natl. Acad. Sci. USA* **88,** 4025–4029.

Sato, Y., Roman, M., Tighe, H., Lee, D., Corr, M., Nguyen, M. D., Silverman, G. J., Lotz, M., Carson, D. A., and Raz, E. (1996). Immunostimulatory DNA sequences necessary for effective intradermal gene immunization. *Science* **273,** 352–354.

Schultz-Cherry, S., Dybing, J. K., Davis, N. L., Williamson, C., Suarez, D. L., Johnston, R., and Perdue, M. L. (2000). Influenza virus (A/HK/156/97) hemagglutinin expressed by an alphavirus replicon system protects chickens against lethal infection with Hong Kong-origin H5N1 viruses. *Virology* **278,** 55–59.

Scott, A. M., and Welt, S. (1997). Antibody-based immunological therapies. *Curr. Opin. Immunol.* **9,** 717–722.

Sharma, J. M. (1999). Introduction to poultry vaccines and immunity. *Adv. Vet. Med.* **41,** 481–494.

Sharp, P. J. (1997). Immunological control of broodiness. *World's Poultry Sci. J.* **53,** 23–31.

Sherman, D. M., Acres, S. D., Sadowski, P. L., Springer, J. A., Bray, B., Raybould, T. J. G., and Muscoplat, C. C. (1983). Protection of calves against fetal colibacillosis by orally administered *E. coli* K99 specific monoclonal antibody. *Infect. Immun.* **42,** 653–658.

Shewan, P. E., and Wilkie, B. N. (1982). Cytotoxin of *Pasteurella haemolytica* acting on bovine leukocytes. *Infect. Immun.* **35,** 91–94.

Shi, Z., Curiel, D. T., and Tang, D. C. (1999). DNA-based non-invasive vaccination onto the skin. *Vaccine* **17,** 2136–2143.

Simons, J. S., and Mikhak, B. (1998). *Ex vivo* gene therapy using cytokine-transduced tumor vaccines: Molecular and clinical pharmacology. *Semin. Oncol.* **35,** 661–676.

Simons, J. S., Mikhak, B., Chang, J. F., DeMarzo, A. M., Carducci, M. A., Lim, M., Weber, C. E., Baccala, A. A., Goemann, M. A., Clift, S. M., Ando, D. G., Levitsky, H. I., Cohen, L. K., Sanda, M. G., Mulligan, R. C., Partin, A. W., Carter, H. B., Piantadosi, S., Marshall, F. F., and Nelson, W. G. (1999). Induction of immunity to prostate cancer antigens: Results of a clinical trial of vaccination with irradiated autologous prostate tumor cells engineered to secrete GM-CSF using *ex vivo* gene transfer. *Cancer Res.* **59,** 5160–5168.

Singh, M., and O'Hagan, D. (1999). Advances in vaccine adjuvants. *Nature Biotechnol.* **17,** 1075–1081.

Smedegaard Madsen, E., Madsen, K. G., Nielsen, J., Holh Jensen, M., Lei, J. C., and Have, P. (1997). Detection of antibodies against porcine parvovirus non-structural protein NS1 may distinguish between vaccinated and infected pigs. *Vet. Microbiol.* **54,** 1–16.

Soiffer, R., Lynch, T., Mihm, M., Jung, K., Rhuda, C., Schmollinger, J. C., Jodi, F. S., Liebster, L, Lam, P., Mentzer, S., Singer, S., Tanabe, K. K., Cosimi, A. B., Duda, R., Sober, A., Bhan, A., Daley, J., Neuberg, D., Parry, G., Rokovich, J., Richards, L., Drayer, J., Berns, A., Clift, S., Cohen, L. K., Mulligan, R. C., and Dranoff, G. (1998). Vaccination with irradiated autologous melanoma cells engineered to secrete human granulocyte-macrophage colony-stimulating factor generates potent anti-tumor immunity in patients with metastatic melanoma. *Proc. Natl. Acad. Sci. USA* **95,** 13141–13146.

Somogyi, P., Frazier, J., and Skinner, M. A. (1993). Fowlpox virus host range restriction: Gene expression, DNA replication, and morphogenesis in non-permissive mammalian cells. *Virology* **197,** 439–444.

Sprent, J., and Surh, C. D. (2001). Generation and maintenance of memory T cells. *Curr. Opin. Immunol.* **13,** 248–254.

Spriggs, D. R., Bronson, R. T., and Fields, B. N. (1983). Hemagglutinin variants of reovirus type 3 have altered central nervous system tropism. *Science* **220,** 505–507.

Stacey, K. J., Sweet, M. J., and Hume, D. A. (1996). Macrophages ingest and are activated by bacterial DNA. *J. Immunol.* **157,** 2116–2122.

Straub, O. C. (1990). Infectious bovine rhinotracheitis. *In* "Virus Infections of Ruminants" (A. Dinter and B. Morein, Eds.), pp. 71–108. Elsevier, Amsterdam.

Streatfield, S. J., Jilka, J. M., Hood, E. E., Turner, D. D., Bailey, M. R., Mayor, J. M., Woodard, S. L., Beifuss, K. K., Horn, M. E., Delaney, D. E., Tizard, I. R., and Howard, J. A. (2001). Plant-based vaccines: Unique advantages. *Vaccine* **19,** 2742–2748.

Subbarao, E. K., Kawaoka, Y., and Murphy, B. R. (1993). Rescue of an influenza A virus wild-type PB2 gene and a mutant derivative bearing a site-specific temperature-sensitive and attenuating mutation. *J. Virol.* **67,** 7223–7228.

Takahashi, H., Takeshita, T., Morcin, B., Putney, S., Germain, R. N., and Berzofsky, J. A. (1990). Induction of CD8+ cytotoxic T cells by immunization with purified HIV-1 envelope protein in ISCOMs. *Nature* **56,** 152–154.

Thapan, M. A., Parr, E. L., Bozzola, J. J., and Parr, M. B. (1991). Secretory immune responses in the mouse vagina after parenteral or intravaginal with an immune stimulating complex (ISCOM). *Vaccine* **9,** 129–133.

Thoelen, S., Van Damme, P., Mathei, C., Leroux-Roels, G., Desombere, I., Safary, A., Vandepapeliere, P., Slaoui, M., and Meheus, A. (1998). Safety and immunogenicity of a hepatitis B vaccine formulated with a novel adjuvant system. *Vaccine* **16,** 708–714.

Thoelen, S., De Clercq, N., and Tornieporth, N. (2001). A prophylactic hepatitis B vaccine with a novel adjuvant system. *Vaccine* **19,** 2400–2403.

Thornton, D. H. (1986). A survey of mycoplasma detection in veterinary vaccines. *Vaccine* **4,** 237–240.

Titomirov, A., Sukharev, S., and Kistanova, E. (1991). *In vivo* electroporation and stable transformation of skin cells of newborn mice by plasmid DNA. *Biochem. Biophys. Acta* **1088,** 131–134.

Tizard, I. (1999). Grease, anthraxgate, and kennel cough: A revisionist history of early veterinary vaccines. *Adv. Vet. Med.* **41,** 7–24.

Tokunaga, T., Yamamoto, H., Shimada, S., Abe, H., Fukuda, T., Fujisawa, Y., Furutani, Y., Yano, O., Kataoka, T., and Sudo, T., *et al.* (1984). Antitumor activity of deoxyribonucleic acid fraction from *Mycobacterium bovis* BCG. I. Isolation, physicochemical characterization, and antitumor activity. *J. Natl. Cancer. Inst.* **72,** 955–962.

Tripp, R. A., Moore, D., Jones, L., Sullender, W., Winter, J., and Anderson, L. J. (1999). Respiratory syncytial virus G and/or SH protein alters Th1 cytokines natural killer cells, and neutrophils responding to pulmonary infection in BALB/c mice. *J. Virol.* **73,** 7099–7107.

Tsang, K. Y., Zaremba, S., Nieroda, C. A., Zhu, M. Z., Hamilton, J. M., and Schlom, J. (1995). Generation of human cytotoxic T cells specific for human carcino-embryonic antigen epitopes from patients immunized with recombinant vaccinia-CEA vaccine. *J. Natl. Cancer Inst.* **87,** 982–990.

Ulmer, J. B., Deck, R. R., Dewitt, C. M., Donnelly, J. J., and Liu, M. A. (1996). Generation of MHC class I-restricted cytotoxic T lymphocytes by expression of a viral protein in muscle cells: Antigen presentation by non-muscle cells. *Immunology* **89,** 59–67.

Ulrich, J. T., and Myers, K. R. (1995). Monophosphoryl lipid A as an adjuvant. Past experiences and new directions. *Pharmaceut. Biotechnol.* **6,** 495–524.

Valenzuela, P., Medina, A., Rutter, W. J., Ammerer, G., and Rutter, B. D. (1982). Synthesis and assembly of hepatitis B surface antigen particles in yeast. *Nature* **298,** 347–350.

van den Eynde, B., and Brichard, V. G. (1995). New tumor antigens recognized by T cells. *Curr. Opin. Immunol.* **7,** 674–681.

van der Bruggen, P., Traversari, C., Chomez, P., Lurquin, C., Deplaen, E., van den Eynde, B., Knuth, A., and Boon, T. (1991). A gene encoding an antigen recognized by cytolytic T lymphocytes on a human melanoma. *Science* **254,** 1643–1647.

van Donkersgoed, J., Dubeski, P. L., Aalhus, J. L., vanderKop, M., Dixon, S., and Starr, W. N. (1999). The effect of vaccines and antimicrobials on the formation of injection site lesions in subprimals of experimentally injected beef calves. *Can. Vet. J.* **40,** 245–251.

van Drunen Littel-van den Hurk, S., Parker, M., Massie, B., van den Hurk., J., Harland, R., Babiuk, L. A., and Zamb, T. J. (1993). Protection of cattle from BHV-1 infection by immunization with recombinant glycoprotein gIV. *Vaccine* **11,** 25–35.

van Drunen Littel-van den Hurk, S., Braun, R. P., Lewis, P. J., Karvonen, B. C., Babiuk, L. A., and Griebel, P. J. (1999). Immunization of neonates with DNA encoding a bovine herpesvirus glycoprotein is effective in the presence of maternal antibodies. *Viral Immunol.* **12,** 67–77.

van Rooijen, G. J. H., and Moloney, M. M. (1995). Plant seed oil-bodies as carriers for foreign proteins. *Biotechnology* **13,** 72–77.

Vinner, L., Nielsen, H. V., Bryder, K., Corbet, S., Nielsen, C., and Fomsgaard, A. (1999). Gene gun DNA vaccination with Rev-independent synthetic HIV-1 gp160 envelope gene using mammalian codons. *Vaccine* **17,** 2166–2174.

Vogel, M., Cichutek, K., Norley, S., and Kurth, R. (1993). Self-limiting infection by int/nef-double mutants of simian immunodeficiency virus. *Virology* **193,** 115–123.

Weeratna, R. D., McCluskie, M. J., Xu, Y., and Davis, H. L. (2000). CpG DNA induces stronger immune responses with less toxicity than other adjuvants. *Vaccine* **18,** 1755–1762.

Wensvoort, G., and Terpstra, C. (1988). Bovine viral diarrhea virus infection in piglets born to sows vaccinated against swine fever with contaminating virus. *Res. Vet. Sci.* **45,** 143–148.

White, A. C., Cloutier, P., and Coughlin, R. (1991). A purified saponin acts as an adjuvant for a T-independent antigen. *Adv. Exp. Med. Biol.* **303,** 207–210.

Whitley, R. J., Kern, E. R., Chatterjee, S., Chou, J., and Roizman, B. (1993). Replication, establishment of latency, and induced reactivation of herpes simplex virus $\gamma 1$ 34.5 deletion mutants in rodent models. *J. Clin. Invest.* **91,** 2837–2843.

Whitton, L., Sheng, N., Oldstone, M., and McKee, T. A. (1993). A string of beads vaccine comprising linked mini-genes, confers protection from lethal virus challenge. *J. Virol.* **67,** 348–352.

Widera, G., Austin, M., Rabussay, D., Goldbeck, C., Barnett, S. W., Chen, M., Leung, L., Otten, G. R., Thudium, K., Selby, M. J., and Ulmer, J. B. (2000). Increased DNA vaccine delivery and immunogenicity by electroporation *in vivo*. *J. Immunol.* **164,** 4635–4640.

Willadsen, P., and Kemp, D. H. (1988). Vaccination with "concealed" antigens for tick control. *Parasitol. Today* **4,** 196–198.

Wittmann, G., Ohlinger, V., and Hohn, U. (1982). Multiplication of Aujesky's disease virus after experimental infection with high and low doses of virus. *Zbl. Vet. Med. B* **29,** 663–675.

Wolff, J. A., Malone, R. W., Williams, P., Ascadi, G., Jani, A., and Felgner, P. L. (1990). Direct gene transfer into mouse muscle *in vivo*. *Science* **247,** 1465–1468.

Wu, J. Y., Gardner, B. H., Murphy, C. I., Seals, J. R., Kensil, C. R., Recchia, J., Belz, G. A., Newman, G. W., and Newman, M. J. (1992). Saponin adjuvant enhancement of antigen-specific immune responses to an experimental HIV-1 vaccine. *J. Immunol.* **148,** 1519–1525.

Yamaguchi, H. (1992). Immunomodulation by medicinal plants. *Adv. Exp. Med. Biol.* **319,** 287–297.

Yamamoto, S., Yamamoto, T., Iho, S., and Tokunaga, T. (2000). Activation of NK cell (human and mouse) by immunostimulatory DNA sequence. *Springer Semin. Immunopathol.* **22,** 35–43.

Yankauckas, M., Morrow, J. E, Parker, S. E., Rhodes, G. H., Dwarki, V. J., and Gromkowski, S. H. (1993). Long-term anti-NP cellular and humoral immunity is induced by intramuscular injection of plasmid DNA containing NP gene. *DNA Cell Biol.* **12,** 771–776.

Yilma, T., Msu, D., Jones, L., Owens, S., Grubman, M., Mebus, C., Yamanaka, M., and Dale, B. (1988). Protection of cattle against Rinderpest with vaccinia virus recombinants expressing the HA or F gene. *Science* **242,** 1058–1061.

Yokomori, K., Baker, S. C., Stohlman, S. A., and Lai, M. M. C. (1992). Hemagglutinin-esterase-specific monoclonal antibodies alter the neurapathogenicity of mouse hepatitis virus. *J. Virol.* **66,** 2865–2874.

Zakhartchouk, A. N., Pyne, C., Mutwiri, G., Papp, Z., Baca-Estrada, M., Griebel, P., Babiuk, L. A., and Tikoo, S. K. (1999). Mucosal immunization of calves with recombinant bovine adenovirus 3: Induction of protective immunity to bovine herpesvirus-1. *J. Gen. Virol.* **80,** 1263–1269.

Zelphati, O., and Szoka, Jr., F. C., (1996). Intracellular distribution and mechanism of delivery of oligonucleotides mediated by cationic lipids. *Pharmaceut. Res.* **13,** 1367–1372.

Zhai, Y., Yang, J. C., Kawakami, Y., Spiess, P., Wadsworth, S. C., Cardoza, L. M., Couture, L. A., Smith, A. E., and Rosenberg, S. A. (1996). Antigen-specific tumor vaccines. Development and characterization of recombinant adenoviruses encoding Mart 1 or gp100 for cancer therapy. *J. Immunol.* **156,** 700–710.

zur Hausen, H. (1994). Molecular pathogenesis of cancer of the cervix and its causation by specific human papillomavirus types. *Curr. Top. Microbiol. Immunol.* **186,** 131–156.

ADVANCES IN VIRUS RESEARCH, VOL. 58

THE POTENTIAL OF PLANT VIRAL VECTORS AND TRANSGENIC PLANTS FOR SUBUNIT VACCINE PRODUCTION

Peter Awram,* Richard C. Gardner,* Richard L. Forster,† and A. Richard Bellamy*

*School of Biological Sciences
University of Auckland
Auckland, New Zealand

†Genesis Research & Development Corporation Limited
Auckland, New Zealand

I. Current Trends in Vaccine Development
II. An Overview: The Use of Plants for Recombinant Protein Expression
III. Plant Protein-Expression Systems
IV. Stable Transgenic Expression Systems
V. Plant Virus Expression Systems
A. Icosahedral Plant Viruses
B. Helical Plant Viruses
C. Advantages and Disadvantages of Plant Viruses for Protein Expression
VI. Enhancing Protein Yield
VII. Expression of Immunogenic Molecules
A. Expression of Small Epitopes
B. Immunogenic Carrier Molecules
C. Expression of Virus-like Particles in Plants for Use in Oral Vaccination
VIII. Immunogenicity and Vaccine Responses
A. Vaccines and the Immune System
B. Oral Delivery of Expressed Protein for Inducing Mucosal Immune Responses
C. Immune Responses to Plant-Derived Antigens
IX. Edible Vaccines and Human Health
A. The Need for Vaccines
B. Human Trials of Edible Plant Vaccines
X. Conclusion
References

I. Current Trends in Vaccine Development

Vaccination has led to a significant improvement in the health of the world's population. The use of vaccines has reduced the spread and infection of a number of major human diseases, including measles, mumps, rubella, and tetanus. Traditionally, these types of vaccines have

0065-3527/02 $35.00

been developed from attenuated or inactivated forms of the pathogen. These approaches have proven to be very effective for the control of disease and have resulted in the eradication of smallpox and the expected eradication of polio in the next decade by the World Health Organization (http://www.who.int/vaccines-polio/).

With the advent of recombinant technologies, subunit vaccines based on proteins expressed in bacteria and yeast have grown in popularity. Because subunit vaccines do not contain an infectious agent that can revert to a more virulent form or survive the inactivation process, they offer advantages over live vaccines because they are incapable of causing disease (Division of Microbiology and Infectious Diseases, 1998). However, subunit vaccines have tended to be expensive, requiring substantial investment in facilities for production and purification. Moreover, the purified product tends to be subject to heat destabilization, and refrigeration is required for storage and transport, which may present problems in developing countries, which often lack adequate health care facilities. Consequently, there is considerable interest in new methods of producing cheap and stable subunit vaccines.

Over the last decade, plants have become an increasingly popular choice for the production of recombinant protein. Interest in the use of plants as bioreactors for production of protein is a result of advances in plant molecular biology and the development of plant protein-expression systems. Production methods can be scaled up to industrial volumes because agricultural practices are easily adapted to the large-scale production of genetically modified plants. The reader is referred to recent reviews of protein expression in plants, which provide an excellent summary of the techniques involved (Beachy, 1997; Fischer *et al.*, 1999; Goddijn and Pen, 1995; Herbers and Sonnewald, 1999; Kusnadi *et al.*, 1997; Miele, 1997).

Interest in vaccine production in plants has also expanded rapidly. There are considerable advantages in expressing antigenic proteins in plants. Plants can be grown locally and cheaply using standard methods, thus reducing problems with distribution, transport, and storage. Subsequent purification of the protein may not be required if the plant material can be delivered orally. These advantages may significantly reduce cost and promote the distribution of vaccines, matters of particular importance in developing countries, where the high costs of subunit vaccines can often limit their use.

Here we present the case that plants can be used for the production of subunit vaccines and outline the systems that are available for their production. The limits of plant biology will be explored along with the limits of the differing expression systems that are available. It is not the intention of this review to extend too far into the interpretation of results that have been achieved priming the immune system or conferring protection, other than to point out some of the issues and options involved and the successes achieved. Since some readers may not be familiar with plant transgenic technologies, brief background information is included, together with references to a number of recent reviews. For those with experience in plant systems, but less knowledge of immune responses, a brief overview of relevant aspects of immunity is also provided.

II. An Overview: The Use of Plants for Recombinant Protein Expression

Several considerations must be addressed when expressing proteins in plants for the purpose of vaccination. The first is that the protein must retain the immunogenic characteristics of the original protein and be capable of inducing a protective response to the disease. This may often require the correct folding of the polypeptide chain and, in some cases, the correct glycosylation of the protein. The stability of the protein must also be considered because it may be subject to degradation during production. It is also necessary to determine how highly purified the protein must be and the costs and difficulties involved in purification. In addition, there is a need to consider how easily production might be scaled up. A detailed description of the processes involved in manufacture of vaccines has been provided by Ebbert *et al.* (1999).

Recent research has shown that plants are capable of satisfying all of these requirements (for reviews see Arntzen, 1997, 1998; Beachy *et al.*, 1996; Langridge, 2000; Ma, 2000; Walmsley and Arntzen, 2000). A large number of peptides and proteins have been expressed in plants and in several cases they have induced antibody responses in animals and protection against a subsequent challenge by the disease-causing organism. Proteins expressed in plants have proven to be very stable and targeting of proteins to specific organelles can facilitate purification. Plants are easy to propagate and grow and as a consequence scaling up production is simple with minimal changes to current agricultural practices.

The major advantage of plants is that they can be grown in nonsterile conditions without costly nutrients and controlled environmental conditions. As a result, plants compare favorably to other methods of recombinant protein production (Table I). Fermentation, in the case

TABLE I

A COMPARISON OF PRODUCTION METHODS FOR MAMMALIAN-DERIVED PROTEINS AND THE FACTORS THAT MAY IMPACT UPON THE DIFFICULTY AND/OR COST OF PRODUCTION

	Growth Media Costs	Equipment Costs	Levels of Protein Produced	Purification	Sensitivity to Growth Conditions[a]	Posttranslational Processing[b]	Codon Usage[b]
Bacteria	Moderate	Moderate	High	Easy to moderate	Moderate	Poor	Poor
Insect cells	Moderate to high	Moderate	Moderate to high	Easy to moderate	Moderate	Moderate	Good
Animal cells	High	High	Low	Moderate to difficult	High	Excellent	Excellent
Yeast cells	Moderate	Moderate	High	Easy to moderate	Moderate	Moderate to excellent	Poor
Plants	Low	Low	Low	Moderate to difficult	Low	Moderate	Moderate

[a] Shear forces, pH, temperature, oxygen.

[b] When expressing a mammalian protein.

of animal, insect, yeast, and bacterial cells, requires large initial outlays for fermenters and expensive growth media. Cultures can become contaminated with pathogens that can later remain with the isolated protein, resulting in the requirement for careful screening of the end product. It is also generally accepted that bacterial systems often do not provide the appropriate posttranslational modifications required for proper functioning of some eukaryotic proteins. In addition, many proteins form insoluble inclusion bodies when overexpressed in bacterial cells and require solubilization and refolding, which may not be successful in many cases.

Notwithstanding the advantages provided by plants, protein expression in plants is not without its own difficulties. Expression of foreign polypeptides in plants may inhibit plant growth, as has been demonstrated in several instances (Poierier *et al.,* 1992; van der Meer *et al.,* 1994). In some cases this growth inhibition has been alleviated by targeting the recombinant protein to a particular subcellular compartment of the plant cell (Poierier *et al.,* 1992). In addition, purification from plants may be more difficult than in other systems due to the presence of the sturdy cellulose cell wall. Further, improper glycosylation may result in a change in antigenicity. Although plants appear to glycosylate proteins in the same positions as animal cells, the carbohydrate chains are usually shorter than and differ in composition from those synthesized in animal cells (Lerouge *et al.,* 1998, 2000; Sturm *et al.,* 1987). It has also been shown that plants can glycosylate foreign proteins with highly immunogenic plant-specific carbohydrates (Cabanes-Macheteau *et al.,* 1999). A lack of correct glycosylation may prevent proper stimulation of the immune system or an inappropriate carbohydrate chain may stimulate an inappropriate immune response. Despite these difficulties, several glycosylated proteins have been expressed in plants without loss of immunogenicity.

A further major disadvantage of plant systems is that, historically, only low expression levels have been achieved for recombinant protein. Much work has therefore been done to increase expression levels. Considerable success has been achieved by altering codon usage and removing sequences that may adversely affect RNA expression (Mason *et al.,* 1998; Strizhov *et al.,* 1996). Targeting protein expression to specific plant compartments such as the endoplasmic reticulum (Richter *et al.,* 2000) and chloroplast (De Cosa *et al.,* 2001) or to specific storage organs such as potato tubers (Mason *et al.,* 1998; Stark *et al.,* 1992) has also increased yields and this approach may also facilitate purification. The use of plant viruses for expression has also been a strategy adopted successfully to increase protein expression levels.

Despite these potential difficulties, it has been estimated that the production of a protein in plants may be up to 50-fold less expensive than production in *Escherichia coli* (Evangelista *et al.,* 1998).

III. Plant Protein-Expression Systems

Plant protein expression can conveniently be divided into transgenic and viral expression systems. Creation of a transgenic plant by stable integration of a foreign gene into the plant genome allows the production of identical or near-identical offspring which can stably pass on the transgene. The alternative approach is to use plant virus expression vectors based on cauliflower mosaic virus (CaMV) (De Zoeten *et al.,* 1989), tobacco mosaic virus (TMV) (Hamamoto *et al.,* 1993), potato virus X (PVX) (Cruz *et al.,* 1996), tobacco etch virus (TEV) (Dolja *et al.,* 1992), or cowpea mosaic virus (CPMV) (Porta *et al.,* 1994). These viral vectors spread systemically through infected plants expressing their transgene, but the transgene is not inherited by the progeny.

Since the first protein was expressed in plants for use as a vaccine (Curtiss and Cardineau, 1990), the field of protein expression in plants has expanded rapidly (for reviews see Fischer *et al.,* 1999; Beachy, 1997; De Wilde *et al.,* 2000; Koprowski and Yusibov, 2001; Kusnadi *et al.,* 1997; Ma, 2000). The plant of choice has typically been tobacco because it is easy to grow and techniques for manipulating this plant are well developed. However, although tobacco is well suited for genetic manipulation, purification of the resulting proteins is required because tobacco contains numerous toxic alkaloids and phenolic compounds. As a result it is likely that other plants are better hosts for vaccine production. *Arabidopsis*, because of its ease of genetic manipulation, has been used as a model system. Bananas, potatoes, tomatoes, lettuce, rice, wheat, soybeans, and corn are also currently under study for use in producing recombinant proteins.

The choice of plant largely depends on whether the transgenic or the virus infection approach is to be applied. The viral approach obviously requires a suitable host for the virus, thus limiting the choice of candidate plants. However, since some plant viral vectors (e.g., TMV) can infect a large number of plant species, it is possible to examine expression of a construct in a number of different plant hosts without the need for additional genetic manipulations.

Since many plants used in horticulture and agriculture are edible, a number of researchers have reasoned that an antigenic protein expressed in plants may be suitable for use as an oral vaccine, without the requirement for purification. A number of oral immunogenicity studies have been performed using protein expressed in transgenic potatoes by feeding the raw tissue to experimental animals or humans and analyzing the immune response (Haq *et al.,* 1995; Mason *et al.,* 1998; Tacket *et al.,* 1998, 2000). Potato is readily manipulated genetically and foreign proteins can be expressed at high levels in the tubers, allowing for easier purification. Potatoes are very easy to cultivate and yields per hectare are very high. Potato tubers can be eaten raw without significant toxic effects, although they are not always well tolerated in the uncooked state, even by rodents.

If an oral vaccine requiring little purification is desired, it is likely that any plant material must be eaten raw since cooking is likely to lead to the denaturation of the protein. For example, when boiled, the immune response of a hepatitis B vaccine candidate produced in transgenic potatoes dropped 24-fold compared to the unboiled control (Kong *et al.,* 2001). In contrast, heat-labile enterotoxin B subunit (LT-B), expressed in corn, survived cooking and retained the correct conformation (Daniell *et al.,* 2001b). Seed crops, such as corn, show promise for vaccine production since protein can be expressed at high levels in the embryo, which is readily harvested and separated from the rest of the seed, allowing concentration of the foreign protein.

Tomato, another expression candidate, is edible in its raw state and has been used for the expression of the rabies virus glycoprotein (McGarvey *et al.,* 1995). However, some plants have proved more successful for protein expression than others and this use of tomato was unsuccessful because less than 0.001% of the total protein was recombinant. However, the protein was recognized by antibodies raised against the original protein, even though it was not identical to the native form, most probably as a result of improper glycosylation.

Finally, it must be realized that expression levels achieved in plants vary considerably depending on the protein that is expressed and the plant used for expression. Currently, there are no reliable methods for predicting the expression levels that will be achieved in a plant, even though there are a number of methods that have been used successfully to increase the levels of expression.

IV. Stable Transgenic Expression Systems

Protein was first expressed in plants for the purposes of vaccine production in 1990 (Curtiss and Cardineau, 1990). These investigators generated transgenic tobacco plants expressing the cell surface adhesion protein SpaA from *Streptococcus mutans,* one of the main bacterial agents causing tooth decay. Made up of over 1500 amino acids, this protein is larger than most proteins that have been expressed in plants. No more than 0.02% of the soluble protein was SpaA, which proved to be too low for either purification or immunization.

The ability to make stable transgenic plants by integration of DNA into the chromosome is the result of the discovery that *Agrobacterium tumefaciens* can insert foreign genetic material into the plant genome and that the resulting cells can be regenerated into whole plants (Koncz *et al.,* 1984). Although foreign DNA can be introduced into plants by a number of other methods, such as biolistic bombardment, the use of *Agrobacterium* remains the preferred approach because it results in more consistent integration of foreign DNA.

Agrobacterium is a plant pathogen that infects a large range of hosts including plants, fungi, and animal cells. Some plants such as tobacco and petunia, have proven to be easy to work with in the laboratory, whereas others, such as rice and corn, are not as amenable to genetic manipulation. *Agrobacterium*-mediated transformation allows the integration of a gene cassette (T-DNA) into the plant genome. The process requires incubation of plant tissue with a bacterial culture containing the recombinant T-DNA. Plant tissue culture is then used to select for the recombinant cells, usually using an antibiotic resistance marker. The resulting transformed tissue is then regenerated into plants. The time required for this process varies from 6 weeks to over 1 year depending on the plant species involved. Once a transgenic plant is obtained, the integration is stable and can be passed on by propagation using normal methods (i.e., seed and cuttings). For reviews of *Agrobacterium*-mediated transformation see Gallois and Marinho (1995), Gasser and Fraley (1989), Hansen and Chilton (1999), Hooykaas (1989), Newell (2000), Paszkowski *et al.* (1992), and Potrykus (1990).

Expression levels in transgenic plants can vary considerably because the integration of the transgene is essentially random and subsequent protein expression can be affected by copy number, positional effects, and gene silencing. As a result, expression of the foreign DNA will vary depending on where it is integrated into the plant genome. It is therefore necessary to screen a number of different regenerated plants to determine expression levels. Using viruses to express the transgene avoids this problem because the genes are not inserted in the plant genome. In transgenic plants, gene silencing can occur for a number of reasons, including multiple integrations, deleterious effects from adjacent sequences, synthesis of anti-sense mRNA, which results in the formation of double-stranded RNA molecules that cannot be translated, or the presence of homologous sequences or DNA methylation effects (De Wilde *et al.,* 2000; Matzke and Matzke, 1998; Vaucheret *et al.,* 1998; Vaucheret and Fagard, 2001). There are a number of ways of decreasing these effects and thereby increasing protein expression. Plants should be screened for single-copy insertion, repetitive homologous sequences should be avoided, and the foreign DNA can be flanked with scaffold attachment regions (SAR) to prevent the effects of adjacent DNA (De Wilde *et al.,* 2000). Although generally not a common problem, it is also necessary to select for plants that express the protein stably over several generations.

Exceptionally high levels of protein have been achieved by integrating genes into the chloroplast genome (for a review see Heifetz, 2000), where the recombinant protein may represent as much as 46% of the total soluble protein (De Cosa *et al.,* 2001). Each plant cell contains thousands of chloroplasts, resulting in thousands of copies of the transgene that can be expressed (De Cosa *et al.,* 2001; McBride *et al.,* 1995). This method of transformation avoids positional and silencing effects because the foreign DNA is integrated by homologous recombination within a spacer region in the chloroplast genome. Polycistrons can be expressed in the chloroplast, allowing the expression of multiple proteins. Chloroplasts have fewer proteolytic pathways than the remainder of the cell (Adam, 2000) and it has been predicted that, as a result, foreign proteins may not be subject to as much degradation when expressed in chloroplasts (Bock, 2001). Using integration into the chloroplast, the cholera toxin B subunit has been expressed in tobacco resulting in a yield of 4.1% of total plant-soluble protein (Daniell *et al.,* 2001a). In addition to high-level expression, chloroplast integration should reduce the potential for transgene spread via pollen because chloroplasts are maternally inherited in most plants.

Although transgenic technologies have been used for the successful production of a number of experimental plant-based vaccines, the time required to produce transgenic plants represents a major disadvantage, particularly if many different expression cassettes or promoter elements require evaluation. Therefore, as noted below, a number of researchers have sought the more rapid expression systems offered by the use of plant viral expression vectors.

V. Plant Virus Expression Systems

Recent advances in our understanding of the molecular biology of plant viruses has enabled the development of expression systems that use viral vectors to produce large amounts of protein in a short period of time. The general approach adopted is to insert the foreign gene into the viral genome under the control of a strong subgenomic promoter. The resulting recombinant viruses can then be introduced into the appropriate host plant by mechanical inoculation, where the virus can spread systemically throughout the plant. If the construct allows viable virus particles to be produced, the recombinant virus has the potential to spread from plant to plant. This property makes propagation simpler, but may create containment issues. With few exceptions, most plant viruses do not integrate into the host cell genome. As a result, foreign genes are not heritable through seeds, unlike transgenic plants. They also do not suffer from positional effects that affect gene expression levels found in stably transformed transgenic plants.

Expression of recombinant protein using viral vectors is a fast and relatively simple method for examining proteins for desired characteristics since the infected plants can produce high amounts of protein within 1–4 weeks of inoculation. This method of expression requires less investment in time compared to the use of transgenic plants before an expressed protein can be isolated for analysis. Using a viral vector, expression can be induced at an advanced developmental stage and thereby the potential for deleterious effects on the developing plants by the expression of the gene product are avoided.

However, there are a number of difficulties involved with the use of plant virus vectors. Inoculation of plants can be inefficient, in part because of the need to penetrate the rigid cellulose cell wall. However, once plants have been infected, high viral titers can be obtained from plant sap and subsequent plants can be inoculated with the packaged virions at high efficiencies. Most viral vectors are RNA viruses and inoculation of the plant is accomplished by mechanical inoculation of the infectious RNA transcripts onto leaves. These transcripts are susceptible to rapid degradation from nucleases and are preferably used quickly after production. In addition, successful inoculation is temperature-dependent, and for many viruses, conditions above 30°C may reduce infection to virtually zero. Systemic spread of the virus from the point of inoculation is also required and it has been found that some constructs interfere with the subsequent movement of the virus from the initial infectious center (Cruz *et al.,* 1996; Usha *et al.,* 1993).

The major problem associated with the use of plant viral vectors is genetic instability caused by the presence of the inserted foreign gene. RNA viruses mutate at a high frequency because of errors introduced during RNA synthesis. Because the foreign insert is not required for the viral life cycle, and because multiple consecutive replication steps are involved in the process of systemic infection of a plant, the insert is commonly deleted or modified during infection. Despite a range of ingenious strategies used to counter this instability, it remains a major obstacle for viral-based vectors. Another disadvantage of the TMV and PVX vectors, at least in our experience, is that they are often somewhat unstable in *E. coli* plasmids, so that care is needed during propagation to maintain the full-length infectious molecule that contains the intact foreign genes.

Despite these difficulties, plant viral vectors exhibit significant advantages over transgenic plants for the production of foreign proteins. The most important of these is the much shorter time required for expression because grown plants can be infected with virus with minimal manipulation. As a result, transient expression using plant viruses allows for rapid analysis of many different constructs, even if a stable transgenic plant is the desired end product. The fact that these viruses can infect a range of different plant species means that the same construct can be used in several different plants to examine species effects on protein production.

As a result, a number of different plant virus systems have been developed for protein expression. These systems can most conveniently be divided according to the architecture of the virus capsid into helical (rod-shaped) viruses or icosohedral viruses. Helical viruses consist of a single strand of RNA around which is wound the capsid protein. Icosahedral viruses encapsulate their genome and associated proteins in a 20-sided particle of a defined size. As a result of their structures, the icosohedral viruses have a limited ability to incorporate additional DNA or RNA as a result of space constraints. In contrast, the helical viruses appear to be more accommodating of additional genetic material because the rodlike particle can be increased in length simply by winding more capsid proteins around the additional RNA. As a result, icosahedral viruses are best suited to expressing small epitopes, whereas helical viruses can often provide the opportunity for the expression of large proteins.

A. Icosahedral Plant Viruses

1. Cowpea Mosaic Virus-Based Vectors

Cowpea mosaic virus (CPMV) is a positive-strand RNA virus of the family Comoviridae. It has a narrow host range, with most experimental work being carried out in its natural host, the legume *Vigna unguiculata* (cowpea). The genome consists of two RNA molecules, RNA-1 and

RNA-2, with RNA-2 encoding the capsid proteins. A systemic infection can be achieved by mechanical abrasion of the leaf and application of both viral RNAs. The icosahedral capsid consists of 60 copies each of the large (L) and small (S) subunits. Lomonossoff and coworkers have developed this system for epitope expression by examining the S protein for its ability to present epitopes (Usha *et al.,* 1993). The crystal structure of the virus particle has been solved, thus allowing for the precise placement of epitopes in the coat proteins (Lin *et al.,* 1999; Lomonossoff and Johnson, 1991; Stauffacher *et al.,* 1987). One of the loops on the S protein was chosen for insertion due to its location on the surface of the capsid (Usha *et al.,* 1993). The initial insertion of an epitope from the VP1 protein of foot-and-mouth disease virus (FMDV) interfered with viral infectivity and spread. In addition, duplicated sequences in the recombinant RNA-2 molecule allowed homologous recombination to occur, causing deletion of the inserted sequences from the viral genome (Porta *et al.,* 1994; Usha *et al.,* 1993). Repositioning of the epitope insertion site between amino acids 22 and 23 of the S coat proteins resulted in the successful assembly of viral particles that contained additional epitopes from human rhinovirus-14 (NIm-1A) and human immunodeficiency virus 1 (HIV-1) (Porta *et al.,* 1994). The amount of virus produced was comparable to wild-type levels, but when injected into rabbits, the NIm-1A-containing particles failed to generate neutralizing antibodies against human rhinovirus-14. To examine this failure, Lin and coworkers (Lin *et al.,* 1996) solved the structure of the recombinant virus particle, allowing them to compare the conformation of the NIm-1A in the S coat protein with the known structure of human rhinovirus-14 (Rossmann *et al.,* 1985). From this comparison it was found that the position of CPMV amino acids at the ends of the NIm-1A epitope were 10 Å further away in the CPMV variant than in the wild-type protein, resulting in the absence of several intraloop hydrogen bonds. This relaxed conformation resulted from the spontaneous cleavage between the two C-terminal residues of the inserted NIm-1A sequence, a fact confirmed by sodium dodecyl sulfate–polyacrylamide gel electrophoresis (SDS–PAGE) (McLain *et al.,* 1995). This cleavage has been reported to occur with all insertions between amino acids 22 and 23 of the S protein (Porta and Lomonossoff, 1998) and as a result may affect the conformation of any epitopes inserted at this site.

Insertion of the HIV-1 glycoprotein epitope gp41 into the same site in the S protein gave better results. Injection of the recombinant virus into mice produced antibodies that were capable of neutralizing up to 97% of the HIV-1 virus (McLain *et al.,* 1995). Further refinement of epitope location and design has allowed inserts of up to 42 amino acids in size to be inserted successfully into the S protein (Xu *et al.,* 1996).

More than 50 different CPMV constructs have been made. In the majority of these constructs, the virus grows normally and large yields of up to 2 mg/g of virus in leaf tissue can be obtained. However, in some constructs it has been found that insertions result in a virus that is incapable of systemic infection (Porta *et al.,* 1994) probably because the coat protein modifications interfere with virus movement in the plant. Particles of CPMV are stable to pH 1.0, are resistant to pepsin degradation, and are stable to 65°C, suggesting that oral immunization with this particle could be possible and that a vaccine might not require cold storage (Dale, 1949; Porta and Lomonossoff, 1998; Xu *et al.,* 1996).

The CPMV capsid is also a highly immunogenic structure (Stace-Smith, 1981). This may be an added advantage for vaccine production because the particle may act as a mucosal adjuvant causing stimulation of the immune system, thus providing an enhanced response to the inserted epitope.

2. Tomato Bushy Stunt Virus-Based Vectors

Tomato bushy stunt virus (TBSV) is a member of the Tombusviridae and has a monopartite, positive-sense RNA genome (Russo *et al.,* 1994). The structure of the particle has been solved and shows that the C-terminus is exposed on the surface (Harrison *et al.,* 1978; Olson *et al.,* 1983). The icosahedral capsid is formed from 180 copies of the 41-kDa coat protein. The coat protein is not necessary for movement or spread of the virus, making it more amenable to genetic insertion (Scholthof *et al.,* 1993). A 13-amino-acid (aa) peptide from HIV-1 gp120 with a 3-aa linker sequence has been expressed at the C-terminus of the coat protein from TBSV in tobacco plants (Joelson *et al.,* 1997). The resulting virus-like particles (VLPs) resembled the wild-type particles. The HIV epitope was recognized by monoclonal antibodies and immunization of mice resulted in high antibody responses to both the HIV peptide and the virus. The immune response for the coat protein–gp120 fusion was greater than that achieved in mice injected with the synthetic gp120 peptide alone.

3. Caulimovirus-Based Vectors

The 8-kb genome of cauliflower mosaic virus (CaMV) consists of a circular, double-stranded (ds) DNA. The outer capsid contains 420 coat protein subunits that make up the icosahedral virion. Open reading frame II, which codes for a protein required for aphid transmission, can be replaced with a foreign gene (De Zoeten *et al.,* 1989). This system has been used to express medically relevant proteins, but not for the synthesis of antigenic proteins for purposes of vaccination.

B. Helical Plant Viruses

1. Tobamoviridae-Based Vectors

The type member of this family, tobacco mosaic virus (TMV), is one of the best-characterized viruses. TMV infects over 200 plant species. Expression levels of the virus can be as high as 50% of the plant dry weight, providing extraordinary virus yields, as high as 60 mg of virus per gram of fresh plant tissue (Copeman *et al.,* 1969). As a result of these favorable characteristics, considerable effort has been made to use TMV as an expression vector (Hamamoto *et al.,* 1993; Hwang *et al.,* 1994).

There have been two approaches to the expression of antigens using TMV. One method is to express the entire coding sequence of a protein using a duplicated subgenomic promoter. Unfortunately, the TMV genome has proved difficult to engineer because constructs are prone to instability and deletion of the inserted DNA is common (Dawson *et al.,* 1989; Donson *et al.,* 1991). New constructs, using heterologous coat proteins and subgenomic promoters from closely related viruses, have reduced these problems and have enabled foreign proteins to be expressed to levels of 2% of the total soluble plant protein (Shivprasad *et al.,* 1999). Using these methods, it has been possible to express full-length proteins such as the structural protein VP1 from foot-and-mouth disease virus and an epitope from hepatitis C fused to the cholera toxin B subunit (Nemchinov *et al.,* 2000; Wigdorovitz *et al.,* 1999b).

As with the icosahedral viruses, epitopes can be inserted into the coat protein of TMV and the viral particle itself then can be used as a carrier for epitope presentation. The TMV coat protein is 158 amino acids long with a molecular weight of 17.5 kDa. At low pH the TMV coat protein molecules form rods without the prior requirement of viral RNA to nucleate rod formation. At neutral or alkaline pH the coat protein forms bilayers (Butler, 1984). TMV virions can survive passage through the gastrointestinal tract, allowing for the possibility of oral delivery (R. N. Beachy, personal communication). The crystal structure of the coat protein has been resolved, indicating that the amino and carboxy termini of the protein are exposed on the virus particle (Namba *et al.,* 1989; Namba and Stubbs, 1986), making these sites good candidates for epitope insertion.

Unfortunately, fusions of even small peptides to the C-terminus of the TMV coat protein have been found to interfere with both virus assembly and virus spread (Takamatsu *et al.,* 1990). This observation has led to the design of new vectors that allow for the production of both wild-type coat protein and recombinant fusions through the use of "leaky" stop codons (Fig. 1). The resulting mixture of native and recombinant forms

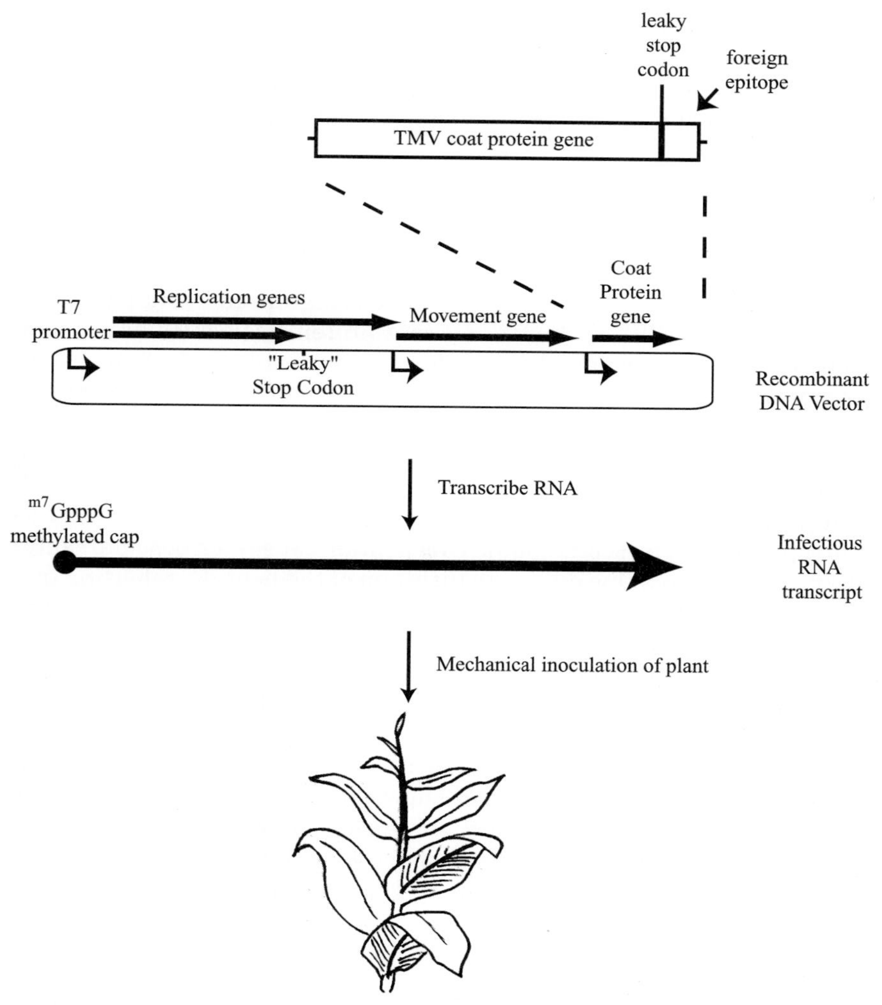

FIG. 1. A generic example of a TMV vector used for expression of an epitope at the C-terminus of the coat protein.

of the coat protein reduces the effect of the perturbations in the recombinant protein (Hamamoto *et al.,* 1993). Placement of the inserted peptide slightly in from the C-terminus (at amino acid 154) avoids the problem created by the disruption of particle assembly (Fitchen *et al.,* 1995) and no leaky terminator is required.

Peptides of up to 25 amino acids in length have been expressed on the C-terminus of the TMV coat protein, with yields of up to 1.2 mg of virus per gram of fresh plant material (Bendahmane *et al.,* 1999; Fitchen *et al.,* 1995; Gilleland *et al.,* 2000; Koo *et al.,* 1999; Staczek *et al.,* 2000; Turpen *et al.,* 1995). In all these cases the virus particles were recognized by antibodies to the appropriate peptides, indicating that the inserted epitopes have the potential to induce an immune response.

2. Potexvirus-Based Vectors

Potato virus X (PVX) is the type member of the monopartite RNA potexviruses. Unlike the rigid rod of TMV, the infectious particle of PVX is a flexuous rod. The 6.4-kb genome is organized similarly to that of TMV and, like TMV, the 237-aa coat protein molecules are assembled around the infectious RNA molecule. Immunological surface mapping of the coat protein indicates that the N-terminus is exposed on the surface of the virions (Baratova *et al.,* 1992)

Santa Cruz and coworkers found that the 27-kDa reporter green fluorescent protein (GFP) could be fused to the N-terminus of the PVX coat protein via the foot-and-mouth disease virus (FMDV) 2A autocatalytic peptide (Cruz *et al.,* 1996). This peptide causes a ribosomal "skip" after its proline residue during translation of the peptide, releasing free PVX coat protein and recombinant GFP with 15 amino acids from the 2a peptide fused to its C-terminus (Donnelly *et al.,* 2001a, 2001b). The peptide does not produce complete cleavage, resulting in the production of recombinant proteins that incorporate both GFP and PVX coat protein. This yields a viral rod with a diameter that is approximately twice that of the wild-type virus because of the presence of the GFP "overcoat." These modified viruses show slower movement through the plant and do not produce infectious virus particles. However, RNA transcripts are infectious via mechanical inoculation, with fluorescent lesions developing after 2–3 days as the virus spreads progressively through the plant.

3. Potyvirus-Based Vectors

Johnsongrass mosaic virus (JGMV) is a flexuous, monopartite RNA potyvirus. As with TMV, structural data indicate that the N- and

C-termini are exposed (Shukla *et al.,* 1988, 1989) and are not required for virus capsid assembly (Jagadish *et al.,* 1991). The N-terminus appears to be amenable to genetic manipulations since peptides up to 62 amino acids long have been substituted (Jagadish *et al.,* 1993). The constructs also express at levels of up to one-fourth of the total cellular protein. As might be anticipated, when protein was assembled into VLPs, the particulate form induced better immune responses in mice than the monomeric protein (Jagadish *et al.,* 1996).

Plum pox potyvirus (PPV) also has been developed into an antigen presentation system (Fernandez-Fernandez *et al.,* 1998). As with JGMV, the N- and C-termini are surface-exposed (Shukla *et al.,* 1988). One or two copies of the 15-amino-acid epitope from canine parvovirus (CPV) were inserted into the N-terminus, replacing the 15 amino acids known to be required for aphid transmission of the virus. Virus yields were equivalent to those of wild type. Two doses up to 50 μg in mice and 500 μg in rabbits were injected intraperitoneally. Neutralizing antibodies were detected against both CPV and PPV; the additional copy of CPV had no observable effect on the antibody titers achieved (Fernandez-Fernandez *et al.,* 1998).

Tobacco etch virus (TEV) is another potyvirus that has been developed for protein expression (Carrington *et al.,* 1993; Dolja *et al.,* 1992, 1998). This system has been used to express reporter genes as well as unrelated viral proteins. To date, TEV has not been used for vaccine production. The recombinant virus has proven to be exceptionally stable, surviving 20 consecutive cycles of infection without modification (Dolja *et al.,* 1992). However, the protein expression levels achieved were low and recovery of protein was only 4 μg/g from fresh tissue (Dolja *et al.,* 1998).

C. Advantages and Disadvantages of Plant Viruses for Protein Expression

Plant viruses show great potential for producing large amounts of protein in short periods of time because the time-consuming processes of integrating foreign DNA into the genome, culturing transgenic material, and growing transgenic progeny are eliminated. Because plants are infected after they are grown, potential problems with toxicity caused by the expressed proteins are reduced. In addition, the virus approach allows great versatility because the plants can be infected with any construct that is desired. As a result, plant viruses allow quick turnaround time for expression and allow numerous constructs to be tested for expression levels and antigenicity. Scale-up is also relatively simple since plants can be grown from existing seed stocks, unlike the "bulking up" required when expressing protein in transgenic plants. For example, Large Scale Biology Corp. (Vaccaville, CA) has obtained U.S. Environmental Protection Agency approval allowing them to grow up to 1000 acres of tobacco per production run and have built a processing plant capable of handling this quantity of material (R. L. Forster, personal communication).

Of course, plant viruses require the extra step of infection that is not required once a transgenic plant is available. Plant viruses have been shown to be unstable, making cloning difficult, as well as making it necessary to check that the inoculating RNA is of the correct sequence. RNA transcripts are susceptible to degradation, although this problem can be alleviated in some systems by techniques using DNA for inoculation (Ahlquist and Janda, 1984; Mori *et al.,* 1991; Shi *et al.,* 1997).

Since some of these viruses have large host ranges, there may be concern about inadvertent spread of the virus to other plants. These concerns can be addressed by the careful choice of virus/host combinations or by the use of host plants engineered to support the replication of defective viruses. For example, Australian researchers have constructed a transgenic tomato plant expressing an ethanol-induced movement protein from tomato mosaic virus (ToMV), a close relative of TMV. A movement-deficient ToMV is used for foreign gene expression, resulting in a virus that will only be viable in the transgenic tomato plant expressing the movement protein (Booth and Rodger, 1999–2000). Approaches of this type can be used to ensure that protein expression can be achieved without the possibility of the virus spreading beyond the target host.

VI. Enhancing Protein Yield

A major limiting factor in the use of plants to express proteins is often the low yield of recombinant protein that is achieved. An expression level of 1% of total soluble protein is generally accepted as a minimal level for it to be economically feasible or practicable to purify protein from a genetically modified plant (Kusnadi *et al.,* 1997). As a result, a number of methods have been developed to increase expression levels. These methods can be applied to transgenic expression and viral expression, although some methods may not work with certain viral systems.

To achieve high levels of protein in plants, several factors must be addressed. The appropriate promoters, enhancers, and leader sequences for the expressed protein must be determined. In addition, it may be necessary to optimize the codon usage and to remove mRNA-destabilizing sequences and polyadenylation signals from the foreign gene. Finally, protein stability is a crucial factor for accumulation of foreign proteins at high levels. These factors in turn depend on the plant system chosen and the method of expression as well as on the stability of the protein.

One of the most effective methods of increasing yields has proven to be the addition of 3′ mRNA-stabilizing sequences. For example, swapping the 3′ region from nopaline synthase with the 3′ regions from either the soybean *vsp*B gene or the potato *pinII* gene resulted in 20- to 50-fold increases in expression of hepatitis B surface antigen in potatoes (Richter *et al.,* 2000). Analysis of mRNA production in a number of different transgenic plants of each construct demonstrated an increase in mRNA levels over the nopaline synthase construct (Richter *et al.,* 2000).

Another effective means of increasing protein levels is by modification of leader or 5′ untranslated regions (UTRs). The sequence of the UTRs greatly affects the levels of protein synthesis by altering ribosome binding and the initiation of protein synthesis (Beachy, 1997). It is not well understood what features of the UTR cause the differences in protein synthesis. The best-studied UTRs are of viral origin, but much work has also been done on characterizing the UTRs of chloroplast mRNAs (Cohen and Mayfield, 1997).

Plants may incorrectly recognize some sequences from foreign genes as mRNA-destabilizing sequences and polyadenylation signals, which will also reduce yields. Such sequences result in lower production of mRNA and reduce the half-life of the mRNA molecule. In addition, codon usage in plants may differ considerably from foreign genes derived from other organisms such as animal pathogens. This may result in reduced protein expression levels of the foreign gene in plants. Use of unusual plant codons in the foreign gene may result in stalling or stopping of translation. Redesigning the coding sequence of the gene and/or removal of mRNA-destabilizing sequences can result in as much as 100-fold increases in expression levels (Strizhov *et al.,* 1996).

Specific leader peptides can be used to target proteins to specific cell compartments such as the endoplasmic reticulum or chloroplast (O'Brien *et al.,* 2000; Richter *et al.,* 2000). It was found that addition of a vacuole-targeting sequence increased the production of hepatitis B surface antigen in potatoes by a factor of three. However, in the same report it was found that a chloroplast-targeting signal reduced expression to undetectable levels (Richter *et al.,* 2000). This result reflects the mixed results that have come from the use of such targeting signals. Other reports have shown no increases in expression levels from targeting (O'Brien *et al.,* 2000). Nevertheless, targeting of protein to specific tissues has been successful in a number of cases. For example, it was found that the *E. coli* ADP glucose pyrophosphorylase gene (glgC16) was more successfully transformed into potato plants and resulted in higher levels of protein when placed under control of the tuber-specific patatin promoter than under the control of the nonspecific 35S CaMV promoter (Stark *et al.,* 1992). Toxic effects of some proteins can be relieved by targeting to specific tissues, as was found for the expression of the heat-labile enterotoxin (LT-B) (Mason *et al.,* 1998). The LT-B codon usage was also optimized for expression in plants, resulting in as much as 14-fold greater expression, but these levels stunted growth of the potato plants. Targeting of the protein to the tubers overcame this growth inhibition (Mason *et al.,* 1998).

Proteins may also be degraded significantly after translation, resulting in reduced yields (Ohtani *et al.,* 1991). A number of methods have been designed to reduce degradation. The protein can be targeted to protein bodies in seeds, chloroplasts, or the endoplasmic reticulum, where there is little proteolytic activity.

VII. Expression of Immunogenic Molecules

A. Expression of Small Epitopes

Although it is possible to express small peptides freely, small peptides, by themselves, tend to be poorly immunogenic. It has been shown that conjugation to a carrier molecule often enhances immunogenicity (Francis, 1991; Kingsman and Kingsman, 1988; Lomonossoff and Johnson, 1996). In addition, small molecules are not taken up as well as larger molecules (Singh-Jasuja *et al.,* 2000). Fusion to a carrier protein can also provide a small peptide with the scaffold necessary to allow for proper folding (Walmsley and Arntzen, 2000). Another advantage of the production of an immunogen as part of a macromolecular structure is that it may make purification simpler because large particles can be separated from other cellular components after cell lysis by centrifugation. The plant virus expression systems that have been developed lend themselves to this form of epitope presentation.

Coat proteins of plant viruses provide regular arrays consisting of macromolecular structures that can present multiple copies of the epitope to the immune system. The advantage of displaying multiple epitopes on a carrier has been demonstrated by Hwang and coworkers (Hwang *et al.,* 1994), who fused an eight-amino-acid epitope from VP1 of poliovirus type 3 to the C-terminal of the TMV coat protein. Isolated protein was then administered to mice in three forms. In one form, VLPs were assembled by lowering the pH to 5.0. In the second form, the VLPs were assembled by the addition of RNA to condense the protein into particles. In the third, the recombinant protein was fully disassembled by raising the pH to 8.0. When injected into mice, both assembled forms elicited antibody responses considerably higher than the disassembled form.

TMV coat protein was the first plant viral protein to be engineered to express antigenic peptides (Hamamoto *et al.,* 1993). Epitopes presented on the virus surface in this manner were capable of eliciting antibody responses in mice (Fitchen *et al.,* 1995). Expressing epitopes on the surface of CPMV also induced antibody responses (McLain *et al.,* 1995). Furthermore, a protective response in mink against the mink enteritis virus was achieved when an epitope from the mink enteritis virus coat protein was presented on the CPMV coat protein surface (Dalsgaard *et al.,* 1997).

One of the obvious limitations encountered when presenting epitopes in the coat protein is that a limited number of amino acids can be inserted without interfering with the proper structure and assembly of the coat protein. For example, the maximum number of amino acids that have been inserted into the TMV coat protein is 25 (Bendahmane *et al.,* 1999; Fitchen *et al.,* 1995; Gilleland *et al.,* 2000; Koo *et al.,* 1999; Staczek *et al.,* 2000; Turpen *et al.,* 1995).

To circumvent the peptide size limitations of TMV and CPMV, Yusibov and coworkers have created a variation on the design of the coat protein epitope carrier utilizing alfalfa mosaic virus (AlMV) coat protein. AlMV has a genome consisting of four RNA molecules of different sizes. Because coat proteins must be bound to each RNA molecule for it to be infectious, this is a difficult virus to work with. The coat protein forms particles of varying shape (spherical, ellipsoid, or bacilliform), depending on the size of the RNA molecule that is being encapsulated (Jaspars, 1985; Shaw, 1996). This indicates that there is considerable flexibility in protein–protein interactions within the capsid. As a result, insertions have been made in the exposed N-terminus without disrupting capsid formation. To avoid the difficulties of working with AlMV, the coat protein has been expressed in the more easily manipulated TMV (Yusibov *et al.,* 1996, 1997). In this way up to 38 amino acids have been inserted into the AlMV coat protein without interfering with particle assembly (Belanger *et al.,* 2000; Modelska *et al.,* 1998).

AlMV coat protein has been used as a carrier to express epitopes derived from rabies virus, HIV, and respiratory syncytial virus. Spinach plants were infected with a TMV vector expressing the AlMV coat protein carrying the rabies epitope. When spinach leaves were fed to mice, serum immunoglobulin G (IgG) as well as mucosal IgA were produced (Modelska *et al.,* 1998). When epitopes from the G protein of human respiratory syncytial virus were expressed in the AlMV coat protein in tobacco plants, a yield of 0.4–0.8 mg/g fresh weight of coat protein was achieved. Doses of up to 1 mg of partially purified viral protein were injected intraperitoneally into mice three times, resulting in complete protection (Belanger *et al.,* 2000).

Icosahedral plant viruses contain a relatively low number of viral proteins that can be used to display epitopes. For example, the CPMV virion contains 60 copies of the S protein, whereas TBSV particles have 180 copies of the coat protein. However, the rod-shaped viruses, such as TMV and PVX, have the potential to display a much higher number of copies. The TMV rod contains over 2000 proteins and thereby could present an equal number of antigenic peptides, provided they are inserted in the surface loop region of the coat protein. However, fusions made to the extreme N-terminus of PVX or the C-terminus of TMV require that the construct also permit the expression of wild-type coat protein because the fusion proteins interfere with the proper functioning of the virus. As a result of this limitation, fewer epitopes are expressed on the particle, although there are many more copies per particle for presentation compared to the icosahedral viruses.

Some epitope vaccines have failed to produce a protective response, even though significant antibody titers have been produced. This failure may merely reflect the general problem sometimes encountered with subunit vaccines where B cells are stimulated (resulting in antibody production) without a corresponding T cell response. It may be possible to solve these difficulties by the insertion of appropriate T cell antigenic peptides into the carrier molecule or by use of an adjuvant to stimulate the immune system. In addition, an appropriate carrier may provide helper T cell epitopes that enhance the immune response.

B. Immunogenic Carrier Molecules

One method for enhancing uptake is to present the antigen as part of a mucosal pathogen (Baron *et al.,* 1987) or toxin (de Aizpurua and Russell-Jones, 1988; Lindner *et al.,* 1994) that has been rendered avirulent. One obvious problem with using such carriers is that an immune response may develop against the carrier, preventing its use in subsequent vaccinations (Cooney *et al.,* 1991; Rooney *et al.,* 1988). Even so, much attention has focused on the use of two mucosal adjuvants, heat-labile enterotoxin (LT) from *E. coli* and cholera toxin (CT) from *Vibrio cholera,* as carriers for vaccine epitopes. These molecules appear to pass through the villous enterocytes as well as the Peyer's patches and stimulate serum antibody responses even at low doses. CT and LT consist of a toxic subunit, A, and a pentameric subunit, B, which binds to the membrane glycolipid ganglioside, Gm1. Without the A subunit, the structure loses its toxicity while retaining the ability to penetrate the intestinal walls, making the pentameric B-subunit complex a good mucosal adjuvant (Dougan *et al.,* 2000; Russell-Jones, 2000; van Ginkel *et al.,* 2000). These subunits are capable of stimulating very high immune responses when administered orally. Although these antigens can be used to stimulate immune responses when presented with other antigens, they are also capable of inducing protection against enterotoxigenic *E. coli* and *V. cholera* (Arakawa *et al.,* 1998a, 1999; Lauterslager *et al.,* 2001; Mason *et al.,* 1998).

Transgenic potatoes, derived from *Agrobacterium*-mediated gene transfer, have been used to express LT-B. Only low levels were recovered, with the highest yield achieved being 0.01% of total protein. Assembly of the pentameric subunit was demonstrated by binding to Gm1 (Haq *et al.,* 1995). Mice fed raw potatoes or crude tobacco plant extract expressing LT-B generated similar serum-specific IgG and IgA responses, although this response was still lower than that seen with oral administration of bacterial LT-B. The lower response may be due to interference by the presence of the plant tissue and the low protein expression levels achieved. More recently, CT-B has been successfully expressed in transgenic potatoes to levels of 0.3% of total protein (Arakawa *et al.,* 1998a). As with LT-B, the plant CT-B was shown to form multimeric complexes and to bind the Gm1 ganglioside.

Further work has shown that fusion of CT-B to other antigenic epitopes can result in enhanced immune responses to the epitopes (Arakawa *et al.,* 1998b). Yu and coworkers have shown that by fusing peptides from rotavirus and enterotoxigenic *E. coli* to CT-B, immune responses can be induced that provide partial protection from challenge (Yu and Langridge, 2001). Carriers such as LT-B and CT-B show great promise for oral vaccines (Estes *et al.,* 1997) because they have been found to work well as oral adjuvants. It is to be hoped that they can be used to enhance the immune response to other plant-expressed antigens.

C. Expression of Virus-Like Particles in Plants for Use in Oral Vaccination

The idea of using a carrier for presentation of the epitope has been further adapted to the design of oral vaccines. A large number of plants are edible and a plant-produced antigen would seem to be well suited for use as an edible vaccine, thereby removing the need for purification of the antigen. A major problem with oral delivery of an antigen is that the gastrointestinal tract is a harsh environment that degrades proteins. Some researchers have noted that the plant cell appears to provide some protection to the antigen (Modelska *et al.,* 1998; Walmsley and Arntzen, 2000). On this basis, the protein may be protected from degradation by the cell wall of the plant until it has passed through the lower intestine, where it can be absorbed through the Peyer's patches and presented by the M cells.

A number of protein complexes have proven to be resistant to degradation during passage through the gut, including the LT and CT subunits, as well as a number of capsid structures from viruses. These proteins tend to be from pathogens transmitted by the fecal–oral route. Presumably as a result, the pathogens have evolved methods of resisting degradation during their passage through the digestive system. A number of capsid proteins from animal viruses fall within this category and have been expressed in plants. These capsid structures have been proven to be immunogenic in other systems and are therefore potential oral vaccinogens.

Norwalk virus causes acute gastroenteritis in humans. A single protein is required for the assembly of a spherical virus capsid capable of resisting the harsh environment of the stomach and intestine. When expressed alone, this protein assembles into virus-like particles (VLPs) that can induce protective immunity. Mason and coworkers expressed the Norwalk capsid protein in tobacco and potato plants (Mason *et al.,* 1996) and found that it assembled into VLPs in plants. When mice were fed raw potato, only 50% developed an IgG response and only 5% developed an IgA response. A further study with humans (Tacket *et al.,* 2000) showed that immune responses could be induced in 19 of 20 volunteers. However, these responses were modest and unlikely to provide complete protection. Low protein yield (0.37% of soluble protein) and the resulting low dose of antigen may have contributed to the poor results achieved.

Rotavirus is another virus that causes gastroenteritis. Unlike Norwalk virus, the inner capsid particle of rotavirus requires two proteins for assembly. When produced using a baculovirus expression

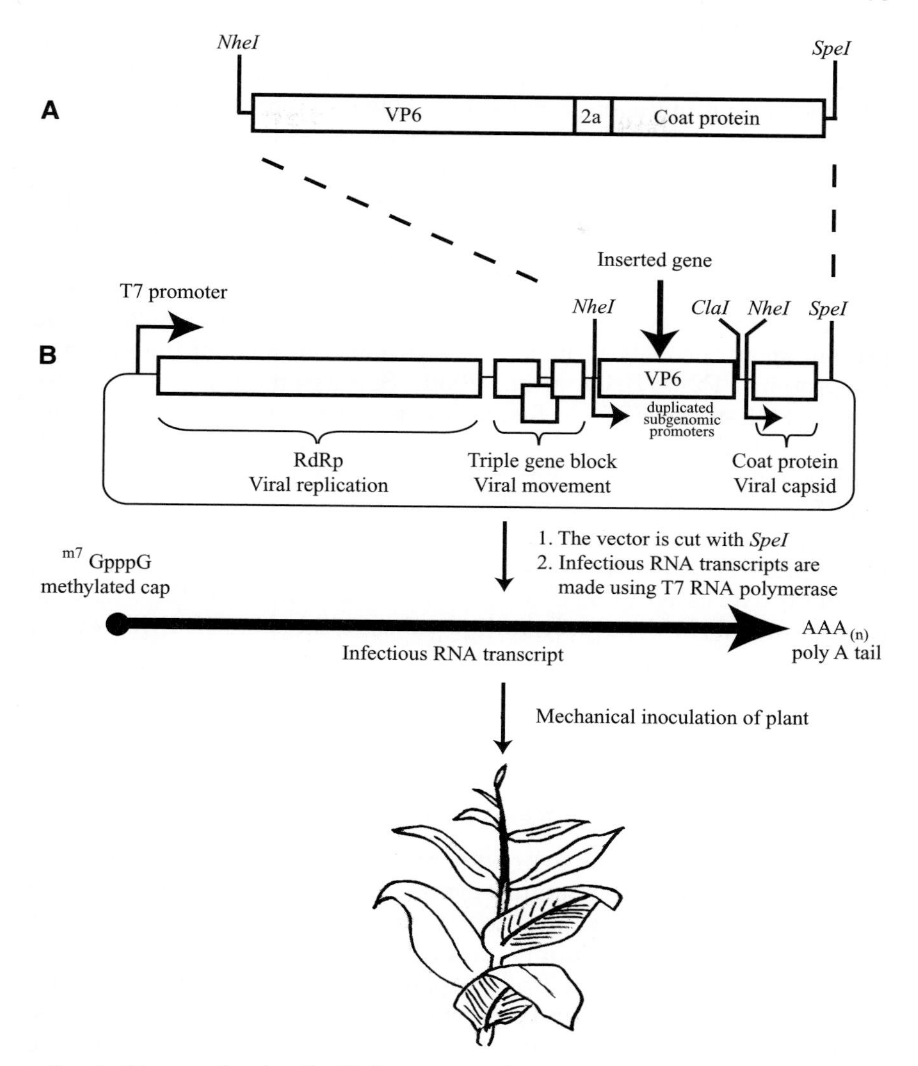

FIG. 2. Diagram showing the PVX vectors used for expression of the rotavirus protein VP6 in tobacco plants. (A) Construct expressing VP6 fused to the PVX coat protein. The FMDV 2a peptide causes a ribosomal "skip" that results in up to 50% of the plant-expressed protein to occur in the form of free VP62a and PVX coat protein peptides. (B) Vector expressing VP6 as free protein.

system, these VLPs are immunogenic when administered orally (O'Neal *et al.,* 1997) and are therefore a likely oral vaccine candidate. VP6 is the major surface protein and represents ~80% of the protein in the capsid particle (Holmes, 1983). In the presence of VP2, the two proteins form a VLP that is identical in shape to the wild-type particle. To express the rotavirus capsid protein, the 44-kDa VP6 protein was fused to the PVX coat protein in the same manner as the GFP–PVX fusion (O'Brien *et al.,* 2000). When expressed in plants, up to 50% of the recombinant proteins cleave after the 2a region, giving free PVX coat protein and VP6 fused to the 2a peptide (Fig. 2). Unexpectedly, the VP6–2a fusion protein spontaneously formed VLPs without the need of the scaffolding protein VP2 (O'Brien *et al.,* 2000). Self-assembly into VLPs also occurred when the VP6–2a fusion was expressed in insect cells using baculovirus (Awram, unpublished results). This self-assembly of VP6 greatly enhances the ease with which VLPs can be produced.

VP6 was also expressed as free protein under a duplicated subgenomic promoter. This construct proved much less stable and therefore was prone to edit out the VP6 gene (O'Brien *et al.,* 2000). In this form, the plant-expressed VP6 behaved in the same manner as the wild-type protein and formed paracrystalline tubes and sheets (O'Brien *et al.,* 2000).

VIII. Immunogenicity and Vaccine Responses

A. Vaccines and the Immune System

The goal of a vaccine is to provide protection from infection. This requires stimulation of the immune system and particularly the T cells for a protective immune response to develop. The immune system can be divided into humoral immunity and cell-mediated immunity (CMI). In general, both types require T cell responses, but these responses are initiated by different means and protect the body by different mechanisms. CMI responses are required for clearing pathogens that are replicating within the cell (i.e., intracellular pathogens such as viruses). Humoral responses are essential for the elimination of extracellular pathogens.

CMI responses are initiated by presentation of antigens in conjunction with major histocompatability complex (MHC) I proteins, which are found on almost all cells. As an intracellular pathogen replicates inside an infected cell, degradation processes by the host result in cytoplasmic proteins being degraded into peptides that are then bound by MHC I molecule. These antigen–MHC I complexes migrate to the surface, where they are recognized by effector T cells, which then kill the infected cell.

Both humoral and CMI responses are dependent on the action of T helper (Th) cells. These involve the recognition of antigens in conjunction with MHC II proteins, which are expressed mainly in antigen-presenting cells (APCs) such as dendritic cells (DCs). APCs internalize antigens, which are then processed into small fragments. Then, in association with MHC II, these fragments are displayed on the cell surface, where they are recognized by Th1 and Th2 cells, which then release cytokines that modulate the immune response. In general, Th1 cells stimulate CMI and neutralizing IgG responses, whereas Th2 cells stimulate B cells to produce antibodies.

When designing a vaccine, the type of immune response that is desired requires consideration because different immune responses work better against different pathogens. In general, Th1 cells are required for the elimination of intracellular pathogens such as *Mycobacterium, Salmonella,* and *Listeria* (Scott *et al.,* 1988), whereas Th2 cells are required for the elimination of extracellular pathogens. As a result, the form of the antigen and delivery method will have important effects on the immune response generated.

The vast majority of pathogens enter the body through mucosal surfaces. As a result, the immune responses at these surfaces are important in preventing infection by these pathogens and emphasis is now being placed on developing vaccines that stimulate mucosal immunity, with the intention of preventing entry of the pathogen. Mucosal immunity results from the presence of antibodies and lymphocytes at the mucosal surfaces that line the respiratory, the digestive, and the urinoreproductive tracts (Fooks, 2000; Ruedl and Wolf, 1995; Yu and Langridge, 2000).

Most vaccines currently in use are administered parenterally and are ineffective for induction of mucosal immunity. Injection usually causes good IgG responses, but smaller or nonexistent mucosal IgA responses. Unfortunately, it is not well understood what makes an effective mucosal vaccine. However, mucosal immunization can stimulate both IgG and IgA responses. The route of mucosal immunization, oral, nasal or rectal, is also important since the best antibody response occurs at the immunizing surfaces. As a result, the most effective method of immunizing against a gastroenteritis pathogen would be expected to be oral delivery.

Antibodies are found in the blood, in extracellular spaces between cells, and on mucosal surfaces, but not inside cells. As a result, antibodies are not effective against an intracellular pathogen once it has entered the cell. However, since many intracellular pathogens enter through mucosal surfaces, antibody defenses at these surfaces can be effective in preventing infection. In addition, when migrating between cells, intracellular pathogens are susceptible to antibody defenses.

However, a mucosal response may not be sufficient to clear an intracellular pathogen from the body and a CMI response may also be required. Although the best way to stimulate CMI is by using a replicating vector capable of cell entry, a number of other methods have been found to stimulate CMI using soluble antigens (Buseyne *et al.*, 2001; McNeela and Mills, 2001; Olszewska and Steward, 2001; Ryan *et al.*, 2001). Examples of this are (1) bacterial toxins that disrupt the membrane structure and enter the cytoplasm (Brunt *et al.*, 1990) and (2) molecules designed to penetrate the membrane (Yewdell *et al.*, 1988). Upon entry of the cytoplasm, these molecules follow the same pathway as internal antigens and are presented in conjunction with MHC I molecules on the cell surface, thereby initiating a CMI response (Albert *et al.*, 1998; Buseyne *et al.*, 2001; Reimann and Schirmbeck, 1999). Some peptides are capable of binding directly to MHC I molecules on the cell surface and therefore can bypass internalization (Townsend *et al.*, 1985). These results indicate that careful antigen selection is required for the design of vaccines effective against intracellular pathogens.

B. Oral Delivery of Expressed Protein for Inducing Mucosal Immune Responses

The basic immune response to an oral antigen is generally thought to occur as outlined in Fig. 3. If an orally delivered antigen survives passage through the stomach, induction of an immune response at the mucosal surfaces starts by the recognition of foreign antigens by M cells. M cells are located in the membranes of structures such as the Peyer's patches. M cells are implicated in uptake, processing of antigen into small fragments, and transport to the underlying B cell follicle (Allan *et al.*, 1993; Neutra *et al.*, 1996). Once the protein is taken up by the M cells, it is delivered to APCs, which cleave the protein into small peptides that are presented on the surface of the cell in association with major histocompatibility complex (MHC) class II molecules. Recognition of these complexes by Th lymphocytes causes their activation. The activation of the Th2 cells leads to the differentiation of secretory IgA+ B cells into IgA plasma-producing cells. After activation, the IgA+ B cells migrate to the mesenteric lymph nodes, where they mature into plasma cells, which then travel to the mucosal membranes. Here, the B cells secrete IgA antibodies, which move through the mucosal epithelial layer, where they bind membrane-bound secretory components and

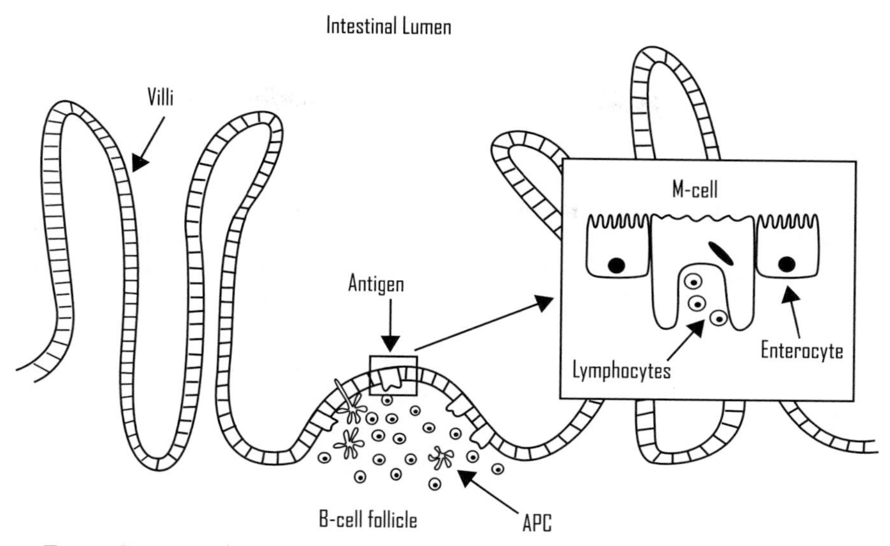

FIG. 3. Cross section of a Peyer's patch, the major inductive tissue of the gastrointestinal tract. Antigen, present in the intestinal lumen, is taken up by the M cells, which process and transport the antigen across the epithelial layer. The M cells then deliver the antigen to antigen-presenting cells (APC) such as dendritic cells. These cells process the antigen and present it to T helper cells and initiate the immune response cascade.

are then exposed to the lumen, where they can bind and neutralize the invading pathogens. There are some indications that some antigens, such as cholera toxin, can be taken up in the gut without going through the Peyer's patches (Russell-Jones, 2000; Yamamoto *et al.,* 2000).

One of the problems encountered with oral delivery is that the gut-associated lymphoid tissue (GALT) is not very efficient at uptake; most of the intestinal epithelium is impermeable to all but the smallest molecules, and although M cells are highly endocytotic, there are few of them and they are not easily accessible due to the presence of villi, which obstruct access. Attempts to alleviate this problem by administration of large amounts of antigen may fail because the body has evolved a mechanism for preventing immune responses to food (oral tolerance) and this mechanism can also affect any delivered antigens (Fooks, 2000; Ruedl and Wolf, 1995; Weiner *et al.,* 1994). Since the gastrointestinal tract is constantly bombarded with foreign molecules, the body has developed mechanisms to prevent immune responses to commonly encountered molecules. In general, most molecules are not taken up and presented to the immune cells in significant quantities and little or no immune response develops. Oral delivery of a plant-derived antigen would require careful limitation of dosages to avoid tolerance effects. However, the induction of tolerance by supplying too much antigen has been used advantageously for vaccination against autoimmune diseases. For example, glutamic acid decarboxylase (GAD), an autoantigen associated with diabetes, has been expressed to high levels in tobacco and potato. When fed to mice that were susceptible to diabetes, the GAD-containing tissue prevented diabetes in 10 of 12 mice compared to 4 of 12 mice fed control plant tissue. These tolerance-related effects are similar to those found when immunizing with GAD derived from other sources (Ma *et al.,* 1997). Another example, also involving diabetes arising from autoimmunity, involved the expression of insulin fused to CT-B. Again, diabetic symptoms were reduced by induction of tolerance in susceptible mice (Arakawa *et al.,* 1998b, 1999).

C. Immune Responses to Plant-Derived Antigens

Initial immune responses to plant-based vaccines were disappointing. For example, hepatitis B surface antigen (HBsAg) was expressed initially in transgenic tobacco and resulted in the formation of virus-like particles (Mason *et al.,* 1992). When a crude extract containing 3% of HBsAg was administered parenterally, an immune response was developed by mice. The response achieved was lower than that induced by the yeast-derived protein, the difference being attributed to the impurity and low concentrations of the plant-derived antigen (Thanavala *et al.,* 1995). Expression of hepatitis B surface antigen in lettuce produced less than 5 ng/g fresh weight of antigen. When injected into humans, only low levels of serum antibodies were induced (Kapusta *et al.,* 1999). Improvements have resulted in the increased expression of hepatitis B surface antigen in potatoes (Richter *et al.,* 2000) and administering the raw potatoes orally induced HBsAg antibody levels greater than those defined as protective (Kong *et al.,* 2001).

More recently, good protection against *E. coli* LT-B and transmissible gastroenteritis virus (TGEV) has been demonstrated using a plant-based expression system (Streatfield *et al.,* 2001). Both LT-B and the spike protein from TGEV, generated via *Agrobacterium*-mediated transformation, were expressed in transgenic corn. Mice were fed 5- or 50-μg doses of recombinant LT-B expressed in corn or 50 μg of wild-type LT-B mixed with their normal feed. All mice receiving LT-B developed increased IgG and IgA antibody responses compared to the controls. In fact, the mucosal IgA response was higher for the recombinant LT-B than for the wild-type LT-B, suggesting that the protection provided by the plant cells allowed better protection during passage through the stomach. When challenged with LT-B, diarrheal symptoms were eliminated. It was also shown that when piglets were fed 2 mg of TGEV protein in corn, they developed significant antibody responses (Streatfield *et al.,* 2001). When challenged with TGEV, the vaccinated animals showed approximately 50% reduction in mortality and severity compared to the control piglets. When compared to the commercial attenuated vaccine, the transgenic-corn-fed piglets showed greater protection, demonstrating the commercial potential of this method.

Plant-based vaccines expressing parvovirus epitopes in the coat protein of CPMV have proven very effective. The 17-amino-acid peptide 3L17 from the S protein of parvoviruses has been shown to provide effective protection against canine parvovirus and mink enteritis virus (MEV) (Dalsgaard *et al.,* 1997; Langeveld *et al.,* 1995). The peptide was expressed on the surface of the CPMV S coat protein and VLPs were purified from the plant material. In one of the experiments (Dalsgaard *et al.,* 1997), yields of 1–1.2 mg of virus particles were obtained per gram of black-eyed bean plants. Subcutaneous injection of the particles (with a saponin adjuvant and adsorbed to an aluminum hydroxide gel) resulted in the production of antibodies and protection from MEV in 11 of 12 of the animals immunized. It was noted that antibodies against CPMV were also present, which could cause problems if multiple vaccinations are required. It was also demonstrated that the CPMV virus could be ultraviolet-inactivated without affecting the immune response (Langeveld *et al.,* 2001).

The structural protein VP1 from foot-and-mouth-disease virus (FMDV) was stably expressed in *Arabidopsis thaliana* (Carrillo *et al.,* 1998) and alfalfa (Wigdorovitz *et al.,* 1999a). Intraperitoneal injection of partially purified protein from *Arabidopsis* resulted in complete protection, whereas protein from alfalfa plants conferred protection in only 12 of 17 mice. In another experiment, the entire VP1 protein was expressed in tobacco plants using TMV. Intraperitoneal injection of crude foliar plant extract elicited antibodies in all mice. When challenged, all 30 immunized mice were protected from infection, whereas all the 24 controls became infected (Wigdorovitz *et al.,* 1999b).

These examples show the potential of plant-expressed vaccines. Successful vaccines have been produced from both transgenic plants and plant viruses, from expression of entire proteins and from small epitopes, and by using injection and oral delivery. Evidence has not shown that any particular method is best and success may depend more on the choice of antigen than on the method of expression and delivery.

IX. Edible Vaccines and Human Health

A. The Need for Vaccines

Twenty percent of the world's children are not immunized for diphtheria, measles, pertussis (whooping cough), polio, tetanus and tuberculosis. These diseases account for the deaths of more than 2 million children every year. Effective vaccines exist for all these diseases. However, in many developing countries, immunization programs are either nonexistent, unreliable, or too costly to maintain. In addition, there are many diseases for which there are no adequate vaccines. For example, it is estimated that rotavirus is responsible for over 870,000 deaths per year worldwide, with most occurring in developing countries (Hoshino and Kapikian, 1994). Over 1 billion dollars is spent in the United States alone on treatment of this disease (Glass *et al.,* 1996).

These examples illustrate the requirement for improved low-cost vaccines. Developing countries currently cannot afford the health care costs required to control these diseases. Even the cost required for vaccination can prove to be excessive because refrigeration methods and the trained personnel required for current vaccines are not available. Development of a plant-based vaccine that is edible could possibly solve some of these problems. Protein expressed in plants is stable at normal temperatures, eliminating the need for refrigeration, and oral vaccines do not require trained personnel for administration. There is also the potential for developing countries to produce their own vaccine. Choice of the appropriate plant would allow the plant to be grown in the country of use. This would reduce the cost of making the vaccine as well as transport costs and reduce the time from production to delivery. In principle, for an edible vaccine, dosages could be controlled by grinding and mixing the edible plant parts to homogeneity and then analyzing the protein content to determine the correct dosage.

B. Human Trials of Edible Plant Vaccines

In 1997 the first human oral vaccination trials of a plant-expressed protein were performed (Tacket *et al.,* 1998). When 50–100 g of raw, transgenic potato expressing LT-B antigen from *E. coli* was fed to participants, antibodies were detected in 10 of 11 participants and in none of the controls. These antibody levels were equivalent to those obtained when volunteers are subjected to 10^6 infectious *E. coli*.

In another trial (Tacket *et al.,* 2000), transgenic potatoes containing Norwalk virus capsid protein were fed to volunteers. Analysis showed that 25–50% of the viral protein assembled into VLPs. Ten individuals were fed 150-g doses of raw potato on days 0, 7, and 21. Doses varied from 215 to 751 μg of VLPs per person depending on the batch of potatoes. Serum IgG increases were modest, as only 35% of the volunteers had serum increases equivalent to those caused by ingestion of 2×250-μg doses of baculovirus-produced VLPs in bicarbonate buffer (Ball *et al.,* 1996). This difference in response may be caused by not all the capsid protein being in the form of VLPs and the free protein having little immunogenicity (Tacket *et al.,* 2000). Another factor may be that the doses used for priming the immune system were too low. In the successful pig feeding trial with TGEV spike protein (Streatfield *et al.,* 2001), 10-day-old piglets were fed two doses of 2 mg, a much higher dose of protein relative to body size than had been previously used. Although there is clearly much research yet to be done, these trials demonstrate the feasibility of developing edible vaccines for human use.

X. Conclusion

Before embarking on the expression of an antigen in plants, careful consideration of the objectives should be made. The intended market, the type of pathogen, the mode of infection, and the requirements of the target group will all have an impact on the design of the vaccine. Clearly, not all antigens are suitable for expression in plants. Plant-based expression is the most suitable route for low-value subunit vaccines that do not require a significant level of purification.

A number of investigators have now demonstrated that it is possible for mucosal delivery of a plant-derived antigen to induce both a mucosal and a systemic response (Arakawa *et al.,* 1998a; Haq *et al.,* 1995; Koo *et al.,* 1999; Mason *et al.,* 1996, 1998). To date, no adverse reactions have been reported during initial clinical studies. In general, nasal vaccination has given better immune responses than has oral vaccination, when the two have been compared directly (Brennan *et al.,* 1999; Durrani *et al.,* 1998). This is likely to be the result of protein degradation in the harsh environment of the alimentary tract. Despite these problems, successful oral vaccines are being generated in plants (Streatfield *et al.,* 2001; Yu and Langridge, 2001). Mucosal vaccines of this type have the advantage of being easier to deliver, requiring less technical expertise and equipment. This is especially true for oral vaccines, explaining the persistence of the researchers, notwithstanding the difficulties encountered.

Obviously, the field of plant vaccines is still in the formative stages. Immunization trials have not involved large numbers of animals and no doubt all the effects of plant vaccines have not yet been discovered. The most successful plant-derived vaccines appear to be those where the greatest yield has been achieved. As a consequence, the importance of achieving high levels of protein expression levels cannot be understated. Expression using viruses or chloroplast–genome integration appear to represent the two most promising methods for achieving high yields of foreign proteins in plants.

Acknowledgment

We acknowledge the valuable comments and suggestions provided by Diana Schubauer, who provided helpful advice on the immunology portions of this manuscript.

References

Adam, Z. (2000). Chloroplast proteases: Possible regulators of gene expression? *Biochimie* **82,** 647–654.

Ahlquist, P., and Janda, M. (1984). cDNA cloning and *in vitro* transcription of the complete brome mosaic virus genome. *Mol. Cell. Biol.* **4,** 2876–2882.

Albert, M. L., Sauter, B., and Bhardwaj, N. (1998). Dendritic cells acquire antigen from apoptotic cells and induce class I-restricted CTLs. *Nature* **392,** 86–89.

Allan, C. H., Mendrick, D. L., and Trier, J. S. (1993). Rat intestinal M cells contain acidic endosomal–lysosomal compartments and express class II major histocompatibility complex determinants. *Gastroenterology* **104,** 698–708.

Arakawa, T., Chong, D. K., and Langridge, W. H. (1998a). Efficacy of a food plant-based oral cholera toxin B subunit vaccine. *Nature Biotechnol.* **16,** 292–297.

Arakawa, T., Yu, J., Chong, D. K., Hough, J., Engen, P. C., and Langridge, W. H. (1998b). A plant-based cholera toxin B subunit–insulin fusion protein protects against the development of autoimmune diabetes. *Nature Biotechnol.* **16,** 934–938.

Arakawa, T., Yu, J., and Langridge, W. H. (1999). Food plant-delivered cholera toxin B subunit for vaccination and immunotolerization. *Adv. Exp. Med. Biol.* **464,** 161–178.

Arntzen, C. J. (1997). High-tech herbal medicine: Plant-based vaccines. *Nature Biotechnol.* **15,** 221–222.

Arntzen, C. J. (1998). Pharmaceutical foodstuffs: Oral immunization with transgenic plants. *Nature Med.* **4**(5 Suppl.), 502–503.

Ball, J. M., Estes, M. K., Hardy, M. E., Conner, M. E., Opekun, A. R., and Graham, D. Y. (1996). Recombinant Norwalk virus-like particles as an oral vaccine. *Arch. Virol. Suppl.* **12,** 243–249.

Baratova, L. A., Grebenshchikov, N. I., Shishkov, A. V., Kashirin, I. A., Radavsky, J. L., Jarvekulg, L., and Saarma, M. (1992). The topography of the surface of potato virus X: Tritium planigraphy and immunological analysis. *J. Gen. Virol.* **73**(Part 2), 229–235.

Baron, L. S., Kopecko, D. J., Formal, S. B., Seid, R., Guerry, P., and Powell, C. (1987). Introduction of *Shigella flexneri* 2a type and group antigen genes into oral typhoid vaccine strain *Salmonella typhi* Ty21a. *Infect. Immun.* **55**(11), 2797–2801.

Beachy, R. N. (1997). Plant biotechnology: The now and then of plant biotechnology. *Curr. Opin. Biotechnol.* **8,** 187–188.

Beachy, R. N., Fitchen, J. H., and Hein, M. B. (1996). Use of plant viruses for delivery of vaccine epitopes. *Ann. N.Y. Acad. Sci.* **792,** 43–49.

Belanger, H., Fleysh, N., Cox, S., Bartman, G., Deka, D., Trudel, M., Koprowski, H., and Yusibov, V. (2000). Human respiratory syncytial virus vaccine antigen produced in plants. *FASEB J.* **14,** 2323–2328.

Bendahmane, M., Koo, M., Karrer, E., and Beachy, R. N. (1999). Display of epitopes on the surface of tobacco mosaic virus: Impact of charge and isoelectric point of the epitope on virus–host interactions. *J. Mol. Biol.* **290,** 9–20.

Berzofsky, J. A., Ahlers, J. D., Derby, M. A., Pendleton, C. D., Arichi, T., and Belyakov, I. M. (1999). Approaches to improve engineered vaccines for human immunodeficiency virus and other viruses that cause chronic infections. *Immunol. Rev.* **170,** 151–172.

Bock, R. (2001). Transgenic plastids in basic research and plant biotechnology. *J. Mol. Biol.* **312,** 425–438.

Booth, B., and Rodger, J. (1999–2000). *Annual Report.* Cooperative Research Centre for the Conservation and Management of Marsupials, Macquarie Centre, NSW, Australia.

Brennan, F. R., Bellaby, T., Helliwell, S. M., Jones, T. D., Kamstrup, S., Dalsgaard, K., Flock, J. I., and Hamilton, W. D. (1999). Chimeric plant virus particles administered nasally or orally induce systemic and mucosal immune responses in mice. *J. Virol.* **73,** 930–938.

Brunt, L. M., Portnoy, D. A., and Unanue, E. R. (1990). Presentation of *Listeria* monocytogenes to CD8+ T cells requires secretion of hemolysin and intracellular bacterial growth. *J. Immunol.* **145,** 3540–3546.

Buseyne, F., Le Gall, S., Boccaccio, C., Abastado, J. P., Lifson, J. D., Arthur, L. O., Riviere, Y., Heard, J. M., and Schwartz, O. (2001). MHC-I-restricted presentation of HIV-1 virion antigens without viral replication. *Nature Med.* **7,** 344–349.

Butler, P. J. (1984). The current picture of the structure and assembly of tobacco mosaic virus. *J. Gen. Virol.* **65**(Part 2), 253–279.

Cabanes-Macheteau, M., Fitchette-Laine, A. C., Loutelier-Bourhis, C., Lange, C., Vine, N. D., Ma, J. K., Lerouge, P., and Faye, L. (1999). N-Glycosylation of a mouse IgG expressed in transgenic tobacco plants. *Glycobiology* **9,** 365–372.

Carrillo, C., Wigdorovitz, A., Oliveros, J. C., Zamorano, P. I., Sadir, A. M., Gomez, N., Salinas, J., Escribano, J. M., and Borca, M. V. (1998). Protective immune response to foot-and-mouth disease virus with VP1 expressed in transgenic plants. *J. Virol.* **72,** 1688–1690.

Carrington, J. C., Haldeman, R., Dolja, V. V., and Restrepo-Hartwig, M. A. (1993). Internal cleavage and trans-proteolytic activities of the VPg-proteinase (NIa) of tobacco etch potyvirus *in vivo*. *J. Virol.* **67**(12), 6995–7000.

Centers for Disease Control and Prevention. (1999). Withdrawal of rotavirus vaccine recommendation. *J. Am. Med. Assoc.* **282,** 2113–2114.

Cohen, A., and Mayfield, S. P. (1997). Translational regulation of gene expression in plants. *Curr. Opin. Biotechnol.* **8,** 189–194.

Cooney, E. L., Collier, A. C., Greenberg, P. D., Coombs, R. W., Zarling, J., Arditti, D. E., Hoffman, M. C., Hu, S. L., and Corey, L. (1991). Safety of and immunological response to a recombinant vaccinia virus vaccine expressing HIV envelope glycoprotein. *Lancet* **337**(8741), 567–572.

Copeman, R. J., Hartman, J. R., and Watterson, J. C. (1969). Tobacco mosaic virus concentration in inoculated and systemically infected tobacco leaves. *Phytopathology* **59,** 1012–1013.

Cruz, S. S., Chapman, S., Roberts, A. G., Roberts, I. M., Prior, D. A., and Oparka, K. J. (1996). Assembly and movement of a plant virus carrying a green fluorescent protein overcoat. *Proc. Natl. Acad. Sci. USA* **93,** 6286–6290.

Curtiss, R. I., and Cardineau, C. A. (1990). *Oral immunisation by transgenic plants.* World Patent Application WO 90/02484.

Dale, W. (1949). Observations on virus disease of cowpea in Trinidad. *Ann. Appl. Biol.* **36,** 327–333.

Dalsgaard, K., Uttenthal, A., Jones, T. D., Xu, F., Merryweather, A., Hamilton, W. D., Langeveld, J. P., Boshuizen, R. S., Kamstrup, S., Lomonossoff, G. P., Porta, C., Vela, C., Casal, J. I., Meloen, R. H., and Rodgers, P. B. (1997). Plant-derived vaccine protects target animals against a viral disease. *Nature Biotechnol.* **15,** 248–252.

Daniell, H., Lee, S. B., Panchal, T., and Wiebe, P. O. (2001a). Expression of the native cholera toxin b subunit gene and assembly as functional oligomers in transgenic tobacco chloroplasts. *J. Mol. Biol.* **311,** 1001–1009.

Daniell, H., Streatfield, S. J., and Wycoff, K. (2001b). Medical molecular farming: Production of antibodies, biopharmaceuticals and edible vaccines in plants. *Trends Plant Sci.* **6,** 219–226.

Dawson, W. O., Lewandowski, D. J., Hilf, M. E., Bubrick, P., Raffo, A. J., Shaw, J. J., Grantham, G. L., and Desjardins, P. R. (1989). A tobacco mosaic virus-hybrid expresses and loses an added gene. *Virology* **172,** 285–292.

de Aizpurua, H. J., and Russell-Jones, G. J. (1988). Oral vaccination. Identification of classes of proteins that provoke an immune response upon oral feeding. *J. Exp. Med.* **167,** 440–451.

De Cosa, B., Moar, W., Lee, S. B., Miller, M., and Daniell, H. (2001). Overexpression of the Bt cry2Aa2 operon in chloroplasts leads to formation of insecticidal crystals. *Nature Biotechnol.* **19,** 71–74.

De Groot, A. S., Bosma, A., Chinai, N., Frost, J., Jesdale, B. M., Gonzalez, M. A., Martin, W., and Saint-Aubin, C. (2001). From genome to vaccine: *In silico* predictions, *ex vivo* verification. *Vaccine* **19,** 4385–4395.

De Wilde, C., Van Houdt, H., De Buck, S., Angenon, G., De Jaeger, G., and Depicker, A. (2000). Plants as bioreactors for protein production: Avoiding the problem of transgene silencing. *Plant Mol. Biol.* **43,** 347–359.

De Zoeten, G. A., Penswick, J. R., Horisberger, M. A., Ahl, P., Schultze, M., and Hohn, T. (1989). The expression, localization, and effect of a human interferon in plants. *Virology* **172,** 213–222.

Division of Microbiology and Infectious Diseases, National Institute of Allergy and Infectious Diseases, National Institutes of Health. (1998). "The Jordan Report: Accelerated Development of Vaccines." Author, Bethesda, Maryland.

Dolja, V. V., McBride, H. J., and Carrington, J. C. (1992). Tagging of plant potyvirus replication and movement by insertion of B-glucuronidase (GUS) into the viral polyprotein. *Proc. Natl. Acad. Sci. USA* **89,** 10280–10212.

Dolja, V. V., Peremyslov, V. V., Keller, K. E., Martin, R. R., and Hong, J. (1998). Isolation and stability of histidine-tagged proteins produced in plants via potyvirus gene vectors. *Virology* **252,** 269–274.

Donnelly, M. L., Hughes, L. E., Luke, G., Mendoza, H., ten Dam, E., Gani, D., and Ryan, M. D. (2001a). The "cleavage" activities of foot-and-mouth disease virus 2A site-directed mutants and naturally occurring "2A-like" sequences. *J. Gen. Virol* **82**(Part 5), 1027–1041.

Donnelly, M. L., Luke, G., Mehrotra, A., Li, X., Hughes, L. E., Gani, D., and Ryan, M. D. (2001b). Analysis of the aphthovirus 2A/2B polyprotein "cleavage" mechanism indicates not a proteolytic reaction, but a novel translational effect: A putative ribosomal "skip." *J. Gen. Virol.* **82**(Part 5), 1013–1025.

Donson, J., Kearney, C. M., Hilf, M. E., and Dawson, W. O. (1991). Systemic expression of a bacterial gene by a tobacco mosaic virus-based vector. *Proc. Natl. Acad. Sci. USA* **88,** 7204–7208.

Dougan, G., Ghaem-Maghami, M., Pickard, D., Frankel, G., Douce, G., Clare, S., Dunstan, S., and Simmons, C. (2000). The immune responses to bacterial antigens encountered *in vivo* at mucosal surfaces. *Phil. Trans. R. Soc. Lond. B Biol. Sci.* **355,** 705–712.

Durrani, Z., McInerney, T. L., McLain, L., Jones, T., Bellaby, T., Brennan, F. R., and Dimmock, N. J. (1998). Intranasal immunization with a plant virus expressing a peptide from HIV-1 gp41 stimulates better mucosal and systemic HIV-1-specific IgA and IgG than oral immunization. *J. Immunol. Meth.* **220,** 93–103.

Ebbert, G. B., Mascolo, E. D., and Six, H. R. (1999). Overview of vaccine manufacturing and quality assurance. *In* "Vaccines," 3rd ed. (S. A. Plotkin and W. A. Orenstein, eds.), pp. 40–46. Saunders, Philadelphia.

Ellis, R. W. (1999). New technologies for making vaccines. *Vaccine* **17,** 1596–1604.

Estes, M. K., Ball, J. M., Crawford, S. E., O'Neal, C., Opekun, A. A., Graham, D. Y., and Conner, M. E. (1997). Virus-like particle vaccines for mucosal immunization. *Adv. Exp. Med. Biol.* **412,** 387–395.

Evangelista, R. L., Kusnadi, A. R., Howard, J. A., and Nikolov, Z. L. (1998). Process and economic evaluation of the extraction and purification of recombinant beta-glucuronidase from transgenic corn. *Biotechnol. Prog.* **14,** 607–614.

Fernandez-Fernandez, M. R., Martinez-Torrecuadrada, J. L., Casal, J. I., and Garcia, J. A. (1998). Development of an antigen presentation system based on plum pox potyvirus. *FEBS Lett.* **427,** 229–235.

Fischer, R., Vaquero-Martin, C., Sack, M., Drossard, J., Emans, N., and Commandeur, U. (1999). Towards molecular farming in the future: Transient protein expression in plants. *Biotechnol. Appl. Biochem.* **30,** 113–116.

Fitchen, J., Beachy, R. N., and Hein, M. B. (1995). Plant virus expressing hybrid coat protein with added murine epitope elicits autoantibody response. *Vaccine* **13,** 1051–1057.

Fooks, A. R. (2000). Development of oral vaccines for human use. *Curr. Opin. Mol. Ther.* **2,** 80–86.

Francis, M. (1991). Enhanced immunogenicity of recombinant and synthetic peptide vaccines. *In* "Vaccines" (G. Gregoriadis, A. Allison, and G. Poste, Eds.), pp. 13–23. Plenum Press, New York.

Gallois, P., and Marinho, P. (1995). Leaf disk transformation using *Agrobacterium tumefaciens*-expression of heterologous genes in tobacco. *Meth. Mol. Biol.* **49,** 39–48.

Gasser, C. S., and Fraley, R. T. (1989). Genetically engineering plants for crop improvements. *Science* **244,** 1293–1299.

Gilleland, H. E., Gilleland, L. B., Staczek, J., Harty, R. N., Garcia-Sastre, A., Palese, P., Brennan, F. R., Hamilton, W. D., Bendahmane, M., and Beachy, R. N. (2000). Chimeric animal and plant viruses expressing epitopes of outer membrane protein F as a combined vaccine against *Pseudomonas aeruginosa* lung infection. *FEMS Immunol. Med. Microbiol.* **27,** 291–297.

Glass, R. I., Kilgore, P. E., and Holman, R. C. (1996). The epidemiology of rotavirus diarrhea in the United States: Surveillance and estimates of disease burden. *J. Infect. Dis.* **174**(Suppl. 1), S5–S11.

Goddijn, O. J. M., and Pen, J. (1995). Plants as bioreactors. *Trends Biotechnol.* **13,** 379–387.

Grandi, G. (2001). Antibacterial vaccine design using genomics and proteomics. *Trends Biotechnol.* **19,** 181–188.

Hamamoto, H., Sugiyama, Y., Nakagawa, N., Hashida, E., Matsunaga, Y., Takemoto, S., Watanabe, Y., and Okada, Y. (1993). A new tobacco mosaic virus vector and its use for the systemic production of angiotensin-I-converting enzyme inhibitor in transgenic tobacco and tomato. *Biotechnology (NY)* **11,** 930–932.

Hansen, G., and Chilton, M. D. (1999). Lessons in gene transfer to plants by a gifted microbe. *Curr. Top. Microbiol. Immunol.* **240,** 21–57.

Haq, T. A., Mason, H. S., Clements, J. D., and Arntzen, C. J. (1995). Oral immunization with a recombinant bacterial antigen produced in transgenic plants. *Science* **268,** 714–716.

Harrison, S. C., Olson, A. J., Schutt, C. E., Winkler, F. K., and Bricogne, G. (1978). Tomato bushy stunt virus at 2.9 Å resolution. *Nature* **276,** 368–373.

Heifetz, P. B. (2000). Genetic engineering of the chloroplast. *Biochimie* **82,** 655–666.

Herbers, K., and Sonnewald, U. (1999). Production of new/modified proteins in transgenic plants. *Curr. Opin. Biotechnol.* **10,** 163–168.

Holmes, I. H. (1983). Rotaviruses. *In* "The Reoviridae" (W. K. Joklik, Ed.), pp. 359–423. Plenum Press, New York.

Hooykaas, P. J. (1989). Transformation of plant cells via *Agrobacterium*. *Plant Mol. Biol.* **13,** 327–336.

Hoshino, Y., and Kapikian, A. Z. (1994). Rotavirus vaccine development for the prevention of severe diarrhea in infants and young children. *Trends Microbiol.* **2,** 242–249.

Hwang, D. J., Roberts, I. M., and Wilson, T. M. (1994). Expression of tobacco mosaic virus coat protein and assembly of pseudovirus particles in *Escherichia coli*. *Proc. Natl. Acad. Sci. USA* **91,** 9067–9071.

Jagadish, M. N., Ward, C. W., Gough, K. H., Tulloch, P. A., Whittaker, L. A., and Shukla, D. D. (1991). Expression of potyvirus coat protein in *Escherichia coli* and yeast and its assembly into virus-like particles. *J. Gen. Virol.* **72**(Part 7), 1543–1550.

Jagadish, M. N., Hamilton, R. C., Fernandez, C. S., Schoofs, P., Davern, K. M., Kalnins, H., Ward, C. W., and Nisbet, I. T. (1993). High level production of hybrid potyvirus-like particles carrying repetitive copies of foreign antigens in *Escherichia coli*. *Biotechnology (NY)* **11,** 1166–1170.

Jagadish, M. N., Edwards, S. J., Hayden, M. B., Grusovin, J., Vandenberg, K., Schoofs, P., Hamilton, R. C., Shukla, D. D., Kalnins, H., McNamara, M., Haynes, J., Nisbet, I. T., Ward, C. W., and Pye, D. (1996). Chimeric potyvirus-like particles as vaccine carriers. *Intervirology* **39,** 85–92.

Jaspars, E. M. J. (1985). Interaction of alfalfa virus nucleic acids and protein. *In* "Molecular Plant Virology," Vol. 1 (J. W. Davies, Ed.), pp. 151–221, CRC Press, Boca Raton, Florida.

Javaherian, K., Langlois, A. J., McDanal, C., Ross, K. L., Eckler, L. I., Jellis, C. L., Profy, A. T., Rusche, J. R., Bolognesi, D. P., Putney, S. D., *et al.* (1989). Principal neutralizing domain of the human immunodeficiency virus type 1 envelope protein. *Proc. Natl. Acad. Sci. USA* **86,** 6768–6772.

Joelson, T., Åkerblom, L., Oxelfelt, P., Strandberg, B., Tomenius, K., and Morris, T. J. (1997). Presentation of a foreign peptide on the surface of tomato bushy stunt virus. *J. Gen. Virol.* **78**(Part 6), 1213–1217.

Kapusta, J., Modelska, A., Figlerowicz, M., Pniewski, T., Letellier, M., Lisowa, O., Yusibov, V., Koprowski, H., Plucienniczak, A., and Legocki, A. B. (1999). A plant-derived edible vaccine against hepatitis B virus. *FASEB J.* **13,** 1796–1799.

Kingsman, S. M., and Kingsman, A. J. (1988). Polyvalent recombinant antigens: A new vaccine strategy. *Vaccine* **6,** 304–306.

Kohler, H., Goudsmit, J., and Nara, P. (1992). Clonal dominance: Cause for a limited and failing immune response to HIV-1 infection and vaccination. *J. Acquired Immune Defic. Syndr.* **5,** 1158–1168.

Koncz, C., Kreuzaler, F., Kalman, Z., and Schell, J. (1984). A simple method to transfer, integrate and study expression of foreign genes, such as chicken ovalbumin and alpha-actin in plant tumors. *EMBO J.* **3,** 1029–1037.

Kong, Q., Richter, L., Yang, Y. F., Arntzen, C. J., Mason, H. S., and Thanavala, Y. (2001). Oral immunization with hepatitis B surface antigen expressed in transgenic plants. *Proc. Natl. Acad. Sci. USA* **98,** 11539–11544.

Koo, M., Bendahmane, M., Lettieri, G. A., Paoletti, A. D., Lane, T. E., Fitchen, J. H., Buchmeier, M. J., and Beachy, R. N. (1999). Protective immunity against murine hepatitis virus (MHV) induced by intranasal or subcutaneous administration of hybrids of tobacco mosaic virus that carries an MHV epitope. *Proc. Natl. Acad. Sci. USA* **96,** 7774–7779.

Koprowski, H., and Yusibov, V. (2001). The green revolution: Plants as heterologous expression vectors. *Vaccine* **19,** 2735–2741.

Kusnadi, A., Nikolov, Z., and Howard, J. (1997). Production of recombinant proteins in transgenic plants: Practical considerations. *Biotechnol. Bioeng.* **56,** 473–484.

Langeveld, J. P., Kamstrup, S., Uttenthal, A., Strandbygaard, B., Vela, C., Dalsgaard, K., Beekman, N. J., Meloen, R. H., and Casal, J. I. (1995). Full protection in mink against mink enteritis virus with new generation canine parvovirus vaccines based on synthetic peptide or recombinant protein. *Vaccine* **13,** 1033–1037.

Langeveld, J. P., Brennan, F. R., Martinez-Torrecuadrada, J. L., Jones, T. D., Boshuizen, R. S., Vela, C., Casal, J. I., Kamstrup, S., Dalsgaard, K., Meloen, R. H., Bendig, M. M., and Hamilton, W. D. (2001). Inactivated recombinant plant virus protects dogs from a lethal challenge with canine parvovirus. *Vaccine* **19,** 3661–3370.

Langridge, W. H. (2000). Edible vaccines. *Sci. Am.* **283,** 66–71.

Lauterslager, T. G., Florack, D. E., van der Wal, T. J., Molthoff, J. W., Langeveld, J. P., Bosch, D., Boersma, W. J., and Hilgers, L. A. (2001). Oral immunisation of naive and primed animals with transgenic potato tubers expressing LT-B. *Vaccine* **19,** 2749–2755.

Lerouge, P., Cabanes-Macheteau, M., Rayon, C., Fischette-Laine, A. C., Gomord, V., and Faye, L. (1998). N-Glycoprotein biosynthesis in plants: Recent developments and future trends. *Plant Mol. Biol.* **38,** 31–48.

Lerouge, P., Bardor, M., Pagny, S., Gomord, V., and Faye, L. (2000). N-Glycosylation of recombinant pharmaceutical glycoproteins produced in transgenic plants: Towards an humanisation of plant N-glycans. *Curr. Pharm. Biotechnol.* **1,** 347–354.

Lin, T., Porta, C., Lomonossoff, G., and Johnson, J. E. (1996). Structure-based design of peptide presentation on a viral surface: The crystal structure of a plant/animal virus chimera at 2.8 Å resolution. *Fold. Des.* **1,** 179–187.

Lin, T., Chen, Z., Usha, R., Stauffacher, C. V., Dai, J. B., Schmidt, T., and Johnson, J. E. (1999). The refined crystal structure of cowpea mosaic virus at 2.8 Å resolution. *Virology* **265,** 20–34.

Lindner, J., Geczy, A. F., and Russell-Jones, G. J. (1994). Identification of the site of uptake of the *E. coli* heat-labile enterotoxin, LTB. *Scand. J. Immunol.* **40,** 564–572.

Lomonossoff, G. P., and Johnson, J. E. (1991). The synthesis and structure of comovirus capsids. *Prog. Biophys. Mol. Biol.* **55,** 107–137.

Lomonossoff, G. P., and Johnson, J. E. (1996). Use of macromolecular assemblies as expression systems for peptides and synthetic vaccines. *Curr. Opin. Struct. Biol.* **6,** 176–182.

Ma, J. K. (2000). Genes, greens, and vaccines. *Nature Biotechnol.* **18,** 1141–1142.

Ma, S. W., Zhao, D. L., Yin, Z. Q., Mukherjee, R., Singh, B., Qin, H. Y., Stiller, C. R., and Jevnikar, A. M. (1997). Transgenic plants expressing autoantigens fed to mice to induce oral immune tolerance. *Nature Med.* **3,** 793–796.

Mason, H. S., Lam, D. M., and Arntzen, C. J. (1992). Expression of hepatitis B surface antigen in transgenic plants. *Proc. Natl. Acad. Sci. USA* **89,** 11745–11749.

Mason, H. S., Ball, J. M., Shi, J. J., Jiang, X., Estes, M. K., and Arntzen, C. J. (1996). Expression of Norwalk virus capsid protein in transgenic tobacco and potato and its oral immunogenicity in mice. *Proc. Natl. Acad. Sci. USA* **93,** 5335–5340.

Mason, H. S., Haq, T. A., Clements, J. D., and Arntzen, C. J. (1998). Edible vaccine protects mice against *Escherichia coli* heat-labile enterotoxin (LT): Potatoes expressing a synthetic LT-B gene. *Vaccine* **16,** 1336–1343.

Matzke, A. J., and Matzke, M. A. (1998). Position effects and epigenetic silencing of plant transgenes. *Curr. Opin. Plant Biol.* **1,** 142–148.

McBride, K. E., Svab, Z., Schaaf, D. J., Hogan, P. S., Stalker, D. M., and Maliga, P. (1995). Amplification of a chimeric *Bacillus* gene in chloroplasts leads to an extraordinary level of an insecticidal protein in tobacco. *Biotechnology (NY)* **13,** 362–365.

McGarvey, P. B., Hammond, J., Dienelt, M. M., Hooper, D. C., Fu, Z. F., Dietzschold, B., Koprowski, H., and Michaels, F. H. (1995). Expression of the rabies virus glycoprotein in transgenic tomatoes. *Biotechnology (NY)* **13,** 1484–1487.

McLain, L., Porta, C., Lomonossoff, G. P., Durrani, Z., and Dimmock, N. J. (1995). Human immunodeficiency virus type 1-neutralizing antibodies raised to a glycoprotein 41 peptide expressed on the surface of a plant virus. *AIDS Res. Hum. Retrovir.* **11,** 327–334.

McNeela, E. A., and Mills, K. H. (2001). Manipulating the immune system: Humoral versus cell-mediated immunity. *Adv. Drug Deliv. Rev.* **51,** 43–54.

Miele, L. (1997). Plants as bioreactors for biopharmaceuticals: Regulatory considerations. *Trends Biotechnol.* **15,** 45–50.

Modelska, A., Dietzschold, B., Sleysh, N., Fu, Z. F., Steplewski, K., Hooper, D. C., Koprowski, H., and Yusibov, V. (1998). Immunization against rabies with plant-derived antigen. *Proc. Natl. Acad. Sci. USA* **95,** 2481–2485.

Mori, M., Mise, K., Kobayashi, K., Okuno, T., and Furusawa, I. (1991). Infectivity of plasmids containing brome mosaic virus cDNA linked to the cauliflower mosaic virus 35S RNA promoter. *J. Gen. Virol.* **72**(Part 2), 243–246.

Namba, K., and Stubbs, G. (1986). Structure of tobacco mosaic virus at 3.6 Å resolution: Implications for assembly. *Science* **231,** 1401–1406.

Namba, K., Pattanayek, R., and Stubbs, G. (1989). Visualization of protein–nucleic acid interactions in a virus. Refined structure of intact tobacco mosaic virus at 2.9 Å resolution by X-ray fiber diffraction. *J. Mol. Biol.* **208,** 307–325.

Nemchinov, L. G., Liang, T. J., Rifaat, M. M., Mazyad, H. M., Hadidi, A., and Keith, J. M. (2000). Development of a plant-derived subunit vaccine candidate against hepatitis C virus. *Arch. Virol.* **145,** 2557–2573.

Neutra, M. R., Frey, A., and Kraehenbuhl, J. P. (1996). Epithelial M cells: Gateways for mucosal infection and immunization. *Cell* **86,** 345–348.

Newell, C. A. (2000). Plant transformation technology. Developments and applications. *Mol. Biotechnol.* **16,** 53–65.

O'Brien, G. J., Bryant, C. J., Voogd, C., Greenberg, H. B., Gardner, R. C., and Bellamy, A. R. (2000). Rotavirus VP6 expressed by PVX vectors in *Nicotiana benthamiana* coats PVX rods and also assembles into viruslike particles. *Virology* **270,** 444–453.

Ohtani, T., Galili, G., Wallace, J. C., Thompson, G. A., and Larkins, B. A. (1991). Normal and lysine-containing zeins are unstable in transgenic tobacco seeds. *Plant Mol. Biol.* **16,** 117–128.

Olson, A. J., Bricogne, G., and Harrison, S. C. (1983). Structure of tomato bushy stunt virus IV. The virus particle at 2.9 Å resolution. *J. Mol. Biol.* **171,** 61–93.

Olszewska, W., and Steward, M. W. (2001). Nasal delivery of epitope based vaccines. *Adv. Drug. Deliv. Rev.* **51,** 161–171.

O'Neal, C. M., Crawford, S. E., Estes, M. K., and Conner, M. E. (1997). Rotavirus virus-like particles administered mucosally induce protective immunity. *J. Virol.* **71,** 8707–8717.

Paszkowski, J., Shillito, R. D., Saul, M., Vandak, V., Hohn, T., Hohn, B., and Potrykus, I. (1992). Direct gene transfer to plants. *Biotechnology* **24,** 387–392.

Phillips, R. E., Rowland-Jones, S., Nixon, D. F., Gotch, F. M., Edwards, J. P., Ogunlesi, A. O., Elvin, J. G., Rothbard, J. A., Bangham, C. R., Rizza, C. R., *et al.* (1991). Human immunodeficiency virus genetic variation that can escape cytotoxic T cell recognition. *Nature* **354,** 453–459.

Poierier, Y., Dennis, D. E., Klomparens, K., and Somerville, C. (1992). Polyhydoxybutyrate, a biodegradable thermoplastic, produced in transgenic plants. *Science* **256,** 520–525.

Porta, C., and Lomonossoff, G. P. (1998). Scope for using plant viruses to present epitopes from animal pathogens. *Rev. Med. Virol.* **8,** 25–41.

Porta, C., Spall, V. E., Loveland, J., Johnson, J. E., Barker, P. J., and Lomonossoff, G. P. (1994). Development of cowpea mosaic virus as a high-yielding system for the presentation of foreign peptides. *Virology* **202,** 949–955.

Potrykus, I. (1990). Gene transfer methods for plants and cell cultures. *Ciba Found. Symp.* **154,** 198–208.

Reimann, J., and Schirmbeck, R. (1999). Alternative pathways for processing exogenous and endogenous antigens that can generate peptides for MHC class I-restricted presentation. *Immunol. Rev.* **172,** 131–152.

Richter, L. J., Thanavala, Y., Arntzen, C. J., and Mason, H. S. (2000). Production of hepatitis B surface antigen in transgenic plants for oral immunization. *Nature Biotechnol.* **18,** 1167–1171.

Rooney, J. F., Wohlenberg, C., Cremer, K. J., Moss, B., and Notkins, A. L. (1988). Immunization with a vaccinia virus recombinant expressing herpes simplex virus type 1 glycoprotein D: Long-term protection and effect of revaccination. *J. Virol.* **62,** 1530–1534.

Rossmann, M. G., Arnold, E., Erickson, J. W., Frankenberger, E. A., Griffith, J. P., Hecht, H.-J., Johnson, J. E., Kamer, G., Luo, M., Mosser, A. G., Rueckert, R. R., Sherry, B., and Vriend, G. (1985). Structure of a human common cold virus and functional relationship to other picornaviruses. *Nature* **317,** 145–153.

Ruedl, C., and Wolf, H. (1995). Features of oral immunization. *Int. Arch. Allergy. Immunol.* **108,** 334–339.

Russell-Jones, G. J. (2000). Oral vaccine delivery. *J. Controlled Release* **65,** 49–54.

Russo, R., Burgyan, J., and Martelli, G. P. (1994). Molecular biology of Tombusviridae. *In* "Advances in Virus Research," Vol. 44 (K. Maramorosch *et al.,* eds.), pp. 381–428. Academic Press, New York.

Ryan, E. J., Daly, L. M., and Mills, K. H. (2001). Immunomodulators and delivery systems for vaccination by mucosal routes. *Trends Biotechnol.* **19,** 293–304.

Scholthof, H. B., Morris, T. J., and Jackson, A. O. (1993). The capsid protein gene of tomato bushy stunt virus is dispensable for systemic movement and can be replaced for localized expression of foreign genes. *Mol. Plant Microbe Interact.* **6,** 309–322.

Scott, P., Natovitz, P., Coffman, R. L., Pearce, E., and Sher, A. (1988). Immunoregulation of cutaneous leishmaniasis. T cell lines that transfer protective immunity or exacerbation belong to different T helper subsets and respond to distinct parasite antigens. *J. Exp. Med.* **168,** 1675–1684.

Shaw, J. G. (1996). Plant viruses. *In* "Fields' Virology" (B. N. Fields, D. M. Knipe, and P. M. Howley, Eds.), Vol. 1, pp. 499–532. Lippincott-Raven, Philadelphia.

Shi, B. J., Ding, S. W., and Symons, R. H. (1997). Plasmid vector for cloning infectious cDNAs from plant RNA viruses: High infectivity of cDNA clones of tomato aspermy cucumovirus. *J. Gen. Virol.* **78**(Part 5), 1181–1185.

Shivprasad, S., Pogue, G. P., Lewandowski, D. J., Hidalgo, J., Donson, J., Grill, L. K., and Dawson, W. O. (1999). Heterologous sequences greatly affect foreign gene expression in tobacco mosaic virus-based vectors. *Virology* **255,** 312–323.

Shukla, D. D., Strike, P. M., Tracy, S. L., Gough, K. H., and Ward, C. W. (1988). The N and C termini of the coat proteins of potyviruses are surface-located and the N terminus contains the major virus-specific epitopes. *J. Gen. Virol.* **69,** 1497–1508.

Shukla, D. D., Tribbick, G., Mason, T. J., Hewish, D. R., Geysen, H. M., and Ward, C. W. (1989). Localization of virus-specific and group-specific epitopes of plant potyviruses by systematic immunochemical analysis of overlapping peptide fragments. *Proc. Natl. Acad. Sci. USA* **86,** 8192–8196.

Singh-Jasuja, H., Toes, R. E., Spee, P., Munz, C., Hilf, N., Schoenberger, S. P., Ricciardi-Castagnoli, P., Neefjes, J., Rammensee, H. G., Arnold-Schild, D., and Schild, H. (2000). Cross-presentation of glycoprotein 96-associated antigens on major histocompatibility complex class I molecules requires receptor-mediated endocytosis. *J. Exp. Med.* **191,** 1965–1974.

Stace-Smith, R. (1981). Comoviruses. *In* "Handbook of Plant Virus Infections and Comparative Diagnosis" (E. Kurstak, Ed.), pp. 171–195. Elsevier/North-Holland Biomedical Press, Amsterdam.

Staczek, J., Bendahmane, M., Gilleland, L. B., Beachy, R. N., and Gilleland, H. E., Jr. (2000). Immunization with a chimeric tobacco mosaic virus containing an epitope of outer membrane protein F of *Pseudomonas aeruginosa* provides protection against challenge with *P. aeruginosa*. *Vaccine* **18,** 2266–2274.

Stark, D. M., Timmerman, K. P., Barry, G. F., Preiss, J., and Kishore, G. M. (1992). Regulation of the amount of starch in plant tissues by ADP glucose pyrophosphorylase. *Science* **258,** 287–291.

Stauffacher, C. V., Usha, R., Harrington, M., Schmidt, T., Hosur, M., and Johnson, J. E. (1987). The structure of cowpea mosaic virus at 3.5 Å. *In* "Crystallography in Molecular Biology" (D. Moras, J. Drenth, B. Strandberg, D. Suck, and K. Wilson, Eds.), pp. 293–308. Plenum Press, New York.

Streatfield, S. J., Jilka, J. M., Hood, E. E., Turner, D. D., Bailey, M. R., Mayor, J. M., Woodard, S. L., Beifuss, K. K., Horn, M. E., Delaney, D. E., Tizard, I. R., and Howard, J. A. (2001). Plant-based vaccines: Unique advantages. *Vaccine* **19,** 2742–2748.

Strizhov, N., Keller, M., Mathur, J., Koncz-Kalman, Z., Bosch, D., Prudovsky, E., Schell, J., Sneh, B., Koncz, C., and Zilberstein, A. (1996). A synthetic cryIC gene, encoding a *Bacillus thuringiensis* delta-endotoxin, confers *Spodoptera* resistance in alfalfa and tobacco. *Proc. Natl. Acad. Sci. USA* **93,** 15012–15017.

Sturm, A., Van Kuik, J. A., Vliegenthart, J. F., and Chrispeels, M. J. (1987). Structure, position, and biosynthesis of the high mannose and the complex oligosaccharide side chains of the bean storage protein phaseolin. *J. Biol. Chem.* **262,** 13392–13403.

Tacket, C. O., Mason, H. S., Losonsky, G., Clements, J. D., Levine, M. M., and Arntzen, C. J. (1998). Immunogenicity in humans of a recombinant bacterial antigen delivered in a transgenic potato. *Nature Med.* **4,** 607–609.

Tacket, C. O., Mason, H. S., Losonsky, G., Estes, M. K., Levine, M. M., and Arntzen, C. J. (2000). Human immune responses to a novel Norwalk virus vaccine delivered in transgenic potatoes. *J. Infect. Dis.* **182,** 302–305.

Takamatsu, N., Watanabe, Y., Yanagi, H., Meshi, T., Shiba, T., and Okada, Y. (1990). Production of enkephalin in tobacco protoplasts using tobacco mosaic virus RNA vector. *FEBS Lett.* **269,** 73–76.

Thanavala, Y., Yang, Y. F., Lyons, P., Mason, H. S., and Arntzen, C. (1995). Immunogenicity of transgenic plant-derived hepatitis B surface antigen. *Proc. Natl. Acad. Sci. USA* **92,** 3358–3361.

Townsend, A. R., Gotch, F. M., and Davey, J. (1985). Cytotoxic T cells recognize fragments of the influenza nucleoprotein. *Cell* **42,** 457–467.

Turpen, T. H., Reinl, S. J., Charoenvit, Y., Hoffman, S. L., Fallarme, V., and Grill, L. K. (1995). Malarial epitopes expressed on the surface of recombinant tobacco mosaic virus. *Biotechnology (NY)* **13,** 53–57.

Usha, R., Rohll, J. B., Spall, V. E., Shanks, M., Maule, A. J., Johnson, J. E., and Lomonossoff, G. P. (1993). Expression of an animal virus antigenic site on the surface of a plant virus particle. *Virology* **197,** 366–374.

van der Meer, I. M., Ebskamp, M. J. M., Visser, R. G. F., Weisbeek, P. J., and Smeekens, S. C. M. (1994). Fructan as a new carbohydrate sink in transgenic potato plants. *Plant Cell* **5,** 561–570.

van Ginkel, F. W., Nguyen, H. H., and McGhee, J. R. (2000). Vaccines for mucosal immunity to combat emerging infectious diseases. *Emerg. Infect. Dis.* **6,** 123–132.

Vaucheret, H., and Fagard, M. (2001). Transcriptional gene silencing in plants: Targets, inducers and regulators. *Trends Genet.* **17,** 29–35.

Vaucheret, H., Beclin, C., Elmayan, T., Feuerbach, F., Godon, C., Morel, J. B., Mourrain, P., Palauqui, J. C., and Vernhettes, S. (1998). Transgene-induced gene silencing in plants. *Plant J.* **16,** 651–659.

Walmsley, A. M., and Arntzen, C. J. (2000). Plants for delivery of edible vaccines. *Curr. Opin. Biotechnol.* **11,** 126–129.

Weiner, H. L., Friedman, A., Miller, A., Khoury, S. J., al-Sabbagh, A., Santos, L., Sayegh, M., Nussenblatt, R. B., Trentham, D. E., and Hafler, D. A. (1994). Oral tolerance: Immunologic mechanisms and treatment of animal and human organ-specific autoimmune diseases by oral administration of autoantigens. *Annu. Rev. Immunol.* **12,** 809–837.

Wigdorovitz, A., Carrillo, C., Dus Santos, M. J., Trono, K., Peralta, A., Gomez, M. C., Rios, R. D., Franzone, P. M., Sadir, A. M., Escribano, J. M., and Borca, M. V. (1999a). Induction of a protective antibody response to foot and mouth disease virus in mice following oral or parenteral immunization with alfalfa transgenic plants expressing the viral structural protein VP1. *Virology* **255,** 347–353.

Wigdorovitz, A., Perez Filgueira, D. M., Robertson, N., Carrillo, C., Sadir, A. M., Morris, T. J., and Borca, M. V. (1999b). Protection of mice against challenge with foot and mouth disease virus (FMDV) by immunization with foliar extracts from plants infected with recombinant tobacco mosaic virus expressing the FMDV structural protein VP1. *Virology* **264,** 85–91.

Xu, F., Jones, T. D., and Rodgers, P. B. (1996). Potential of chimaeric plant virus particles as novel, stable vaccines. *Dev. Biol. Stand.* **87,** 201–205.

Yamamoto, M., Rennert, P., McGhee, J. R., Kweon, M. N., Yamamoto, S., Dohi, T., Otake, S., Bluethmann, H., Fujihashi, K., and Kiyono, H. (2000). Alternate mucosal immune system: Organized Peyer's patches are not required for IgA responses in the gastrointestinal tract. *J. Immunol.* **164,** 5184–5191.

Yewdell, J. W., Bennink, J. R., and Hosaka, Y. (1988). Cells process exogenous proteins for recognition by cytotoxic T lymphocytes. *Science* **239,** 637–640.

Yu, J., and Langridge, W. H. (2000). Novel approaches to oral vaccines: Delivery of antigens by edible plants. *Curr. Infect. Dis. Rep.* **2,** 73–77.

Yu, J., and Langridge, W. H. (2001). A plant-based multicomponent vaccine protects mice from enteric diseases. *Nature Biotechnol.* **19,** 548–552.

Yusibov, V., Kumar, A., North, A., Johnson, J. E., and Loesch-Fries, L. S. (1996). Purification, characterization, assembly and crystallization of assembled alfalfa mosaic virus coat protein expressed in *Escherichia coli*. *J. Gen. Virol.* **77**(Part 4), 567–573.

Yusibov, V., Modelska, A., Steplewski, K., Agadjanyan, M., Weiner, D., Hooper, D. C., and Koprowski, H. (1997). Antigens produced in plants by infection with chimeric plant viruses immunize against rabies virus and HIV-1. *Proc. Natl. Acad. Sci. USA* **94,** 5784–5788.

Zavala, F., Cochrane, A. H., Nardin, E. H., Nussenzweig, R. S., and Nussenzweig, V. (1983). Circumsporozoite proteins of malaria parasites contain a single immunodominant region with two or more identical epitopes. *J. Exp. Med.* **157,** 1947–1957.

ADVANCES IN VIRUS RESEARCH, VOL. 58

TREATMENT OF ARENAVIRUS INFECTIONS: FROM BASIC STUDIES TO THE CHALLENGE OF ANTIVIRAL THERAPY

Elsa B. Damonte and Celia E. Coto

Laboratorio de Virología
Departamento de Química Biológica
Facultad de Ciencias Exactas y Naturales
Universidad de Buenos Aires
1428 Buenos Aires, Argentina

I. Introduction
II. Classification
III. Arenaviruses as Agents of Emerging Diseases
IV. The Virus
V. Virus Infection and the Host Cell
VI. The Replicative Cycle and Possible Targets for Therapeutic Agents
A. Viral Entry
B. RNA Transcription and Replication
C. Protein Maturation and Exocytic Transport
D. Assembly and Budding
VII. Present Treatment of Human Disease
VIII. Concluding Remarks
References

I. Introduction

Arenaviruses are enveloped, single-stranded, ambisense RNA viruses with a segmented genome consisting of two segments, designated large (L) and small (S). They are rodent-associated viruses with one exception, Tacaribe virus (TCRV), which infects bats of the genus *Artibeus*. They are zoonotic agents that can cause severe human diseases, known primarily as the hemorrhagic fevers, occurring in regions of South America (Bolivia: Machupo virus, MACV; Argentina: Junín virus, JUNV; Brazil: Sabiá virus, SABV; Venezuela: Guanarito virus, GTOV) and in Africa (Lassa virus, LASV). Murine lymphocytic choriomeningitis virus (LCMV), the prototype virus of Arenaviridae, is the only member with a worldwide distribution and is associated with acute aseptic meningitis in humans (Baird and Rivers, 1938).

Of the characterized pathogenic arenaviruses, LASV is the most dangerous: It is estimated that it causes 100,000–300,000 infections and

0065-3527/02 $35.00

approximately 5000 deaths annually (McCormick *et al.,* 1987; Bowen *et al.,* 2000).

Although arenaviruses constitute a real menace to human beings, knowledge of their singular biological properties has greatly contributed to the development of several scientific fields. Lassa fever outbreaks among hospitalized persons and nurses (Carey *et al.,* 1972) have shown how dangerous it is to take insufficient precautions in the laboratory or the hospital, generating a demand that perilous viruses should be handled under Biosafety Level 4 containments. Study of the peculiar ambisense gene arrangement of arenaviruses led to a new notion of how viruses could express their genetic makeup. Finally, studies of LCMV in mice provided a model for investigating persistent infection and brought to light many of the principles on which modern-day viral immunology and immunopathology are based.

The danger of arenaviruses for human health and their increased emergence during recent years has led to efforts to develop vaccines or effective antiviral agents against them. Attenuated and recombinant vaccines have been tried against JUNV and LASV (Auperin, 1993; Maiztegui *et al.,* 1998), but this subject will not be discussed in this chapter. Instead, we have chosen to review current knowledge on arenavirus replication steps in order to provide clues concerning the molecular targets useful for the development of specific antiviral agents for improving present therapy.

II. Classification

The Arenaviridae family is made up of 20 antigenically related members grouped in the *Arenavirus* genus. Their appearance in the electron microscope (Dalton *et al.,* 1968; Murphy *et al.,* 1970) and serological cross-reactivity in complement fixation and immunofluorescence tests between them have played a key role in the identification of the arenaviruses as a distinct virus family (Rowe *et al.,* 1970b; Casals, 1975; Wulff *et al.,* 1978).

Arenaviruses were classified into two groups: the New World or Tacaribe complex and the Old World or lymphocytic choriomeningitis–Lassa complex, according to their geographical region of isolation and serological cross-reactivity (Wulff *et al.,* 1978; Buchmeier *et al.,* 1981). The results of phylogenetic analyses of S RNA nucleotide sequence data are consistent with the New World–Old World division of the family (Buchmeier *et al.,* 1995; Bowen *et al.,* 1996b). The members of the Old World group are LCMV (Armstrong and Lillie, 1934), LASV (Frame *et al.,* 1970), Mopeia virus (MOPV) (Wulff *et al.,* 1977), Mobala virus

TABLE I
OLD WORLD ARENAVIRUSES (LCMV–LASSA COMPLEX): RESERVOIR, LOCATION, AND DISEASE

Virus	Reservoir	Location	Habitat	Human Disease
LCMV	*Mus musculus*	Europe, Asia, and the Americas	Peridomestic, grasslands	Febrile syndrome/ aseptic meningitis
LASV	*Mastomys* spp.	West Africa	Savannah, forest clearings	Lassa fever
MOPV	*Mastomys natalensis*	Southern Africa	Savannah	No
MOBV	*Praomys jacksoni*	Central African Republic	Savannah	No
IPPYV	*Arvicanthis* spp.	Central African Republic	Grassland, savannah	No

(MOBV) (González *et al.,* 1983), and Ippy virus (IPPYV) (Swanepoel *et al.,* 1985); the last three members are apparently not associated with human disease (Table I). Lassa fever is endemic in several African countries and a recent phylogenetic analysis of LASV strains showed that they comprise four lineages, three of which are found in Nigeria and the fourth in Guinea, Liberia, and Sierra Leone (Bowen *et al.,* 2000).

The New World arenaviruses comprise three phylogenetic lineages, designated A, B, and C. Lineage A includes Tamiami (TAMV) (Jennings *et al.,* 1970), Flexal (FLEV) (Pinheiro *et al.,* 1977), Whitewater Arroyo (WWAV) (Fulhorst *et al.,* 1996), Paraná (PARV) (Webb *et al.,* 1970), and Pichinde viruses (PICV) (Trapido and San Martín, 1971), and a new arenavirus, recently isolated in Venezuela, named Pirital virus (PIRV) (Fulhorst *et al.,* 1997). Lineage B includes Amaparí (AMAV) (Pinheiro *et al.,* 1966), Guanarito (GTOV) (Tesh *et al.,* 1994), Junín (JUNV) (Parodi *et al.,* 1958), Machupo (MACV) (Johnson *et al.,*1965), Sabiá (SABV) (Coimbra *et al.,* 1994; González *et al.,* 1996), and Tacaribe viruses (TCRV) (Downs *et al.,* 1963). Lineage C contains Latino (LATV) (Webb *et al.,* 1973) and Oliveros viruses (OLVV) (Bowen *et al.,* 1996a) (Table II).

There is no obvious correlation between New World arenavirus phylogeny and their geographical distribution. For example, the ranges of JUNV and OLVV overlap in Argentina (Bowen *et al.,* 1996a; Mills *et al.,* 1996), but these viruses are quite distinct and occupy different phylogenetic lineages. Moreover, several studies have shown that LCMV also overlaps the same region of JUNV (Sabattini *et al.,* 1974; Ambrosio *et al.,* 1994) and OLVV (Mills *et al.,* 1996). A similar observation is applicable to PIRV (nonpathogenic) and GTOV (pathogenic), viruses isolated from the same rodents in the same region of Venezuela

TABLE II
NEW WORLD ARENAVIRUSES (TACARIBE COMPLEX): LINEAGE ASSIGNMENT, RESERVOIR, LOCATION, AND DISEASE[a]

Lineage	Virus	Reservoir	Location	Habitat	Human Disease
A	FLEV	*Oryzomys* spp.	Brazil	Tropical forest	Yes[b]
	PARV	*Oryzomys buccinatus*	Paraguay	Tropical forest savannah	No
	PICV	*Oryzomys albigularis*	Colombia	Tropical forest	No
	TAMV	*Sigmodon hispidus*	Florida	Grasslands marsh	No No
	WWAV	*Neotoma albigula*	Western U.S.	Grasslands	Hemorrhagic fever
	PIRV	*Sigmodon alstoni/ Zygodontomys brevicauda*	Venezuela	Grasslands	No
B	AMAV	*Oryzomys gaeldi, Neacomys guianae*	Brazil	Tropical forest	No
	GTOV	*Sigmodon alstoni/ Zygodontomys brevicauda*	Venezuela	Grasslands	Venezuelan hemorrhagic fever (VHF)
	JUNV	*Calomys musculinus*	Argentina	Grasslands, cultivated fields	Argentinian hemorrhagic fever (AHF)
	MACV	*Calomys callosus*	Bolivia	Grasslands, peridomestic	Bolivian hemorrhagic fever (BHF)
	SABV	Unknown	Brazil	—	Hemorrhagic fever
	TCRV	*Artibeus* spp.	Trinidad	—	Yes[b]
C	LATV	*Calomys callosus*	Bolivia	Grasslands, peridomestic	No
	OLVV	*Bolomys obscurus*	Argentina	Grasslands	No

[a] Cupixi is a virus isolated in Brazil from *Oryzomis capito* which belongs to the Tacaribe complex, but is not assigned to a lineage yet.
[b] Associated only with single, nonfatal laboratory-acquired infection.

(Fulhorst *et al.,* 1997; Weaber *et al.,* 2000). Likewise, LATV (nonpathogenic) and MACV (pathogenic) are parasites of the same host, *Calomys callosus*, although they were isolated from two different zones of Bolivia (Webb *et al.,* 1973).

III. ARENAVIRUSES AS AGENTS OF EMERGING DISEASES

The emergence of infectious diseases transmitted by animals is an issue of great concern for public health agencies and a continuing threaten to the way of life of persons living in endemic areas. Arenaviruses are typical agents of endemic emerging diseases, which can be harmless or dreadful producers of hemorrhagic fever in humans. As stated, most arenaviruses are confined to particular geographical regions in intimate association with rodents. Those that are known have come to light either because their passing to humans caused an alarming disease or as a result of a systematic survey for the presence of virus or specific antibodies in native rodents. Under these circumstances one can expect the number of Arenaviridae members to grow and this is indeed the case. It has been estimated that in recent decades a new arenavirus was recognized every 3 years.

Over a period of 40 years (1956–1996), at least 13 members of the Tacaribe complex were found in South America. Only one, TAMV, was discovered in North America, in South Florida, in cotton rats in 1970 (Jennings *et al.,* 1970). Twenty-six years later, a study of the prevalence of antibodies to arenaviruses in indigenous rodents from the southern and western United States revealed the presence of a new arenavirus associated with *Neotoma* spp. (Kosoy *et al.,* 1996). Further studies by Fulhorst *et al.* (1996) led to the isolation of virus from two arenavirus antibody-positive *N. albigula*, collected at Whitewater Arroyo, New Mexico. Phylogenetic analysis of the nucleocapsid protein gene sequence of the agent, named Whitewater Arroyo, showed that the virus was closely related to TAMV. At the time of WWAV isolation, its potential pathogenicity remained unknown, but in April 2000 a 14-year-old girl living in the western United States died after suffering from hemorrhagic fever (Enserink, 2000). According to the California Department of Health Services, the agent responsible was WWAV carried by wood rats. During a rodent survey, Fulhorst also found a third North American arenavirus present in deer mice; the virus, whose discovery has not been published yet, has been named Bear Canyon virus (Enserink, 2000).

The emergence of arenavirus is highly related to the ecological characteristics of these viruses, which are able to establish chronic viremic infections in specific rodent hosts. Chronically infected animals shed virus in their urine, which contaminates their environment and then spreads to humans by contact facilitated by skin abrasions or cuts, or via aerosols. The natural history of the human diseases is determined by the pathogenicity of the virus, the geographical distribution, the habitat and the habits of the rodent reservoir host, and the nature of the human–rodent interaction (Childs and Peters, 1993).

LCMV is focally distributed throughout the world due to its association with the common house mouse (*Mus musculus*). As a consequence, the distribution of human cases is focal and also seasonal, probably because mice move into houses and barns in winter. Feral mice may also introduce the virus in laboratory and commercial mouse, rat, hamster, guinea pig, and rhesus monkey colonies (Childs and Peters, 1993).

LASV is enzootic in *Mastomys natalensis,* which is a peridomestic rodent that lives in or near human dwellings and transmits the virus vertically to its offspring and horizontally to humans (McCormik, 1987; ter Meulen *et al.,* 1996). Uniquely among the arenaviruses, LASV person-to-person spread is also common. Lassa fever initially came to medical attention in a setting of nosocomial spread and a hospital-based Nigerian epidemic occurring in 1969 (Carey *et al.,* 1972). Since the late 1980s urban outbreaks have occurred in Nigerian cities with a mortality rate of 30%.

JUNV is enzootic in *Calomys musculinus,* the main reservoir, but can also infect other wild rodents (Mills *et al.,* 1991). Sixty years ago human infection with the virus was limited to sporadic cases characterized by a flu-like syndrome of variable severity (Arribalzaga, 1955). The introduction of widespread planting of maize in the rural areas of the pampas in the late 1940s favored the reproduction of *C. musculinus* over other indigenous rodents and increased its contacts with humans. When JUNV infects humans it causes a severe disease, called Argentine hemorrhagic fever; the virus was first isolated from human cases during a huge outbreak of the disease in 1958 (Parodi *et al.,* 1958). Since then, the virus has spread to over 100,000 km^2 of what is now rich farmland of the humid pampas. Argentine hemorrhagic fever has always been an occupational disease. Initially, adult males harvesting grain crops by hand were affected. Now, after mechanization, both the harvesters and the grain truck drivers are the main individuals exposed to infection (Maiztegui, 1975; Maiztegui *et al.,* 1986). Since the disease was first recognized annual outbreaks have occurred without interruption, with a total of 26,000 cases notified to April 2001 (D. A. Enría, personal communication).

MACV emerged in Bolivia in 1952 when a revolution forced the people living in the plains to attempt subsistence agriculture at the forest edge. *Calomys callosus* is a forest rodent well-adapted to human contact and the reservoir of MACV (Johnson *et al.,* 1966; Johnson, 1981). Human outbreaks of Bolivian hemorrhagic fever resulted from rodent invasion of villages, with peak occurrence between 1962 and 1964 and 20% mortality. Trapping of *C. callosus* in the surroundings of human houses resulted in the control of the disease (Johnson, 1981). However, in 1994 a fatal familiar outbreak occurred in Beni Department, indicating that the virus is still present (Centers for Disease Control [CDC], 1994).

Among the pathogenic arenaviruses, SABV is the only one with origin, geographical distribution, natural maintenance cycle, and epidemiology unknown. This virus emerged in 1990 when it was isolated from the serum of a fatal case of hemorrhagic fever in Brazil (Coimbra *et al.,* 1994; González *et al.,* 1996). Later, two nonfatal laboratory infections occurred (Barry *et al.,* 1995).

IV. The Virus

The Arenaviridae are pleomorphic particles, with a size range from 50 to 300 nm (Murphy *et al.,* 1970). Virions possess a thick unit membrane with external surface projections or spikes, seen in negatively stained preparations. In their interior, one or more electron-dense granules, 20–25 nm in diameter, are observed in most particles (Compans, 1993). These granules were identified as host-cell-derived ribosomes and their characteristic granular internal structure was the basis for the name given to the family (arena, from the Latin for "sandy") (Rowe *et al.,* 1970a). However, these packaged ribosomes do not appear to be essential for virus replication (Leung and Rawls, 1977). The genomic L and S RNAs (approximately 7.2 and 3.4 kb, respectively) are within virions as filamentous nucleocapsids, 5–15 nm in diameter, in circular configurations (Vezza *et al.,* 1978; Young and Howard, 1983). A short sequence is conserved at the 3′ termini of L and S RNAs in arenaviruses, and this sequence is complementary to the 5′ termini of the viral RNAs, producing a panhandle structure responsible for the circular forms of nucleocapsids observed by electron microscopy. The ambisense open reading frames in S and L RNAs are separated by noncoding intergenic regions, which present a strong secondary structure, which, depending on the virus, can be arranged in one or two hairpins (reviewed in Franze-Fernández *et al.,* 1993). In addition to S and L RNAs, abundant 28S and 18S RNAs of ribosomal origin as well as heterogeneous host- and virus-derived RNA species of 4–6 S are found in virion RNA preparations.

The S RNA encodes the major structural proteins of the virion, which are the viral nucleocapsid protein NP, tightly associated with genome RNAs, and the two envelope glycoproteins, GP1 and GP2, which are derived by posttranslational cleavage of a precursor polypeptide GPC. GP1 is a peripheral membrane glycoprotein, whereas GP2 is an integral glycoprotein. For LCMV, both glycoproteins were shown to be oligomerized as homotetramers to form the spike complex, with GP1 exposed at the top and noncovalently linked to the stalk of GP2 (Burns and

Buchmeier, 1993). The L segment encodes for a 200-kDa protein named L, thought to be the viral RNA polymerase (Fuller-Pace and Southern, 1989), and an 11-kDa protein called Z, with a RING finger domain and proposed structural/regulatory functions (Garcin *et al.,* 1993; Salvato, 1993). Both Z and NP are able to bind zinc through specific sequences (Salvato *et al.,* 1992; Tortorici *et al.,* 2001b), but the significance of this zinc-binding activity is unknown.

V. Virus Infection and the Host Cell

Arenaviruses infect a wide variety of cell types from different species, with a variable efficiency of multiplication. In general, for most arenaviruses replication occurs without disturbing host cell macromolecular synthesis and without producing cytopathogenicity. Only under certain conditions, dependent either on the virus strain or the proportion of defective interfering particles in the virus population, have an effect on cellular biosynthesis and a cytolytic response been observed (Friedlander *et al.,* 1984; López and Franze-Fernández, 1985; Candurra *et al.,* 1990). The monkey kidney Vero cell line is one of the few cell types in which a typical cytopathic effect is observed and consequently it is the cell system used to titrate virus by plaque assay.

Noncytolytic infection with LCMV results in alterations of specialized functions of differentiated cells. Several reports have described the disruption in the transcription of genes for growth hormone, thyroid hormone, and other products, a property related to the virus tropism for differentiated cells (Klavinskis and Oldstone, 1987; De la Torre and Oldstone, 1992; Teng *et al.,* 1996). A very subtle and selective effect on cell translation also has been demonstrated for the LCMV Z protein. In infected and transfected cells, Z was found associated with the eukaryotic translation initiation factor 4E (eIF-4E) and through this interaction Z represses the production of certain proteins, for example, cyclin D1, at the posttranscriptional and the post-RNA transport levels (Campbell Dwyer *et al.,* 2000). During infection, Z translational repression may be reduced by NP, providing a mechanism for modulating the viral effects on the host cell by the interplay between viral proteins.

The infection of cells with arenaviruses is characterized by two distict phases. After an initial period of active virus production, which may or may not be associated with cytopathic effects, cultures readily progress toward a state of long-term persistent infection. The molecular mechanisms responsible for the establishment and maintenance of arenavirus persistence are very complex and obscure. The main properties of these persistently infected cells can be summarized as follows:

1. Morphology and growth characteristics are similar to those of uninfected cells.
2. There is a cyclic pattern of release of reduced levels of infectious virus, with long periods without virus production (Weber *et al.*, 1983; D'Aiutolo and Coto, 1986; Bruns *et al.*, 1990).
3. Virus recovered from persistent infections in general are variants from the original virus, such as thermosensitive mutants, plaque mutants, virulence-altered variants, or slow-growth virus (Coto *et al.*, 1981, 1993; Weber *et al.*, 1985; Bruns *et al.*, 1990).
4. There is an accumulation of defective interfering particles and presence of deleted and/or truncated RNAs (Welsh and Buchmeier, 1979; Giménez and Compans, 1980; Van der Zeijst *et al.*, 1983; D'Aiutolo and Coto, 1986; Iapalucci *et al.*, 1994; Meyer and Southern, 1997).
5. There is complete resistance to superinfection with homologous virus, but, depending on the virus, susceptibility to heterologous arenaviruses (Damonte *et al.*, 1983).

VI. The Replicative Cycle and Possible Targets for Therapeutic Agents

Understanding the viral life cycle provides key information for the rational design of antiviral drugs. Many aspects of the arenavirus multiplication cycle have not been studied. In this section, knowledge of the different stages of the arenavirus cycle and the antiviral possibilities investigated for each step are presented.

A. Viral Entry

1. Attachment

The viral replicative cycle starts with the attachment of the virus to the surface of the host cell. A virion protein, named viral attachment protein (VAP), binds to components of the plasma membrane acting as virus receptors. Recently, α-dystroglycan (α-DG), a high molecular weight glycoprotein, was identified as a cellular receptor for some arenaviruses using the virus overlay protein blot assay (VOPBA) procedure with purified α-DG protein and also determining the virus binding to mutant mouse cells bearing a null mutation of the gene encoding DG (Cao *et al.,* 1998). DG is encoded by a single gene and posttranslationally processed into α and β chains, peripheral and integral membrane glycoproteins, respectively, which form a complex spanning the plasma membrane (Henry and Campbell, 1999). The DG complex is highly conserved and expressed in a variety of tissues and cells, which explains the wide host range of arenaviruses. LCMV as well as the Old World arenaviruses LASV and MOBV, and OLVV, a group C New World arenavirus, were found to bind to α-DG, but GTOV, a group B New World arenavirus, failed to recognize α-DG (Cao *et al.,* 1998). The proteinaceous nature of the cell receptor for JUNV, another group C arenavirus, was also demonstrated by enzymatic treatment, but the cellular protein was not identified (Raiger Iustman *et al.,* 1995).

With respect to the identification of VAP, several lines of evidence suggest that presumably GP1 is the envelope glycoprotein responsible for arenavirus adsorption to the host cell. First, GP1-specific monoclonal antibodies blocked LCMV binding to cells *in vitro,* whereas antibodies against the transmembrane glycoprotein GP2 failed to affect virus binding (Borrow and Oldstone, 1992). Second, only antibodies directed against GP1 can also neutralize LCMV infectivity (Parekh and Buchmeier, 1986). Third, Scolaro *et al.* (1990) reported that a host range mutant of JUNV unable to bind to murine cells showed an altered GP1 peptide mapping. Interestingly, Smelt *et al.* (2001) demonstrated that LCMV variant strains differing in their pathogenic potential for mice and presenting point mutations at GP1 could be divided into two functional groups with respect to binding affinity to α-DG. The first group of LCMV strains exhibited a high affinity of binding to α-DG and a marked dependence on this protein for cell entry, and invariably established a persistent infection. In contrast, the other group of strains showed a low level of or no binding to α-DG and a reduced dependence on this protein for cell entry, and mouse infection was rapidly cleared, changing the pathology of infection from persistent to acute. These studies established an association between receptor usage and pathogenesis, and furthermore, they are indicative that additional, unidentified cell surface receptors or cofactors allowing an α-DG-independent viral uptake must exist, and can be used for certain arenaviruses, such as GTOV.

Two approaches have been reported on arenavirus inhibitors which have been attributed as interfering with virus adsorption. As reported for several viruses, the addition of the soluble receptor α-DG blocked LCMV and LASV infection *in vitro* (Cao *et al.,* 1998). Different types of polysulfates, including sulfated polysaccharides (dextran sulfate, heparin, pentosan polysulfate), polyacetal polysulfate, and polyvinylalcohol sulfate and its copolymer with acrylic acid, were found to be highly selective inhibitors of JUNV and TCRV replication (Andrei and De Clercq, 1990; Witvrouw *et al.,* 1994). The potential clinical application of both kinds of agents, soluble receptors and polyanionic substances, has not been investigated.

2. Internalization and Uncoating

After virions are bound to the cell receptor, arenavirus entry involves an endocytic process, which includes virion uptake into vesicles followed by a low-pH-dependent fusion of viral and endosome membranes and the release of the nucleocapsid into the cytoplasm. The inhibition of arenavirus infection by lysosomotropic agents was the first evidence of an endosomal entry route. Two classes of compounds which raise the pH of endosomes were found to be effective in blocking internalization of several arenaviruses, including LASV, MOPV, and PICV (Glushakova and Lukashevich, 1989), JUNV (Castilla *et al.,* 1994), and LCMV (Borrow and Oldstone, 1994). Weak bases such as ammonium chloride and chloroquine diffuse across membranes in nonprotonated form, but become protonated in acidic compartments such as endosomes and accumulate, raising the vesicular pH, whereas carboxylic ionophores, such as nigericin and monensin, raise vesicular pH by exchanging H^+ for Na^+ or K^+. The inhibitory action of these compounds was exerted at early times during arenavirus infection without affecting attachment, but preventing penetration. That arenavirus entry occurred by endocytosis was also demonstrated by other experimental evidence: (1) Virion entry in vesicles was visualized for LCMV by immunoelectronmicroscopy (Borrow and Oldstone, 1994); (2) the ammonium chloride-induced blockade of JUNV infection was overcome by buffering the extracellular medium at a pH below 6.0, conditions that lead to direct fusion of virus envelope with the cell membrane (Castilla *et al.,* 1994); and (3) specific inhibitors of the vacuolar-proton ATPase (bafilomycin A1 and concanamycin A), the enzyme responsible for maintaining the low pH of endosomes, inhibited simultaneously JUNV replication and vesicle acidification (Castilla *et al.,* 2001).

Further studies have shown that the fusion should be catalyzed by the spike glycoprotein complex, which after exposure to acid pH undergoes conformational changes characterized by a greatly reduced antibody binding to GP1, a concomitant increased binding to GP2, and finally the irreversible dissociation of GP1 from the virions (Di Simone *et al.,* 1994; Castilla *et al.,* 1994; Di Simone and Buchmeier, 1995). It may be assumed that the acid pH triggered a change in the interaction between GP1 and GP2 leading probably to the exposure of a hidden fusion peptide. Accordingly, a previous report on LASV described the acid-induced fusion activity of a synthetic amphiphilic peptide homologous to a sequence localized in the internal GP2 (Glushakova *et al.,* 1992). In addition, after a brief acid treatment JUNV-infected cells expressing viral glycoproteins on their surface could mediate the formation of syncytia by fusion with the adjacent cells (Castilla and Mersich, 1996).

As mentioned above, classical lysosomotropic compounds were effective inhibitors of arenavirus replication, but as they interfere with vital cellular processes, they do not have therapeutic perspectives. However, a wide spectrum of pharmacologically active substances licensed for clinical use has been assayed and found to be active against arenaviruses blocking the early stages of the viral cycle. One of the first compounds studied three decades ago was amantadine, known to succesfully block the uncoating of influenza virus. Amantadine proved effective *in vitro* against LCMV and several members of the Tacaribe complex, but its action *in vivo* has been discouraging because amantadine medication shortened the life span of guinea pigs and mice infected with JUNV or LCMV, respectively (Coto *et al.,* 1969; Pfau *et al.,* 1972; Pfau, 1975). A series of diverse amines with a wide pharmacological use, including anesthetics (procaine), antihistaminics (chlorpheniramine), and neuroleptic drugs (trifluoperazine and chlorpromazine), exerted a concentration-dependent inhibition of *in vitro* JUNV, TCRV, and PICV multiplication at doses not affecting cell viability (Castilla *et al.,* 1994; Candurra *et al.,* 1996). From time of addition and removal experiments, it could be concluded that all these compounds inhibited an early stage in the replicative cycle, viral entry, although trifluoperazine also interfered with a later stage of viral maturation. Both phenotiazines, trifluoperazine and chlorpromazine, acted on JUNV multiplication through their interaction with calmodulin, which participates in various cellular processes as a modulator of many Ca^{2+}-dependent enzymes and is directly involved as a structural protein in the cytoskeleton. Other substances that affect the cytoskeletal networks, such as the Ca^{2+}-chelator EGTA, nifedipine, a Ca^{2+}-channel blocker, colchicine, and nocodazole, were also demonstrated as inhibitors of JUNV multiplication at early times (Candurra *et al.,* 1999), suggesting that arenavirus entry is dependent on the integrity of the cytoskeleton. Caffeine, a methylxanthine with a stimulating effect in the central nervous system, also achieved inhibition of JUNV production *in vitro* at an early stage, previous to protein expression (Candurra and Damonte, 1999). The possibility of testing any of these substances widely applied for medical use in an experimental model of hemorrhagic fever deserves to be considered, either alone or combined with ribavirin (see Sections VI.B and VII), to reduce the undesirable collateral actions of both agents.

The search for compounds with antiviral activity in natural sources is another interesting strategy not fully explored for arenaviruses. Positive results have been obtained with purified extracts from leaves of *Melia azedarach* L., with antiviral properties against several animal viruses. *In vitro* JUNV multiplication was inhibited by preventing virus uncoating due to interference with vacuolar acidification (Castilla *et al.*, 1998). Interestingly, extracts of *M. azedarach* have also shown effectiveness in preventing TCRV encephalitis in mice, with a degree of protection from 66% to 100%, depending on the virus dose (Andrei *et al.*, 1986).

B. RNA Transcription and Replication

The molecular mechanism of transcription and replication in arenaviruses is still a matter of speculation, but the hallmark of the process is the ambisense coding strategy of both RNA fragments (Auperin *et al.*, 1984). The S RNA encodes NP protein at its 3′ portion in the genome complementary sense and the glycoprotein precursor GPC at the 5′ portion in the genome sense.

Similarly, the L RNA encodes the virus polymerase L protein at the 3′ portion and the Z protein at the 5′ end in opposite orientations. Although S and L genome segments contain protein-coding sense sequences at their 5′ regions, they are not directly translated and thus arenaviruses behave at this point like true negative-strand viruses with transcription as the first biosynthetic process after uncoating.

The ambisense arrangement and the structural characteristics of genomes provide a mechanism for temporal regulation of transcription and replication. Northern blot analyses of arenavirus-infected cells revealed the presence of full-length S and L genomes and antigenomes, both found as nucleocapsids tightly bound to NP, and four subgenomic unencapsidated mRNAs (Raju *et al.*, 1990; Franze-Fernández *et al.*, 1993; Romanowski, 1993). Primary transcription of the mRNAs for NP and L proteins complementary to the 3′ portion of S and L fragments is the initial event. The hairpin configuration in the intergenic region is thought to serve as a transcription termination signal, leading to the release of the viral polymerase. The switch from transcription to replication involves the bypass of the hairpin-dependent termination signal and is probably mediated at least by the intracellular level of NP. Very recently, *in vitro* interaction of NP with the intergenic S RNA sequence and *in vivo* antiterminator activity of NP were reported for

JUNV (Tortorici *et al.*, 2001a). The observation that inhibition of protein synthesis blocks TCRV and JUNV S RNA replication while allowing primary transcription of the NP mRNA (Franze-Fernandez *et al.*, 1987; Tortorici *et al.*, 2001a) supports the function of NP as a transcriptional antiterminator in arenaviruses. A more complex scheme involving interactions with other viral or cellular proteins cannot be discarded. After NP synthesis has occurred, replication would proceed via the synthesis of full-length antigenomic RNAs. These antigenomes serve both as replication intermediates in the formation of newly synthesized genomes and as templates for the transcription of S and L subgenomic mRNAs for GPC and Z, respectively.

Besides the above-summarized regulation of gene expression with discordant transcription of NP and L mRNAs relative to GPC and Z, there are some distinctive features in arenavirus transcription:

1. The 5′ ends of the S-derived mRNAs of TCRV, LCMV, and PICV were found to extend 1–7 nucleotides beyond the end of the genome template and were capped (Garcin and Kolakofsky, 1990; Meyer and Southern, 1993; Polyak *et al.*, 1995). To account for the origin of these nontemplated nucleotides, a "cap-snatching" mechanism for mRNA initiation as found previously for influenzavirus and bunyaviruses was proposed, but there is not yet information to exclude or confirm this hypothesis. In fact, the evidence for a role of nuclear functions in arenavirus infection is contradictory. Enucleation of cells resulted in inhibition of PICV multiplication (Banerjee *et al.*, 1976). Conflicting results were reported on the effects of actinomycin D: Inhibition of PICV late in the cycle was reported by Rawls *et al.* (1976), whereas López *et al.* (1986) did not observe any inhibition on virus yields. JUNV multiplication was inhibited in the presence of α-amanitine, an inhibitor of cellular DNA-dependent RNA polymerase II, and this enzyme activity was increased in nuclei of infected cells (Mersich *et al.*, 1981). Finally, NP-related nuclear inclusions were detected in PICV-infected Vero cells (Young *et al.*, 1987).
2. Termination of transcription occurs at multiple different sites within the intergenic region of TCRV, LCMV, and JUNV (Iapalucci *et al.*, 1991; Meyer and Southern, 1993; Tortorici *et al.*, 2001a) leading to heterogeneity at the 3′ ends of mRNAs, which are not polyadenylated.
3. Experiments based on *in vitro* transcription combined with immunodepletion suggest that the Z protein of TCRV is required for genome replication and, to a lesser degree, for mRNA synthesis (Garcin *et al.*, 1993). However, Lee *et al.* (2000), using a reverse

genetics system, observed that only NP and L proteins of LCMV are necessary for efficient RNA transcription and replication. More research is required to elucidate the precise participation of each virion protein in the biosynthetic pathway.

Although incompletely known, RNA transcription and replication represent an attractive target for antiviral action. In fact, the only drug partially effective for treatment of human arenavirus infections (see Section VII) is ribavirin (1-β-D-ribofuranosyl-1,2,4-triazole-3-carboxamide), a guanosine analogue shown to be inhibitory against arenaviruses in studies performed *in vitro* and in animal models (Jahrling *et al.,* 1980; Weissenbacher *et al.,* 1986; McKee *et al.,* 1988; Lucia *et al.,* 1989). Ribavirin has a broad spectrum of antiviral activity against several viruses other than arenaviruses. It has been approved for clinical use as an aerosol in the treatment of acute lower respiratory tract infections caused by respiratory syncytial virus (Hall *et al.,* 1983). The mechanism of action of ribavirin has not been completely elucidated. After phosphorylation to ribavirin 5′-monophosphate, the primary interaction is with cellular inosine monophosphate (IMP) dehydrogenase, producing a reduction in GTP pools (Streeter *et al.,* 1973). Ribavirin is also phosphorylated to its 5′-triphosphate and in this form can affect the initiation and elongation of viral mRNAs by competitive inhibition of mRNA-capping enzymes and viral polymerases (Goswami *et al.,* 1979; Gilbert and Knight, 1986). Thus, ribavirin action is not highly specific and the several disadvantages recorded for human treatment (see Section VII) have prompted the screening of several other nucleoside analogues to obtain more selective agents targeted to RNA synthesis (Andrei and De Clercq, 1993).

Compounds chemically related to ribavirin, such as its 3-carboxamide derivative ribamidine and the C-nucleoside analogues tiazofurin and selenazofurin, have shown comparable antiarenavirus activity *in vitro* (Huggins *et al.,* 1984; Burns *et al.,* 1988), all of them acting as inhibitors of cellular IMP dehydrogenase. Ribamidine also provided protection against a lethal challenge of PICV in a hamster model, showing an *in vivo* potency between 1/3 and 1/10 that of ribavirin (Smee *et al.,* 1993). Pyrazofurin, another C-nucleoside analogue, had a marked inhibitory effect on JUNV and TCRV *in vitro* replication, but this compound was an inhibitor of the enzyme orotidylic acid monophosphate (OMP) decarboxylase, interfering with the conversion of OMP to UMP (Andrei and De Clercq, 1990).

A series of acyclic and carbocyclic adenosine analogues known to be inhibitors of *S*-adenosylhomocysteine (SAH) hydrolase, a central enzyme in the transmethylation reactions required for the 5′ capping of viral mRNAs, also affected the replication of JUNV and TCRV (Andrei and De Clercq, 1990). A different kind of adenosine analogue, 3′-fluoro-3′-deoxyadenosine, was found to be active against LCMV, but not PICV (Smee *et al.,* 1992). This compound did not act as an inhibitor of SAH hydrolase. Some analogues of the nucleoside cytidine were also assayed against arenaviruses. Cyclopentylcytosine and cyclopentenylcytosine, cytosine analogues targeted to the CTP synthetase that catalyzes the final step in the biosynthesis of CTP converting UTP to CTP, showed a high inhibitory action against JUNV and TCRV multiplication in Vero cells (Andrei and De Clercq, 1990). However, the selectivity indices of these compounds, as also occurred for other of the above-mentioned analogues, were not very promising and were highly dependent on the method used to determine cellular toxicity, that is, DNA synthesis, cell growth, or cell morphology.

In order to enhance antiviral specifity, the development of antisense technologies is a more recent strategy targeted to the viral nucleic acid. Using the LCMV model, selective ribozymes, RNA molecules which cleave RNA in a sequence-specific manner, were designed. In tissue culture, these ribozymes were shown to diminish LCMV RNA levels and reduce infectious virus yields by approximately 100-fold (Xing and Whitton, 1993). However, many aspects concerning the optimization of ribozyme structure, delivery system, and antiviral effect remain unresolved.

C. Protein Maturation and Exocytic Transport

The expression of the two genome segments gives rise to a total of five mature proteins, all of them structural components of the virion. Some aspects of the processing and modification of viral proteins are known and may aid in finding new antiviral targets. The two glycoproteins are synthesized as the precursor GPC at the endoplasmic reticulum and migrate via the Golgi complex. The posttranslational processing involves as a first event the transition of the GPC oligosaccharide chains from the high-mannose type to the complex form, with loss of glucose and mannose residues and acquisition of glucosamine, galactose, and fucose. Then the cleavage of GPC to generate GP1 at the amino-terminal end and GP2 at the carboxy-terminal end occurs late in transit through or exit from the *trans*-Golgi, and finally the mature glycoproteins are inserted at the plasma membrane. This maturation pathway of arenavirus glycoproteins and its influence on virion production was demonstrated using a series of drugs that inhibit sequential steps of glycosylation and intracellular exocytosis. Inhibition of the first step of N-glycosylation, the addition of the mannose-rich chains, with tunicamycin blocked the cleavage and transport of the unglycosylated GPC and reduced the release of infectious virions (Padula and Martinez Segovia, 1984; Wright *et al.,* 1989, 1990). The use of trimming glucosidase and mannosidase inhibitors such as 1-deoxynojirimycin, 1-deoxymannojirimycin, castanospermine, and swainsonine demonstrated that, although the addition of the oligosaccharide chains was essential for glycoprotein cleavage, transport, and virion infectivity, the acquisition of a complex structure of the carbohydrate chains was not required for these events to occur (Wright *et al.,* 1990; Silber *et al.,* 1993). Studies with monensin and brefeldin A, which blocked glycoprotein processing and transport at progressively later stages through the Golgi cisternae, allowed the conclusion that proteolytic cleavage of GPC together with the subsequent transport of GP1 and GP2 to the cell surface is a requisite for the formation of infectious arenaviruses (Damonte *et al.,* 1994; Candurra and Damonte, 1997).

Using synthetic peptides, Buchmeier *et al.* (1987) showed that the cleavage site of LCMV GPC was located within a stretch of nine amino acids that contained the dibasic residues Arg–Arg at positions 262–263. It was proposed that the GPC cleavage was mediated by a Golgi-associated protease acting at or following the dibasic residues, which are highly conserved among most arenaviruses (Burns and Buchmeier, 1993). More precisely, Lenz *et al.* (2000) identified the N-terminal tripeptide when GP2 was isolated from purified virions and demonstrated that the cleavage site of LASV GPC is located after the tetrapeptide sequence Arg–Arg–Leu–Leu. This tetrapeptide is conserved in all LASV isolates published, and LCMV and the New World arenaviruses have similar ones. A systematic mutational analysis revealed that the essential motif for LASV cleavage is Arg–X (Leu, Isoleu, or Val)–Leu, homologous to the consensus sequence recognized by a novel class of cellular endoproteases (Lenz *et al.,* 2000). The glycosylation inhibitors mentioned above are not very selective in their antiviral action, but the elucidation of the cleavage motif of LASV offers the possibility of designing substrate analogues to block GPC cleavage and may have meaningful therapeutic potential for treatment of Lassa fever.

Formation of infectious virions depends not only on glycosylation and proteolytic cleavage of the glycoproteins, but also on myristoylation of GPC, a protein modification detected on JUNV-infected cells by incorporation of [^{3}H]myristic acid (S. M. Cordo, personal communication). The enzyme *N*-myristoyltransferase links myristic acid to the penultimate glycine residue in the N-terminal corresponding consensus sequence, previously reported in JUNV S RNA (Romanowski, 1993). Myristic acid analogues, such as 2-hydroxymyristic acid and 13-oxamyristic acid, were found to inhibit JUNV and TCRV production without apparent toxicity to the cells (Cordo *et al.*, 1999). The cleavage and cell membrane expression of JUNV glycoproteins were not affected by the analogues, suggesting that myristoylation is not essential for the intracellular exocytic transport of the envelope proteins from the site of synthesis to the cell surface, but it may have an important role in their interaction with the plasma membrane during virion assembly and/or budding.

The exocytic pathway of viral glycoproteins may also be affected by agents producing alterations in the properties of the cell membrane. On this basis, compounds disturbing the lipid composition have been analyzed as potential antivirals. Lauric acid, a saturated fatty acid with 12 carbons, was the most effective inhibitor of JUNV and TCRV multiplication due to a blockade in the insertion of the viral glycoproteins into the plasma membrane (Bartolotta *et al.*, 2001). This antiviral activity appeared to be correlated with a stimulation of the triacylglycerol cell content because both effects were dependent on the continued presence of the fatty acid.

D. Assembly and Budding

Assembly and budding are the least understood stages in the arenavirus replicative cycle. Since the very early reports showing electron microscopy images of virions budding from the plasma membrane (Murphy *et al.*, 1970), very limited progress has been attained. The presence of host ribosomes in the virions and the reported packaging of more than one copy of any genome segment in the particles (Romanowski, 1993) suggest that assembly is not a very accurate process.

Arenaviruses appear to lack an internal matrix protein, which in other enveloped viruses plays an important role in the organization of viral compounds during assembly. The lack of a matrix protein may lead to unusual protein interactions during virus assembly, and thus a closer interaction with cytoskeletal components may be required to target viral proteins to a common location and to attain virus maturation. In fact, cell fractionation studies have shown the association with the cellular cytoskeleton of JUNV proteins as well as infectious virus particles (Candurra *et al.*, 1999), suggesting the involvement of the cytoskeleton in the initiation of the assembly and budding processes.

Cross-linking studies indicated that NP may interact with GP2 and Z (Burns and Buchmeier, 1991; Salvato *et al.*, 1992), both proteins likely to be associated with viral membrane because they segregated as hydrophobic proteins upon detergent extraction of virions. On this basis, alternative mechanisms have been proposed for arenavirus assembly: (1) a direct interaction between NP in the nucleocapsids and the cytoplasmic tail of the glycoproteins inserted in the membrane, and (2) the assignment to Z of a role analogous to that of the matrix protein of other viruses, acting as a bridge between nucleocapsids and glycoproteins at the internal face of the envelope (Compans, 1993; Salvato, 1993). No conclusive experimental evidence is available to support either mechanism. However, it is interesting that many known antiretroviral compounds with diverse chemical structures targeted to the Zn-finger motifs in the HIV nucleocapsid protein NCp7 have shown a very potent virucidal activity against the arenaviruses JUNV, TCRV, and PICV (García *et al.,* 2000). It remains to prove whether the inactivation of arenavirus particles by these inhibitors, which included azoic compounds, hydrazide derivatives, aromatic and aliphatic disulfides, and dithianes, is effectively due to an alteration on Z or NP that affects its interaction with other viral proteins in the virion leading to destabilization of the infectious virus.

From time-related inhibition experiments, several natural and synthetic compounds have been found to be inhibitors of late stages in the replicative cycle, but their precise target has not been elucidated. Some examples of these active compounds include: a natural brassinosteroid and a series of synthetic derivatives with plant growth-promoting properties that affected the multiplication of JUNV, PICV, and TCRV (Wachsman *et al.,* 2000); sulfated polyhydroxysteroids isolated from marine organisms and their synthetic derivatives and analogues, which were active aginst JUNV (Comin *et al.,* 1999); and extracts from *Melia azedarach,* which inhibited two events of the JUNV replicative cycle requiring cell membrane participation, uncoating (as mentioned in Section VI.A.2) and budding (Castilla *et al.,* 1998).

VII. Present Treatment of Human Disease

Several arenaviruses induce hemorrhagic fever in humans; however, we will focus only on the diseases produced by LASV and JUNV because they generate periodic outbreaks of hemorrhagic fever with high mortality rate.

Lassa fever is very variable in its presentation, making it difficult to diagnose either in endemic areas or in returning travelers because clinical signs may be confused with yellow fever, malaria, or typhoid. Lassa fever must be suspected if temperatures remain elevated for 1 week or more (Monath *et al.,* 1973). It may present insidious development of fever, headache, and malaise, progressing to pharyngitis, pains in the back, chest, and joins, vomiting, and proteinuria. In severe cases, conjuctivitis, pneumonitis, hepatitis, encephalopathy, nerve deafness, and/or hemorrhages are seen, and death occurs in about 20% of hospitalized cases, usually following cardiovascular collapse. Mortality is higher during the third trimester of pregnancy, and fetal loss is almost invariable. Although some similarities exist with persons infected with the South American arenaviruses, Lassa fever is much more severe and results in a high mortality rate. Despite these observations, it is clear that there is a spectrum of disease associated with LASV in humans from a mild, almost asymptomatic condition to the serious and often fatal hemorrhagic illness (Howard, 1986).

An effective therapy was described using LASV immune plasma administrated to a patient with a laboratory-acquired infection in 1969 (Leifer *et al.,* 1970). Later, studies performed in animals showed that the benefit of using immune plasma was relative, besides the difficulties of obtaining, testing, controlling, and storing the serum. In a study performed in Sierra Leone, West Africa, the efficacy of ribavirin and convalescent plasma was evaluated in the treatment of LASV infections (McCormick *et al.,* 1986). The drug was most effective when administered intravenously during the first 6 days after the onset of illness, significantly decreasing case fatality rates from 50% to 5–9%. Ribavirin is now the preferred method of treatment for patients diagnosed with Lassa fever and it is also recommended as a prophylactic agent in cases of possible exposure to LASV.

Argentine hemorrhagic fever (AHF) is a more typical hemorrhagic fever than Lassa fever. The pathology is largely confined to the circulatory system. After penetrating the skin or mucosa, the virus undergoes a 1- or 2-week incubation period, and initial symptoms are quite nonspecific. Among the first findings are marked asthenia, muscular pain, dizziness, skin and mucosal rashes, lymph node enlargement, cutaneous petechiae, and retrocular pain. At 6–10 days after onset, symptomatology tends to worsen in most patients, and cardiovascular, digestive, renal, or neurological involvement becomes more severe, together with hematologic and clotting alterations. In each individual case, findings are mainly hemorrhagic or neurological and at 10–15 days, over 80% of the patients improve noticeably, whereas the remainder are prone to worsen. Total mortality reaches 16% in the absence of early convalescence plasma treatment. Regardless of the clinical form, convalescence is quite lengthy, but total recovery takes place without sequelae, except for those patients presenting the so-called neurological syndrome (Weissenbacher *et al.,* 1987).

Treatment of AHF patients with immune plasma administered within 8 days of overt clinical disease reduces mortality to 1–2% (Maiztegui *et al.,* 1979). Ribavirin therapy for experimental AHF includes studies performed in guinea pigs (Kenyon *et al.,* 1986), the marmoset, *Callitrhix jacchus* (Weissenbacher *et al.,* 1986), and rhesus macaques (McKee *et al.,* 1988). Rivabirin was not effective in guinea pigs, although viral replication was delayed and mean time to death was prolonged. Better results were obtained in *C. jacchus* because ribavirin administration lowered viremia and increased survival, although late neurological alterations were observed in infected animals. A very important result was obtained with rhesus macaques: Treatment with ribavirin at the time of infection protected the animals from clinical disease. Based on these promising results a clinical evaluation of ribavirin was performed in AHF patients (Enría and Maiztegui, 1994). For that purpose ribavirin was clinically evaluated in a double-blind trial in which patients with more than 8 days of evolution received either intravenous ribavirin or placebo. The results of this study demonstrated that rivabirin had an antiviral effect, but did not show efficacy in reducing mortality. The effect of the drug remains to be seen if it is administered at an earlier step of the disease. It must be remarked that side effects associated with ribavirin treatment such as thrombocytosis and anemia have been recorded (Weissenbacher *et al.,* 1986; McKee *et al.,* 1988). Current specific therapy consists in the early administration of immune plasma in defined doses of specific neutralizing antibodies per kilogram of body weight (Enría *et al.,* 1984).

VIII. Concluding Remarks

In this review, we have attempted to summarize the present state and perspectives of antiviral therapy for arenavirus infections in relation to the dramatic increase in virus emergence during the last decade. Because of the very high case-fatality rates in patients with Lassa fever and the continuing appearance of new viruses in North and South America, including agents responsible for severe human infections, control through chemotherapy warrants special attention.

The replicative cycle of arenaviruses comprises a number of steps that could be considered adequate targets for chemotherapeutic intervention. As reviewed here, numerous compounds have been reported to inhibit the replication of arenaviruses acting at the early stages of attachment and entry, the biosynthetic processes of replication and transcription, or the late steps of maturation, exocytosis, and budding.

Although these studies have brought a better understanding of the mechanisms of virus multiplication and the interaction with the host cell, very few successes on true therapeutic possibilities have been obtained. As occurs with other viruses, the most explored approach of antiviral development has been the utilization of different kinds of nucleoside analogues to block viral RNA transcription and/or replication. However, all the nucleoside analogues tested against arenaviruses up to now have the serious drawback of a considerable level of toxicity and, in considering the equilibrium risk–benefit ratio, ribavirin is still accepted at present as the most valuable chemotherapeutic agent for arenavirus treatment. It must be remarked that so far the best therapy for Argentine hemorrhagic fever patients is the administration of convalescent plasma.

The methods used to discover antiviral drugs have evolved considerably over recent years, particularly driven by the medical need for effective compounds for the treatment of diseases associated with HIV. The traditional methodology of *in vitro* screening of compounds against a particular target has been the only strategy employed against arenaviruses. An alternative approach based on using high-resolution structural data on target biomolecules to design compounds has not be applied in Arenaviridae due to the lack of studies on the crystal structure of the virus component candidates as substrates.

Numerous efforts have also been devoted to obtaining safe vaccines to protect the population against hemorrhagic fever agents. In fact, an attenuated live vaccine named Candid 1 has been developed for Argentine hemorrhagic fever (Maiztegui *et al.,* 1998) and has been evaluated in the human population of the endemic area. However, vaccines probably will never be the complete answer to the control of arenavirus infections. Even with an effective vaccine, occasional outbreaks are expected to occur due to the characteristics of the viruses, for example, because of ecological changes in the habits of the natural rodent reservoir. In addition, the appearance of novel strains or virus species not cross-reacting with vaccine components may lead to the production of isolated cases, outbreaks, or epidemics, all health-threatening situations requiring the administration of an effective chemotherapy.

ACKNOWLEDGMENTS

Research in the authors' laboratory was supported by Agencia Nacional de Promoción Científica y Tecnológica (ANPCyT), Consejo Nacional de Investigaciones Científicas y Técnicas (CONICET), and Universidad de Buenos Aires, Argentina. E.B.D. and C.E.C. are members of the Research Career of CONICET.

References

Ambrosio, A. M., Feuillade, M. R., Gamboa, G. S., and Maiztegui, J. I. (1994). Prevalence of lymphocytic choriomeningitis virus infection in a human population of Argentina. *Am. J. Trop. Med. Hyg.* **50,** 381–386.

Andrei, G., and De Clercq, E. (1990). Inhibitory effect of selected antiviral compounds on arenavirus replication *in vitro*. *Antiviral Res.* **14,** 287–300.

Andrei, G., and De Clercq, E. (1993). Molecular approaches for the treatment of hemorrhagic fever virus infections. *Antiviral Res.* **22,** 45–75.

Andrei, G. M., Lampuri, J. S., Coto, C. E., and de Torres, R. A. (1986). An antiviral factor from *Melia azedarach* L. prevents Tacaribe virus encephalitis in mice. *Experientia* **42,** 843–845.

Armstrong, C. R., and Lillie, R. D. (1934). Experimental lymphocytic choriomeningitis of monkeys and mice produced by a virus encountered in studies of the 1933 St Louis encephalitis epidemic. *Public Health Rep.* **50,** 831–842.

Arribalzaga, R. A. (1955). Una nueva enfermedad epidémica a germen desconocida: Hipertermia nefrotóxica, leucopenia y enantémica. *Día Médico* **27,** 1204–1210.

Auperin, D. D. (1993). Construction and evaluation of recombinant virus vaccines for Lassa fever. *In* "The Arenaviridae" (M. S. Salvato, Ed.), pp. 259–280. Plenum Press, New York.

Auperin, D. D., Romanowski, V., Galinski, M., and Bishop, D. H. L. (1984). Sequencing studies of Pichinde arenavirus S RNA indicate a novel coding strategy, an ambisense viral S RNA. *J. Virol.* **52,** 897–904.

Baird, R. D., and Rivers, T. M. (1938). Relationship of lymphocytic choriomeningitis to acute aseptic meningitis (Wallgren). *Am. J. Public Health* **28,** 47.

Banerjee, S. N., Buchmeier, M. J., and Rawls, W. E. (1976). Requirement of the cell nucleus for the replication of an arenavirus. *Intervirology* **6,** 190–196.

Barry, M., Russi, M., Armstrong, L., Geller, D. L., Tesh, R., Dembry, L., González, J. P., Khan, A., and Peters, C. J. (1995). Treatment of laboratory-acquired Sabiá virus infection. *N. Engl. J. Med.* **333,** 294–296.

Bartolotta, S., García, C. C., Candurra, N. A., and Damonte, E. B. (2001). Effect of fatty acids on arenavirus replication: Inhibition of virus production by lauric acid. *Arch. Virol.* **146,** 777–790.

Borrow, P., and Oldstone, M. B. A. (1992). Characterization of lymphocytic choriomeningitis virus-binding protein(s): A candidate cellular receptor for the virus. *J. Virol.* **66,** 7270–7281.

Borrow, P., and Oldstone, M. B. A. (1994). Mechanism of lymphocytic choriomeningitis virus entry into cells. *Virology* **198,** 1–9.

Bowen, M. D., Peters, C. J., Mills, J. M., and Nichol, S. T. (1996a). Oliveros virus: A novel arenavirus from Argentina. *Virology* **217,** 362–366.

Bowen, M. D., Peters, C. J., and Nichol, S. T. (1996b). The phylogeny of New World (Tacaribe complex) arenaviruses. *Virology* **219,** 285–290.

Bowen, M. D., Rollin, P. E., Ksiazek, T. G., Hustad, H. L., Bausch, D. G., Demby, A. H., Bajani, M. D., Peters, C. J., and Nichol, S. T. (2000). Genetic diversity among Lassa virus strains. *J. Virol.* **74,** 6992–7004.

Bruns, M., Kratzberg, T., Zeller, W., and Lehmann-Grube, F. (1990). Mode of replication of lymphocytic choriomeningitis virus in persistently infected cultivated mouse L cells. *Virology* **177,** 615–624.

Buchmeier, M. J., Lewicki, H. A., Tomori, O., and Oldstone, M. B. A. (1981). Monoclonal antibodies to lymphocytic choriomeningitis and Pichinde viruses: Generation, characterization and cross-reactivity with other arenaviruses. *Virology* **113,** 73–85.

Buchmeier, M. J., Southern, P. J., Parekh, B. S., Wooddell, M. K., and Oldstone, M. B. A. (1987). Site-specific antibodies define a cleavage site conserved among arenavirus GP-C glycoproteins. *J. Virol.* **61,** 982–985.

Buchmeier, M. J., Clegg, J. C. S., Franze-Fernandez, M. T., Kolakofsky, D., Peters, C. J., and Southern, P. J. (1995). Family Arenaviridae. *In* "Virus Taxonomy: Classification and Nomenclature of Viruses" (F. A. Murphy, C. M. Fauquet, D. H. L. Bishop, S. A. Ghabrial, A. W. Jarvis, G. P. Martelli, M. A. Mayo, and M. D. Summers, Eds.), pp. 319–323. Springer-Verlag, New York.

Burns, J. W., and Buchmeier, M. J. (1991). Protein–protein interactions in lymphocytic choriomeningitis virus. *Virology* **183,** 620–629.

Burns, J. W., and Buchmeier, M. J. (1993). Glycoproteins of the arenaviruses. *In* "The Arenaviridae" (M. S. Salvato, Ed.), pp. 17–35. Plenum Press, New York.

Burns, N. J., III, Barnett, B. B., Huffman, J. H., Dawson, M. Y., Sidwell, R. W., De Clercq, E., and Kende, M. (1988). A newly developed immunofluorescent assay for determining the Pichinde virus-inhibitory effects of selected nucleoside analogues. *Antivir. Res.* **10,** 89–98.

Campbell Dwyer, E. J., Lai, H., MacDonald, R. C., Salvato, M. S., and Borden, K. L. B. (2000). The lymphocytic choriomeningitis virus RING protein Z associates with eukaryotic initiation factor 4E and selectively represses translation in a RING-dependent manner. *J. Virol.* **74,** 3293–3300.

Candurra, N. A., and Damonte, E. B. (1997). Effect of inhibitors of the intracellular exocytic pathway on glycoprotein processing and maturation of Junin virus. *Arch. Virol.* **142,** 2179–2193.

Candurra, N. A., and Damonte, E. B. (1999). Acción inhibitoria de la cafeína sobre la multiplicación del virus Junín. *Rev. Arg. Microbiol.* **31,** 135–141.

Candurra, N. A., Scolaro, L. A., Mersich, S. E., Damonte, E. B., and Coto, C. E. (1990). A comparison of Junin virus strains: Growth characteristics, cytopathogenicity and viral polypeptides. *Res. Virol.* **141,** 505–515.

Candurra, N. A., Maskin, L., and Damonte, E. B. (1996). Inhibition of arenavirus multiplication *in vitro* by phenotiazines. *Antivir. Res.* **31,** 149–158.

Candurra, N. A., Lago, M. J., Maskin, L., and Damonte, E. B. (1999). Involvement of the cytoskeleton in Junin virus multiplication. *J. Gen. Virol.* **80,** 147–156.

Cao, W., Henry, M. D., Borrow, P., Yamada, H., Elder, J. H., Ravkov, E. V., Nichol, S. T., Compans, R. W., Campbell, K. P., and Oldstone, M. B. A. (1998). Identification of α-dystroglycan as a receptor for lymphocytic choriomeningitis virus and Lassa fever virus. *Science* **282,** 2079–2081.

Carey, D. E., Kemp, G. E., White, H. A., Pinneo, L., Addy, R. F., Fom, A., Stroh, G., Casals, J., and Henderson, B. E. (1972). Lassa fever. Epidemiological aspects of the 1970 epidemic, Jos, Nigeria. *Trans. R. Soc. Trop. Med. Hyg.* **66,** 402–408.

Casals, J. (1975). Arenaviruses. *Yale J. Biol. Med.* **48,** 115–140.

Castilla, V., and Mersich, E. E. (1996). Low-pH-induced fusion of Vero cells infected with Junin virus. *Arch. Virol.* **141,** 1307–1317.

Castilla, V., Mersich, S. E., Candurra, N. A., and Damonte, E. B. (1994). The entry of Junin virus into Vero cells. *Arch. Virol.* **136,** 363–374.

Castilla, V., Barquero, A. A., Mersich, S. E., and Coto, C. E. (1998). *In vitro* anti-Junin virus activity of a peptide isolated from *Melia azedarach* L. leaves. *Int. J. Antimicrob. Agents* **10,** 67–75.

Castilla, V., Palermo, L. M., and Coto, C. E. (2001). Involvement of vacuolar proton ATPase in Junin virus multiplication. *Arch. Virol.* **146,** 251–263.

Centers for Disease Control. (1994). Bolivian hemorrhagic fever—El Beni Department, Bolivia, 1994. *MMWR* **43,** 943–946.

Childs, J. E., and Peters, C. J. (1993). Epidemiology and ecology of arenaviruses and their hosts. *In* "The Arenaviridae" (M. S. Salvato, Ed.), pp. 331–384. Plenum Press, New York.

Coimbra, T. L. M., Nassar, E. S., Burattini, M. N., de Souza, L. T. M., Ferreira, I. B., Rocco, I. M., Travassos da Rosa, A. P., Vasconcelos, P. F. C., Pinheiro, F. P., Le Duc, J. W., Rico-Hesse, R., González, J.-P., Jahrling, P. B., and Tesh, R. B. (1994). New arenavirus isolated in Brazil. *Lancet* **343,** 391–392.

Comin, M. J., Maier, M. S., Roccatagliata, A. J., Pujol, C. A., and Damonte, E. B. (1999). Evaluation of the antiviral activity of natural sulfated polyhydroxysteroids and their synthetic derivatives and analogs. *Steroids* **64,** 335–340.

Compans, R. W. (1993). Arenavirus ultrastructure and morphogenesis. *In* "The Arenaviridae" (M. S. Salvato, Ed.), pp. 3–16. Plenum Press, New York.

Cordo, S. M., Candurra, N. A., and Damonte, E. B. (1999). Myristic acid analogs are inhibitors of Junin virus replication. *Microbes Infection* **1,** 609–614.

Coto, C. E., Calello, M. A., and Parodi, A. S. (1969). Efecto de la amantadina-HCl sobre la infectividad del virus Junín (FHA) *in vitro* e *in vivo*. *Rev. Asoc. Arg. Microbiol.* **1,** 3–8.

Coto, C. E., Vidal., M. C., D'Aiutolo, A. C., and Damonte, E. B. (1981). Selection of spontaneous ts mutants of Junin and Tacaribe viruses in persistent infections. *In* "The Replication of Negative Strand Viruses" (D. H. L. Bishop and R. W. Compans, Eds.), pp. 11–14. Elsevier, New York.

Coto, C. E., Damonte, E. B., Alché, L. E., and Scolaro, L. A. (1993). Genetic variation in Junin virus. *In* "The Arenaviridae" (M. S. Salvato, Ed.), pp. 85–101. Plenum Press, New York.

D'Aiutolo, A. C., and Coto, C. E. (1986). Vero cells persistently infected with Tacaribe virus: Role of interfering particles in the establishment of the infection. *Virus Res.* **6,** 235–244.

Dalton, A. J., Rowe, W. P., Smith, G. H., Wilsnack, R. E., and Pugh, W. E. (1968). Morphological and cytochemical studies on lymphocytic choriomeningitis virus. *J. Virol.* **2,** 1465–1478.

Damonte, E. B., Mersich, S. E., and Coto, C. E. (1983). Response of cells persistently infected with arenaviruses to superinfection with homotypic and heterotypic viruses. *Virology* **129,** 474–478.

Damonte, E. B., Mersich, S. E., and Candurra, N. A. (1994). Intracellular processing and transport of Junin virus glycoproteins influences virion infectivity. *Virus Res.* **34,** 317–326.

De la Torre, J. C., and Oldstone, M. B. A. (1992). Selective disruption of growth-hormone transcription machinery by viral infection. *Proc. Natl. Acad. Sci. USA* **89,** 9939–9943.

Di Simone, C., and Buchmeier, M. J. (1995). Kinetics and pH dependence of acid-induced structural changes in the lymphocytic choriomeningitis virus glycoprotein complex. *Virology* **209,** 3–9.

Di Simone, C., Zandonatti, M. A., and Buchmeier, M. J. (1994). Acidic pH triggers LCMV membrane fusion activity and conformational change in the glycoprotein spike. *Virology* **198,** 455–465.

Downs, W. G., Anderson, C. R., Spence, L., Aitken, T. H. G., and Greenhall, A. H. (1963). Tacaribe virus, a new agent isolated from *Artibeus* bats and mosquitoes in Trinidad, West Indies. *Am. J. Trop. Med. Hyg.* **12,** 640–642.

Enría, D. A., and Maiztegui, J. I. (1994). Antiviral treatment of Argentine hemorrhagic fever. *Antivir. Res.* **23,** 23–31.

Enría, D. A., Briggiler, A. M., Fernández, N. J., Levis, S. C., and Maiztegui, J. I. (1984). Importance of dose of neutralizing antibodies in treatment of Argentine hemorrhagic fever with immune plasma. *Lancet* **ii**(8397), 255–256.

Enserink, M. (2000). Emerging diseases. New arenavirus blamed for recent deaths in California. *Science* **289,** 842–843.
Frame, J. D., Baldwin, Jr., J. M., Gocke, D. J., and Troup, J. M. (1970). Lassa fever, a new virus disease of man from West Africa. I. Clinical description and pathological findings. *Am. J. Trop. Med. Hyg.* **19,** 670–676.
Franze-Fernández, M. T., Zetina, C., Iapalucci, S., Lucero, M. A., Bouissou, C., Lopez, R., Rey, O., Daheli, M., Cohen, G. N., and Zakin, M. M. (1987). Molecular structure and early events in the replication of Tacaribe arenavirus S RNA. *Virus Res.* **7,** 309–324.
Franze-Fernández, M. T., Iapalucci, S., López, N., and Rossi, C. (1993). Subgenomic RNAs of Tacaribe virus. *In* "The Arenaviridae" (M. S. Salvato, Ed.), pp. 113–132. Plenum Press, New York.
Friedlander, A. M., Jahrling, P. B., Merrill, J. P., and Tobery, S. (1984). Inhibition of mouse peritoneal macrophage DNA synthesis by infection with the arenavirus Pichinde. *Infect. Immun.* **43,** 283–288.
Fulhorst, C. F., Bowen, M. D., Ksiazek, T. G., Rollin, P. E., Nichol, S. T., Kosoy, M. Y., and Peters, C. J. (1996). Isolation and characterization of Whitewater Arroyo virus, a novel North American arenavirus. *Virology* **224,** 114–120.
Fulhorst, C. F., Bowen, M. D., Salas, R. A., de Manzione, N. M. C., Duno, G., Utrera, A., Ksiazek, T. G., Peters, C. J., Nichol, S. T., de Miller, E., Tovar, D., Ramos, B., Vasquez, C., and Tesh, R. B. (1997). Isolation and characterization of Pirital virus, a newly discovered South American arenavirus. *Am. J. Trop. Med. Hyg.* **56,** 558–553.
Fuller-Pace, F. V., and Southern, P. J. (1989). Detection of virus-specific RNA-dependent RNA polymerase activity in extracts from cells infected with lymphocytic choriomeningitis virus: *In vitro* synthesis of full-length viral RNA species. *J. Virol.* **63,** 1938–1944.
García, C. C., Candurra, N. A., and Damonte, E. B. (2000). Antiviral and virucidal activities against arenaviruses of zinc-finger active compounds. *Antiviral Chem. Chemother.* **11,** 231–238.
Garcin, D., and Kolakofsky, D. (1990). A novel mechanism for the initiation of arenavirus genome replication. *J. Virol.* **64,** 6196–6203.
Garcin, D., Rochat, S., and Kolakofsky, D. (1993). The Tacaribe arenavirus small zinc finger protein is required for both mRNA synthesis and genome replication. *J. Virol.* **67,** 807–812.
Gilbert, B. E., and Knight, V. (1986). Biochemistry and clinical application of ribavirin. *Antimicrob. Agents Chemother.* **30,** 201–205.
Giménez, H. B., and Compans, R. W. (1980). Defective interfering Tacaribe virus and persistently infected cells. *Virology* **107,** 229–239.
Glushakova, S. E., and Lukashevich, I. S. (1989). Early events in arenavirus replication are sensitive to lysosomotropic compounds. *Arch. Virol.* **194,** 157–161.
Glushakova, S. E., Omelyanenko, V. G., Lukashevich, I. S., Bogdanov, A. A., Moshnikova, A. B., Kozytch, A. T., and Torchilin, V. P. (1992). The fusion of artificial lipid membranes induced by the synthetic arenavirus fusion peptide. *Biochim. Biophys. Acta* **1110,** 202–208.
González, J. P., McCormick, J. B., Saluzzo, J. F., Herve, J. P., Georges, A. J., and Johnson, K. M. (1983). An arenavirus isolated from wild-caught rodents (*Praomys* species) in the central African republic. *Intervirology* **19,** 105–112.
González, J. P., Bowe, M. D., Nichol, S. T., and Rico-Hesse, R. (1996). Genetic characterization and phylogeny of Sabiá virus, an emergent pathogen in Brazil. *Virology* **221,** 318–324.
Goswami, B. B., Borek, E., Sharma, O. K., Fujitaki, J., and Smith, R. A. (1979). The broad spectrum antiviral agent ribavirin inhibits capping of mRNA. *Biochem. Biophys. Res. Commun.* **89,** 830–836.

Hall, C. B., Walsh, E. E., Hruska, J. F., Betts, R. F., and Hall, W. J. (1983). Ribavirin treatment of experimental respiratory syncytial viral infection. *J. Am. Med. Assoc.* **249,** 2666–2670.

Henry, M. D., and Campbell, K. P. (1999). Dystroglycan inside and out. *Curr. Opin. Cell Biol.* **11,** 602–607.

Howard, C. R. (1986). Human arenavirus infections. *In* "Arenaviruses" (A. J. Zuckerman Ed.), pp. 47–69. Elsevier, Amsterdam.

Huggins, J. W., Robins, R. K., and Canonico, P. G. (1984). Synergistic antiviral effects of ribavirin and the C-nucleoside analogs tiazofurin and selenazofurin against togaviruses, bunyaviruses, and arenaviruses. *Antimicrob. Agents Chemother.* **26,** 476–480.

Iapalucci, S., López, N., and Franze-Fernández, M. T. (1991). The 3′ end termini of Tacaribe arenavirus subgenomic RNAs. *Virology* **182,** 269–278.

Iapalucci, S., Cherñavsky, A., Rossi, C., Burgín, M. J., and Franze-Fernández, M. T. (1994). Tacaribe virus gene expression in cytopathic and non-cytopathic infections. *Virology* **200,** 613–622.

Jahrling, P. B., Hesse, R. A., Eddy, G. A., Johnson, K. M., Callis, R. T., and Stephen, E. L. (1980). Lassa virus infection of rhesus monkeys: Pathogenesis and treatment with ribavirin. *J. Infect. Dis.* **141,** 580–589.

Jennings, W. L., Lewis, A. L., Sather, G. E., Pierce, L. V., and Bond, J. O. (1970). Tamiami virus in the Tampa Bay area. *Am. J. Trop. Med. Hyg.* **19,** 527–536.

Johnson, K. M. (1981). Arenaviruses in rodents. *In* "Comparative Diagnosis of Viral Diseases" (E. Kurtak and C. Kurstak, Eds.), pp. 511–525. Academic Press, New York.

Johnson, K. M., Mackenzie, R. B., Webb, P. A., and Kuns, M. L. (1965). Chronic infection of rodents by Machupo virus. *Science* **150,** 1618–1619.

Kenyon, R. H., Canonico, P. G., Green, D. G., and Peters, C. J. (1986). Effect of ribavirin and tributylribavirin on Argentine hemorrhagic fever (Junin virus) in guinea pigs. *Antimicrob. Agents Chemother.* **29,** 521–523.

Klavinskis, L. S., and Oldstone, M. B. A. (1987). Lymphocytic choriomeningitis virus can persistently infect thyroid epithelial cells and perturb thyroid hormone production. *J. Gen. Virol.* **68,** 1867–1873.

Kosoy, M. Y., Elliott, L. H., Ksiazek, T. G., Fulhorst, C. F., Rollin, P. E., Childs, J. E., Mills, J. N., Maupin, G. O., and Peters, C. J. (1996). Prevalence of antibodies to arenaviruses in rodents from the southern and western United States: Evidence for an arenavirus associated to the genus *Neotoma*. *Am. J. Trop. Med. Hyg.* **54,** 570–576.

Lee, K. J., Novella, I. S., Teng, M. N., Oldstone, M. B. A., and de la Torre, J. C. (2000). NP and L proteins of lymphocytic choriomeningitis virus (LCMV) are sufficient for efficient transcription and replication of LCMV genomic RNA analogs. *J. Virol.* **74,** 3470–3477.

Leifer, E., Gocke, D. J., and Bourne, H. (1970). Lassa fever, a new virus disease of man from West Africa. II. Report of a laboratory-acquired infection treated with plasma from a person recently recovered from the disease. *Am. J. Trop. Med. Hyg.* **19,** 677–679.

Lenz, O., ter Meulen, J., Feldmann, H., Klenk, H.-D., and Garten, W. (2000). Identification of a novel consensus sequence at the cleavage site of the Lassa virus glycoprotein. *J. Virol.* **74,** 11418–11421.

Leung, W. C., and Rawls, W. E. (1977). Virion-associated ribosomes are not required for the replication of Pichinde virus. *Virology* **81,** 174–176.

López, R., and Franze-Fernández, M. T. (1985). Effect of Tacaribe virus infection on host cell protein and nucleic acid synthesis. *J. Gen. Virol.* **66,** 1753–1761.

López, R., Grau, O., and Franze-Fernández, M. T. (1986). Effect of actinomycin D on arenavirus growth and estimation of the generation time for a virus particle. *Virus Res.* **5,** 213–220.

Lucia, H. L., Coppenhaver, D. H., and Baron, S. (1989). Arenavirus infection in the guinea pig model: Antiviral therapy with recombinant interferon-α, the immunomodulator CL246,738 and ribavirin. *Antivir. Res.* **12,** 279–292.

Maiztegui, J. I. (1975). Clinical and epidemiological patterns of Argentine hemorrhagic fever. *Bull. WHO* **55,** 567–575.

Maiztegui, J. I., Fernández, N., and Damilano, A. J. (1979). Efficacy of immune plasma in treatment of Argentine hemorrhagic fever and association between treatment and a late neurological syndrome. *Lancet* **ii**(8154), 1216–1217.

Maiztegui, J. I., Feuillade, M., and Briggiler, A. (1986). Progressive extension of the endemic area and changing incidence of Argentine hemorrhagic fever. *Med. Microbiol. Immunol.* **175,** 149–152.

Maiztegui, J. I., McKee, K. T., Barrera Oro, J. G., Harrison, L. H., Gibbs, P. H., Feuillade, M. R., Enria, D. A., Briggiler, A. M., Levis, S. C., Ambrosio, A. M., Halsey, N. A., and Peters, C. J., and the AHF Study Group (1998). Protective efficacy of a live attenuated vaccine against Argentine hemorrhagic fever. *J. Infect. Dis.* **177,** 277–283.

McCormick, J. B. (1987). Epidemiology and control of Lassa fever. *Curr. Top. Microbiol. Inmunol.* **134,** 69–78.

McCormick, J. B., King, I. J., Webb, P. A., Scribner, C. L., Craven, R. B., Johnson, K. M., Elliot, L. H., and Belmont-Williams, R. (1986). Lassa fever. Effective therapy with ribavirin. *N. Engl. J. Med.* **314,** 20–26.

McCormick, J. B., Webb, P. A., Krebs, J. W., Johnson, K. M., and Smith, E. S. (1987). A prospective study of the epidemiology and ecology of Lassa fever. *J. Infect. Dis.* **155,** 437–444.

McKee, K. T., Huggins, J. W., Trahan, C. J., and Mahlandi, B. G. (1988). Ribavirin prophylaxis and therapy for experimental Argentine hemorrhagic fever. *Antimicrob. Agents Chemother.* **32,** 1304–1309.

Mersich, S. E., Damonte, E. B., and Coto, C. E. (1981). Induction of RNA polymerase II activity in Junin virus-infected cells. *Intervirology* **16,** 123–127.

Meyer, B. J., and Southern, P. J. (1993). Concurrent sequence analysis of 5′ and 3′ RNA termini by intramolecular circularization reveals 5′ nontemplated bases and 3′ terminal heterogeneity for lymphocytic choriomeningitis virus mRNAs. *J. Virol.* **67,** 2621–2627.

Meyer, B. J., and Southern, P. J. (1997). A novel type of defective viral genome suggests a unique strategy to establich and maintain persistent lymphocytic choriomeningitis virus infections. *J. Virol.* **71,** 6757–6764.

Mills, J. N., Ellis, B. A., McKee, K. T., Jr., Ksiazek, T. G., Barrera Oro, J. G., Maiztegui, J. I., Calderon, G. E. Peters, C. J., and Childs, J. E. (1991). Junin virus activity in rodents from endemic and nonendemic loci in central Argentina. *Am. J. Trop. Med. Hyg.* **44,** 589–597.

Mills, J. N., Barrera Oro, J. G., Bressler, D. S., Childs, J. E., Tesh, R. B., Smith, J. F., Enría, D. A., Geisbert, T. W., McKee, K. T., Jr., Bowen, M. D., Peters, C. J., and Jahrling, P. B. (1996). Characterization of Oliveros virus, a new member of the Tacaribe complex (Arenaviridae: Arenavirus). *Am. J. Trop. Med. Hyg.* **54,** 399–404.

Monath, T. P., Mertens, P. E., Patton, R., Moser, C. R., Baum, J. J., and Pinneo, L. (1973). A hospital epidemic of Lassa fever in Zorzor, Liberia, March–April 1972. *Am. J. Trop. Hyg.* **23,** 1140–1149.

Murphy, F. A., Webb, P. A., Johnson, K. M., Whitfield, S. G., and Chappell, W. A. (1970). Arenaviruses in Vero cells: Ultrastructural studies. *J. Virol.* **6,** 507–518.

Padula, P. J., and Martinez Segovia, Z. M. (1984). Replication of Junin virus in the presence of tunicamycin. *Intervirology* **22,** 227–231.

Parekh, B. S., and Buchmeier, M. J. (1986). Proteins of lymphocytic choriomeningitis virus: Antigenic topography of the viral glycoproteins. *Virology* **153,** 168–178.

Parodi, A. S., Greenway, D. J., Rugiero, H. R., Rivero, E., Frigerio, M. J., Mettler, W. E., Garzon, F., Boxaca, M., Guerrero, L. B., and Nota, N. R. (1958). Sobre la etiología del brote epidémico de Junín. *Día Médico* **30,** 2300–2302.

Pfau, C. J. (1975). Arenavirus chemotherapy: Retrospect and prospect. *Bull. WHO* **52,** 737–744.

Pfau, C. J., Trowbridge, R. S., Welsh, R. M., Staneck, L. D., and O'Connell, C. M. (1972). Arenaviruses: Inhibition by amantadine hydrochloride. *J. Gen. Virol.* **14,** 209–211.

Pinheiro, F. P., Shope, R. E., de Andrade, A. H. P., Bensabath, G., Cacios, G. V., and Casals, J. (1966). Amapari, a new virus of the Tacaribe group from rodents and mites of Amapa territory, Brazil. *Proc. Soc. Exp. Biol. Med.* **122,** 531–535.

Pinheiro, F. P., Woodall, J. P., Da Rosa, A. P. A. T., and Da Rosa, J. F. T. (1977). Studies of Arenaviruses in Brazil. *Medicina* (Buenos Aires) **37**(Suppl. 3), 175–181.

Polyak, S. J., Zheng, S., and Harnish, D. G. (1995). 5′ Termini of Pichinde arenavirus S RNAs and mRNAs contain nontemplated nucleotides. *J. Virol.* **69,** 3211–3215.

Raiger Iustman, L. J., Candurra, N., and Mersich, S. E. (1995). Influencia del tratamiento enzimático sobre la interacción virus Junín-células Vero. *Rev. Arg. Microbiol.* **27,** 28–32.

Raju, R., Raju, L., Hacker, D., Garcin, D., Compans, R. W., and Kolakofski, D. (1990). Nontemplated bases at the 5′ ends of Tacaribe virus mRNAs. *Virology* **174,** 53–59.

Rawls, W. E., Banerjee, N. S., McMillan, C. A., and Buchmeier, M. J. (1976). Inhibition of Pichinde virus replication by actinomycin D. *J. Gen. Virol.* **33,** 421–434.

Romanowski, V. (1993). Genetic organization of Junin virus, the etiological agent of Argentine hemorrhagic fever. *In* "The Arenaviridae" (M. S. Salvato, Ed.), pp. 51–84. Plenum Press, New York.

Rowe, W. P., Murphy, F. A., Bergold, G. H., Casals, J., Hotchin, J., Johnson, K. M., Lehmann-Grube, F., Mims, C. A., Traub, E., and Webb, P. A. (1970a). Arenoviruses: Proposed name for a newly defined virus group. *J. Virol.* **5,** 651–652.

Rowe, W. P., Pugh, W. E., Webb, P. A., and Peters, C. J. (1970b). Serological relationship of the Tacaribe complex of viruses to lymphocytic choriomeningitis virus. *J. Virol.* **5,** 289–292.

Sabattini, M. S., Barrera Oro, J. G., Maiztegui, J. I., and de Ferradas, B. R. (1974). Actividad del virus de la coriomeningitis linfocítica en el área endémica de fiebre hemorrágica argentina. II aislamiento a partir de un *Mus musculus* campestre capturado en el sudeste de Córdoba. *Medicina* (Buenos Aires) **37**(Suppl. 3), 149–161.

Salvato, M. S. (1993). Molecular biology of the prototype arenavirus, lymphocytic choriomeningitis virus. *In* "The Arenaviridae" (M. S. Salvato, Ed.), pp. 133–156. Plenum Press, New York.

Salvato, M. S., Schweighofer, K. J., Burns, J., and Shimomaye, E. M. (1992). Biochemical and immunological evidence that the 11-kDa zinc-binding protein of lymphocytic choriomeningitis virus is a structural component of the virus. *Virus Res.* **22,** 185–198.

Scolaro, L. A., Mersich, S. E., and Damonte, E. B. (1990). A mouse attenuated mutant of Junin virus with an altered envelope glycoprotein. *Arch. Virol.* **111,** 257–262.

Silber, A. M., Candurra, N. A., and Damonte, E. B. (1993). The effects of oligosaccharide trimming inhibitors on glycoprotein expression and infectivity of Junin virus. *FEMS Microbiol. Lett.* **109,** 39–44.

Smee, D. F., Morris, J. L. B., Barnard, D. L., and Van Aerschot, A. (1992). Selective inhibition of arthropod-borne and arenaviruses by 3′-fluoro-3′-deoxyadenosine. *Antivir. Res.* **18,** 151–162.

Smee, D. F., Gilbert, J., Leonhardt, J. A., Barnett, B. B., Huggins, J. H., and Sidwell, R. W. (1993). Treatment of lethal Pichinde virus infections in weanling LVG/Lak hamsters with ribavirin, ribamidine, selenazofurin, and ampligen. *Antivir. Res.* **20,** 57–70.

Smelt, S. C., Borrow, P., Kunz, S., Cao, W., Tishon, A., Lewicki, H., Campbell, K. P., and Oldstone, M. B. A. (2001). Differences in affinity of binding of lymphocytic choriomeningitis virus strains to the cellular receptor α-dystroglycan correlate with viral tropism and disease kinetics. *J. Virol.* **75,** 448–457.

Streeter, D. G., Witkowski, J. T., Khare, G. P., Sidwell, R. W., Bauer, R. J., Robins, R. K., and Simon, L. N. (1973). Mechanism of action of 1-β-D-ribofuranosyl-1,2,4-triazole-3-carboxamide (Virazole), a new broad-spectrum antiviral agent. *Proc. Natl. Acad. Sci. USA* **70,** 1174–1178.

Swanepoel, R., Leman, P. A., Shepherd, A. J., Shepherd, S. P., Kiley, M. P., and McCormick, J. B. (1985). Identification of Ippy virus as a Lassa-fever related virus. *Lancet* **1,** 639.

Teng, M. N., Borrow, P., Oldstone, M. B. A., and de la Torre, J. C. (1996). A single amino acid change in the glycoprotein of lymphocytic choriomeningitis virus is associated with the ability to cause growth hormone deficiency syndrome. *J. Virol.* **70,** 8438–8443.

ter Meulen, J. I., Lukashevich, I., Sidibe, K., Inapogui, A., Marx, M., Dorlemann, A., Yansane, M. L., Koulemou, K., Chang-Claude, J., and Schmitz, H. (1996). Hunting of peridomestic rodents and consumption of their meat as possible risk factors for rodent-to-human transmission of Lassa virus in the Republic of Guinea. *Am. J. Trop. Med. Hyg.* **55,** 661–666.

Tesh, R. B., Jarhling, P. B., Salas, R. A., and Shope, R. E. (1994). Description of Guanarito virus (Arenaviridae: Arenavirus), the etiologic agent of Venezuelan hemorrhagic fever. *Am. J. Trop. Med. Hyg.* **50,** 452–459.

Tortorici, M. A., Albariño, C. G., Posik, D. M., Ghiringhelli, P. D., Lozano, M. E., Rivera Pomar, R., and Romanowski, V. (2001a). Arenavirus nucleocapsid protein displays a transcriptional antitermination activity *in vivo*. *Virus Res.* **73,** 41–55.

Tortorici, M. A., Ghiringhelli, P. D., Lozano, M. E., Albarino, C. G., and Romanowski, V. (2001b). Zinc-binding properties of Junin virus nucleocapsid protein. *J. Gen. Virol.* **82,** 121–128.

Trapido, H., and San Martín, C. (1971). Pichinde virus. A new virus of the Tacaribe group from Colombia. *Am. J. Trop. Med. Hyg.* **20,** 631–641.

Van der Zeijst, B. A. M., Bleumink, N., Crawford, L. V., Swyryd, E. A., and Stark, G. R. (1983). Viral proteins and RNAs in BHK cells persistently infected by lymphocytic choriomeningitis virus. *J. Virol.* **48,** 262–270.

Vezza, A. C., Clewley, J. P., Gard, G. P., Abraham, N. Z., Compans, R. W., and Bishop, D. H. L. (1978). Virion RNA species of the arenaviruses Pichinde, Tacaribe and Tamiami. *J. Virol.* **26,** 485–495.

Wachsman, M. B., López, E. M. F., Ramírez, J. A., Galagovsky, L. R., and Coto, C. E. (2000). Antiviral effect of brassinosteroids against herpes virus and arenaviruses. *Antivir. Chem. Chemother.* **11,** 71–77.

Weaber, S. C., Salas, R. A., de Manzione, N., Fulhorst, C. F., Duno, G., Utrera, A., Mills, J. N., Ksiazek, T. G., Tovar, D., and Tesh, R. B. (2000). Guanarito virus (Arenaviridae) isolates from endemic and outling localities in Venezuela: Sequence comparisons among and within strains isolated from Venezuelan hemorrhagic fever patients and rodents. *Virology* **266,** 189–195.

Webb, P. A., Johnson, K. M., Hibbs, J. B., and Kuns, M. L. (1970). Parana, a new Tacaribe complex virus from Paraguay. *Arch. Ges. Virusforsch.* **32,** 379–388.

Webb, P. A., Johnson, K. M., Peters, C. J., and Justines, G. (1973). Behaviour of Machupo and Latino viruses in *Calomys callosus* from two geographic areas of Bolivia. *In*

"Lymphocytic Choriomeningitis Virus and Other Arenaviruses" (F. Lehmann-Grube, Ed.), pp. 313–321. Springer-Verlag, Berlin.

Weber, C., Martínez Peralta, L., and Lehmann-Grube, F. (1983). Persistent infection of cultivated cells with lymphocytic choriomeningitis virus. *Arch. Virol.* **77,** 271–276.

Weber, E. L., Guerrero, L. B. de, and Boxaca, M. C. (1985). MRC5 cells: A model for Junin virus persistent infection. *J. Gen. Virol.* **66,** 1179–1183.

Weissenbacher, M. C., Avila, M. M., Calello, M. A., Merani, M. S., McCormick, J. B., and Rodriguez, M. (1986). Effect of ribavirin and immune serum on Junin virus infected primates. *Med. Microbiol. Immunol.* **175,** 183–186.

Weissenbacher, M. C., Laguens, R. P., and Coto, C. E. (1987). Argentine hemorrhagic fever. *Curr. Top. Microbiol. Immunol.* **134,** 79–116.

Welsh, R. M., and Buchmeier, M. J. (1979). Protein analysis of defective interfering virus and persistently infected cells. *Virology* **96,** 503–515.

Witvrouw, M., Desmyter, J., and De Clercq, E. (1994). Antiviral portrait series. 4. Polysulfates as inhibitors of HIV and other enveloped viruses. *Antivir. Chem. Chemother.* **5,** 345–359.

Wright, K. E., Salvato, M. S., and Buchmeier, M. J. (1989). Neutralizing epitopes of lymphocytic choriomeningitis virus are conformational and require both glycosylation and disulfide bonds for expression. *Virology* **171,** 417–426.

Wright, K. E., Spiro, R. C., Burns, W., and Buchmeier, M. J. (1990). Post-translational processing of the glycoproteins of lymphocytic choriomeningitis virus. *Virology* **177,** 175–183.

Wulff, H., McIntosh, B. M., Hammer, D. B., and Johnson, K. M. (1977). Isolation of an arenavirus closely related to Lassa virus from *Mastomys natalensis* in south-east Africa. *Bull. WHO* **55,** 441–444.

Wulff, H., Lange, J. V., and Webb, P. A. (1978). Interrelationships among arenaviruses measured by indirect immunofluorescence. *Intervirology* **9,** 344–350.

Xing, Z., and Whitton, L. (1993). An anti-lymphocytic choriomeningitis virus ribozyme expressed in tissue culture cells diminishes viral RNA levels and leads to a reduction in infectious virus yield. *J. Virol.* **67,** 1840–1847.

Young, P. R., and Howard, C. R. (1983). Fine structure analysis of Pichinde virus nucleocapsids. *J. Gen. Virol.* **64,** 833–842.

Young, P. R., Chanas, A. C., Lee, S. R., Gould, E. A., and Howard, C. R. (1987). Localization of an arenavirus protein in the nuclei of infected cells. *J. Gen. Virol.* **68,** 2465–2470.

ADVANCES IN VIRUS RESEARCH, VOL. 58

EVALUATION OF DRUG RESISTANCE IN HIV INFECTION

Benedikt Weissbrich, Martin Heinkelein,
and Christian Jassoy

Institute for Virology and Immunobiology
Julius Maximilians University
97078 Würzburg, Germany

I. Introduction
II. Targets of Antiviral Drug Therapy in HIV Infection
   A. Reverse Transcriptase
   B. Protease
   C. Envelope Glycoprotein and Virus Entry
   D. Integrase
III. Mechanisms of Antiviral Drug Resistance
   A. General Aspects
   B. Resistance to Reverse Transcriptase Inhibitors
   C. Resistance to Protease Inhibitors
   D. Resistance to Fusion and Integrase Inhibitors
IV. Technologies for Measuring Drug Sensitivity
   A. Overview
   B. Phenotypic Sensitivity Testing Using Primary HIV Isolates
   C. Phenotypic Sensitivity Testing with Recombinant HIV
   D. Replication-Incompetent HIV Particles for Phenotypic Sensitivity Testing
   E. Enzymatic Assays for the Analysis of Drug Resistance
   F. Genotyping
   G. Measuring Sensitivity to Inhibitors of Virus
      Entry and the Integrase
V. Clinical Implications
   A. Interpretation of HIV Sensitivity Assays
   B. Clinical Use of HIV Drug Sensitivity Assays
VI. Conclusions
   References

I. Introduction

Loss of helper T lymphocytes and destruction of lymphatic tissues are the major pathological sequelae of infection with the human immunodeficiency virus (HIV). They ultimately result in the development of the acquired immunodeficiency syndrome (AIDS). The precise mechanisms that underly the pathological processes are unresolved and remain a matter of debate. However, it is clear that virus replication constitutes the driving force of the disease process because disease

0065-3527/02 $35.00

progression is correlated with the concentration of virus in the blood (Mellors *et al.,* 1996).

Since 1995, antiviral therapy has significantly improved the quality of life and extended the life expectancy of infected persons in regions of the world where therapy is available. As far as is known, current therapeutic regimens are unable to cure the disease. Therefore, drugs have to be taken for many years, probably for life. Antiviral treatment for an extended time period raises several new issues such as the problems of patient motivation, long-term toxicity of the treatment, costs, and therapeutic failure due to the development of drug resistance (O. Cohen and Fauci, 2001).

Resistance to antiviral therapy reflects the ability of HIV to replicate in the presence of anti-HIV drugs and this is associated with an accelerated disease progression. Acquired resistance to chemotherapy has been a challenge with both bacterial and parasitic infectious diseases for some time. Mechanisms for avoiding the development of therapeutic failure due to resistance such as maintenance of sufficiently high drug levels and combination therapy consisting of two, three, or more drugs at a time are being employed for the treatment of these infections. In an analogous fashion, combination drug therapy now represents the basic principle of the treatment of HIV infection (Carpenter *et al.,* 2000). Nevertheless, clinical experience shows that the development of resistance to single or multiple antiviral drugs is not a rare event. Due to the availability of an increasing number of anti-HIV drugs, alternative therapeutic options for controling drug-resistant virus strains often exist. Ideally, appropriate adjustment of treatment is based on the demonstration of sensitivity to alternative drugs.

Several methods of drug sensitivity testing are routinely used for the clinical management of bacterial diseases. Assays such as the disk diffusion, the broth dilution, and the E test are quick, relatively cheap, and simple to perform. In contrast, testing of the sensitivitiy to antiviral drugs is routinely available only in a few highly specialized laboratories, for several reasons. First, drug sensitivity testing was rarely required in the pre-AIDS era because only a very limited range of virus infections was treatable and generally few potential alternative drugs were available (Crumpacker, 2001; Hirsch *et al.,* 1996). In addition, even if detected, cases of resistance to antiviral drugs were rare and usually limited to particular circumstances of disease-accompanying or iatrogenic immunosuppression (Balfour, 1999). Finally, since viruses replicate intracellularly, tests that rely on proliferation of the pathogen, analogous to those routinely used for bacteria, require sophisticated cell culture techniques.

At present, two enzymes, the reverse transcriptase (RT) and the protease (PR), are targets for antiviral therapy of HIV infection. Inhibitors of virus entry and of the viral enzyme integrase are currently in preclinical and clinical development. Several methods have been developed to diagnose drug resistance of HIV to RT and PR inhibitors and to identify drugs to which the virus remains sensitive. In general, drug sensitivity testing seems to offer clinical benefit to the infected individual under therapy. Beginning with an introduction to the structure and function of the viral enzymes, the process of virus–cell fusion, and the mechanisms that underly viral resistance, this chapter presents an overview of the methods developed and used for the analysis of drug sensitivity. The experience with sensitivity testing in HIV infection may bear implications for other chronic viral infections.

II. Targets of Antiviral Drug Therapy in HIV Infection

A. Reverse Transcriptase

1. Reverse Transcription of the HIV Genome

A characteristic feature of HIV and other retroviruses is transcription of the viral RNA genome in a double-stranded complementary DNA (cDNA) by a viral enzyme called reverse transcriptase (RT). Reverse transcription of the viral genome takes place in the cytoplasm after entry of the nucleocapsid. Retrovirus nucleocapsids contain two identical copies of viral RNA. Like DNA polymerases, RTs need a primer to be able to start DNA synthesis. A cellular tRNA for lysin ($tRNA^{Lys3}$) contained in the virion serves as primer for the HIV RT. It binds to a complementary stretch of 18 nucleotides approximately 100 bases downstream of the 5′ end of the genome. This region, the primer-binding site (PBS), is essential to virus replication. Synthesis of negative-strand DNA occurs in the 5′ direction complementary to the 5′ U5 and R region of the viral genome up to the 5′ terminus of the template strand and results in a short stretch of a DNA–RNA hybrid. RNAse H activity, located likewise on the RT molecule, subsequently removes the RNA template.

The viral RNA contains identical R regions on both the 5′ and the 3′ ends of the genome. This is essential to the continuation of cDNA synthesis. The single-stranded DNA copied from the 5′ R region hybridizes with the 3′ R domain either on the same template or on the other RNA genome packaged in the virion, a process called strand transfer or template exchange. Using the U5–R DNA strand as primer, the RT begins to synthesize the negative-strand DNA from the 3′ end of the viral genome all the way to the PBS. The RNAse H removes most of the RNA from the resulting DNA/RNA hybrid except for two purine-rich stretches of RNA in the middle of the genome and at the 3′ U3. The positive-strand DNA is synthesized using the purine-rich stretches of RNA as primers. Oligoribonucleotides produced upon RNAse H digestion of the viral RNA may serve as primers for the generation of additional stretches of positive-strand DNA.

A second, intramolecular template exchange anneals the positive-strand DNA comprising the U3, the R, and the U5 regions and the copy of the PBS with the original PBS site. This may lead to a circularly closed copy of proviral DNA. However, to remain functional, the annealed DNA strands have to be partially displaced, possibly through a helicase activity of the RT, and DNA synthesis continued to the end of the DNA template strands. This leads to a double-stranded, linear, blunt-ended DNA facsimile of the viral genome that contains at the termini identical copies of the U3, R, and U5 regions, the long terminal repeats (LTR; Fig. 1) (Goff, 2001).

2. *Structure and Function*

The HIV RT is encoded by the pol open reading frame of the viral genome (di Marzo-Veronese *et al.,* 1986) (Fig. 1). It is a heterodimer of two polypeptides of 66 and 51 kDa, respectively. The p66 peptide contains two catalytic sites for polymerase and RNAse H activity, respectively, which are both required for the generation of the proviral DNA. The polypeptide p51 is generated from a p66 precursor by proteolytic removal of the carboxy-terminal end. Although its protein sequence is identical to the respective part of p66, p51 exhibits a different conformation and function. It interacts with p66 at several regions (Becerra *et al.,* 1991; Wang *et al.,* 1994) and confers a structural component to the enzyme that stabilizes the p66 subunit, enables optimal binding to the tRNA primer, and is required for the RNAse H function (Harris *et al.,* 1998; Jacques *et al.,* 1994; Tasara *et al.,* 1999). As with other polymerases, the structure of the RT has been compared to a right hand with thumb, palm, and fingers (Jacobo-Molina *et al.,* 1993; Kohlstaedt *et al.,* 1992). The active site of the RT polymerase is located in the palm domain. Fingers, palm, and thumb form what is known as the primer grip to accommodate the primer/template complex. The thumb is connected to the RNAse H activity, which is located at the carboxyl-terminal end of the p66 polypeptide. The distance between the catalytic sites is approximately 60 Å, or 18 nucleotides of a nucleic acid double helix (Gopalakrishnan *et al.,* 1992; Huang *et al.,* 1998).

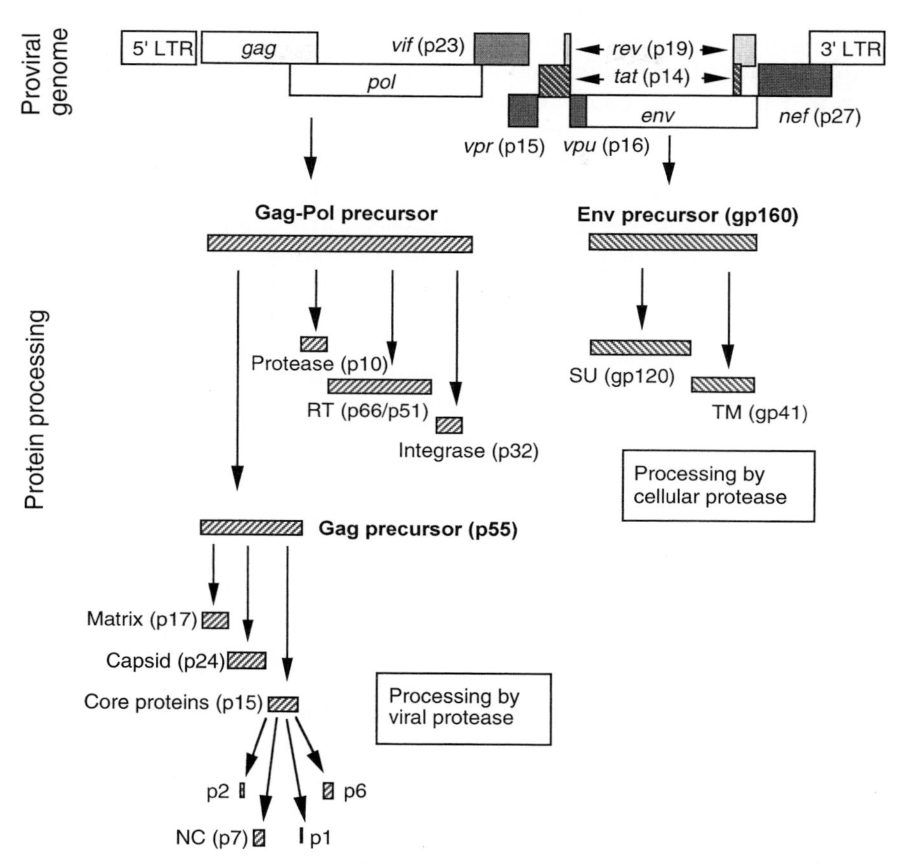

FIG. 1. Genomic organization of HIV and processing of viral proteins. The viral proteins encoded by gag, pol, and env are generated by processing of protein precursors. RT, reverse transcriptase; NC, nucleocapsid; SU, surface glycoprotein; TM, transmembrane glycoprotein.

During reverse transcription the primer/template strand extends along the p66 subunit to the catalytic site of the RNAse H in a molecular cleft.

The velocity of strand elongation by the RT, the processivity of the enzyme, is comparably low (Majumdar *et al.,* 1988). The slow processivity is due to the fact that during strand synthesis, the RT may pause or temporarily dissociate from the template when facing certain primer/template sequences (Bebenek *et al.,* 1993; Hsieh *et al.,* 1993; Klarmann *et al.,* 1993). Individual steps of strand elongation, such as binding of the primer/template complex, binding of a dideoxy nucleotide triphosphate (dNTP), addition of the dNTP to the primer, release of a pyrophosphate molecule, and translocation of the RNA/DNA strand, are associated with conformational changes of the RT molecule. For instance, binding of the primer/template causes the thumb to move away to allow contact between the fingers and the primer end. Upon binding of dNTPs, the fingers bend closer to the palm in such a way that two fingertips form part of the dNTP-binding site. After covalent linkage of the nucleotide to the primer, the grip loosens to release pyrophosphate and for translocation of the primer/template (Huang *et al.*, 1998).

Both polymerase and RNAse H activity require two divalent cations such as Mg^{2+}. In the case of the polymerase active site, the cations are associated with aspartates at positions 110, 185, and 186 (Patel *et al.*, 1995). They function by generating and stabilizing a state of transition of the dNTP and the primer required for covalent bonding of the dNTP to the primer and release of a pyrophosphate. These aspartate residues thus represent critical components of the catalytic site (Larder *et al.*, 1987). The aspartates at positions 185 and 186 are part of the "Tyr–Met–Asp–Asp motive" (amino acids 183–186) of the HIV RT. Analogous sequences (Tyr–X–Asp–Asp, where X stands for a hydrophobic amino acid) are a constitutive component of the catalytic site of other RTs and DNA and RNA polymerases as well (Kamer and Argos, 1984).

Like other chemical reactions, the enzymatic process that catalyzes strand elongation is not a one-way street, but it is generally a reversible reaction. Covalent bonding of dNTP to the primer and release of a pyrophosphate compete with the addition of a pyrophosphate moiety to the terminal nucleotide monophosphate and release of the dNTP from the primer. Under normal conditions, the reverse reaction, the pyrophosphorylysis, may be neglected. However, pyrophosphorylysis plays a role in some of the processes that underly drug resistance.

3. *Mechanisms of RT Inhibition*

Drugs that inhibit reverse transcription exert their effect by either one of two different mechanisms. The first class of inhibitors of the HIV reverse transcriptase comprises molecules that bind to the active site of the enzyme and compete with the authentic substrate for the catalytic site. In enzymatic reactions, binding of one of these drugs reduces the availability of free enzyme for the catalytic process. Competitive inhibitors of the HIV RT are derivatives of deoxyribonucleosides such as zidovudine (AZT), zalcitabine (ddC), didanosine (ddI), lamivudin (3TC), stavudine (d4T), or abacavir or of deoxyribonucleotides such as tenofovir. These drugs are processed in the cell to di- and triphosphate nucleotides by host cell kinases (Furman *et al.,* 1986). The phosphorylated drugs compete for the active site in the primer grip of the RT.

Although competition for the active site of reverse transcriptase may be relevant to inhibition by the drugs to some degree, the main mechanism of action of this class of inhibitors is termination of strand elongation. Nucleoside and nucleotide inhibitors lack a 3′ hydroxyl group that is required for the synthesis of the DNA chain. Thus, incorporation of one of the drug-derived nucleotides into the primer/template prevents subsequent strand elongation (Furman *et al.,* 1986; St Clair *et al.,* 1987). Action of the drugs on the viral RT in the cell is relatively selective compared to that on cellular polymerases. This is mainly due to a 100-fold greater affinity for the viral RT compared to cellular polymerases and the observation that the intracellular concentration is higher than the K_i value for RT, but lower than the K_i for the cellular polymerases alpha and beta (Furman *et al.,* 1986).

A second group of HIV RT inhibitors is not composed of derivatives of nucleosides or nucleotides, but comprises a wide variety of different chemical classes. They have therefore been designated nonnucleoside reverse transcriptase inhibitors (NNRTI). These drugs, including nevirapine, delavirdine, efavirenz, and several others in preclinical and clinical development, share a common mode of action. Instead of competing with the nucleoside triphosphate substrate, these substances bind to a hydrophobic pocket close to the catalytic site of the polymerase (Ding *et al.,* 1995a, b; Kohlstaedt *et al.,* 1992; Ren *et al.,* 1995). The pocket is not present in the original enzyme, but is newly created by displacement of the polypeptide segment connecting palm and thumb upon interaction with the inhibitor (Huang *et al.,* 1998; Rodgers *et al.,* 1995). The noncompetitive inhibitors of the reverse transcriptase lock the polymerase in an inactive conformation by inducing structural changes at the catalytic site or the primer grip that alter the shape or restrict the mobility of these elements (Ding *et al.,* 1995a; Esnouf *et al.,* 1995; Tantillo *et al.,* 1994).

Another mechanism of RT inhibition not utilized by any of the drugs licensed for HIV but demonstrable *in vitro* is blockade of strand elongation by pyrophosphate analogues. Molecules like pyrophosphate and foscarnet compete with the β and γ phosphates of the deoxynucleotide triphosphate substrate for the pyrophosphate-binding pocket in the primer grip. This prevents positioning of the pyrophosphate moiety of the substrate (Hizi *et al.,* 1991; Majumdar *et al.,* 1988). Although it is not utilized in HIV infection, this pharmacological mechanism is being exploited for inhibition of the viral DNA-directed DNA polymerase in the treatment of cytomegalovirus infection.

B. Protease

1. Proteolysis in HIV Replication

The HIV protease is required for maturation of the virus. During HIV replication full-length viral RNA transcripts are translated into Gag/Pol precursor polyproteins. The precursor proteins are targeted to the cell membrane, where they assemble into virion particles. For maturation to infectious virus, cleavage of the polyproteins is required. The Gag and Gag/Pol precursors are cleaved into six structural proteins and three enzyme products yielding the matrix (p17) and capsid protein (p24), the core proteins p2, nucleocapsid (p7), p1, and p6, the protease, reverse transcriptase heterodimer, and integrase (Fig. 1). After excision, the structural proteins are capable of self-assembly. It is not totally clear how the cleavage process starts. According to one model, protease subunits of the polyprotein dimerize and exert some activity in regard to processing of Gag/Pol precursor molecules. Excised protease monomers spontaneously hybridize to the functional dimer and the enzyme catalyzes the subsequent polyprotein cleavage process (Swanstrom and Wills, 1997).

Maturation of HIV occurs extracellularly after budding of the virus from the cell membrane. Pharmacological inhibition of the protease does not prevent production and release of virus particles. However, because cleavage of the polyprotein precursor into the individual proteins is essential for infectivity of the virus, the virus particles generated in the presence of a protease inhibitor are noninfectious. Ultramicroscopically, the virions exhibit characteristic structural differences of which the lack of the conical core structure is the most prominent. Instead, the virions present with a thick layer consisting of uncleaved Gag and Gag/Pol polyproteins beneath the lipid membrane (Wilk and Fuller, 1999).

Proteases, including the HIV protease, catalyze the cleavage of peptide bonds by stabilizing the substrate–product intermediate in the transition state. During transition, the proton and the oxygen of a dissociated water molecule are linked to the amino and the carboxyl groups of the substrate, respectively. Subsequent electron transfer leads to dissolution of the peptide bond and to the more stable formation of the cleaved end products (Babine and Bender, 1997). The HIV protease preferentially cleaves peptide bonds flanked by regions of increased hydrophobicity. In particular, the amino acid residue at the first position upstream of the scissile bond (P1) is hydrophobic and unbranched (Pettit *et al.,* 1991). This is relevant to the chemical design of a protease inhibitor.

2. Structure and Function

The HIV protease belongs to the aspartic protease class of enzymes, which comprises proteases with such diverse functions as pepsin, renin, and cathepsin D (Babine and Bender, 1997; Seelmeier *et al.,* 1988). The enzyme is encoded by the 5′ end of the pol open reading frame of the virus and translated from full-length transcripts of the viral cDNA as part of the Gag–Pol fusion protein precursor. The functional protease is a homodimer of two polypeptides each of 99 amino acids and approximately 11 kDa (Lapatto *et al.,* 1989; Navia *et al.,* 1989; Wlodawer *et al.,* 1989). The polypeptide chains are positioned in a twofold rotational C_2 symmetry and form a cleft for binding of the substrate. The bottom of the cleft, close to the center of the enzyme dimer, is formed by two peptide loops from either protein subunit, which contains the amino acid sequence Asp–Thr–Gly (amino acids 25–27) and represents the active site of the enzyme. The two aspartate residues bind a molecule of water. These elements are critical for generating the transition state of the substrate–product intermediate, hence the designation aspartic protease. The surface of the substrate-binding cleft is shaped by β sheets from each of the polypeptides. The amino acid residues interacting with the substrate (e.g., Arg8, Leu23, Ala28, Asp29, Asp30, Val32, Ile47, Ile50, Pro81, Val82, Ile84 on both polypeptide chains) are primarily hydrophobic (Babine and Bender, 1997). Mobile β-sheet "flaps" cover the binding area. The flaps take up a closed conformation and interact at the tips when substrate is bound and open to allow access of substrate and release of the cleaved product (Miller *et al.,* 1989).

3. Mechanism of Inhibition

Inhibitors of the HIV protease are peptidomimetics, that is, synthetic derivatives of short peptides (type I mimetics) or chemically unrelated substances (type III mimetics), which bind to the catalytic site of the enzyme and act as competitive inhibitors with regard to the natural substrates. All of the drugs currently licensed for clinical use (saquinavir, ritonavir, indinavir, nelfinavir, amprenavir, lopinavir) and many of those in clinical and preclinical development are type I mimetics. These substances represent transition state analogues of the enzymatic process that mimic a reaction pathway intermediate (Ripka and Rich, 1998). In peptide-derived transition state analogues the scissile peptide bond has been replaced by a nonhydrolyzable structure of comparable geometry. A hydroxyl group occupies the place of the water in the catalytic space and engages the aspartate residues. Chemically, peptidomimetic transition state analogues are based on hydroxyethylamine (saquinavir, nelfinavir), hydroxyethylamino sulfonamide (amprenavir), or hydroxylamine pentanamide (indinavir), or represent derivatives of C_2-symmetric molecules (ritonavir, lopinavir) (Babine and Bender, 1997). The peptidomimetic drugs retain hydrophobic side chains. Otherwise much of the original peptide backbone has been substituted to accommodate the requirements of target specificity, solubility, stability, and oral bioavailability (Ripka and Rich, 1998; Swanstrom and Eron, 2000; Wlodawer and Erickson, 1993).

Other candidate drugs represent type III mimetics derived from nonpeptide template structures. For instance, the dihydropyrone tipranavir is a warfarin derivative and DMP-450 is based on a seven-membered cyclic urea structure. The compounds contain specific elements that interact with the enzyme in a topographical and functional manner reminiscent of a peptide-derived transition state analogue (Ripka and Rich, 1998). In addition, the drugs carry components to displace the "flap water" that usually links the HIV protease flaps (Ile50, Ile50′) with the substrate during the transition state. Because of this distinctive feature, the substances have also been labeled "flap water mimics" (Babine and Bender, 1997).

C. Envelope Glycoprotein and Virus Entry

1. The Mechanism of Virus Entry

Infection of susceptible cells by HIV involves a series of steps that have not yet been elucidated in their entirety. Nevertheless, several elements of the infection process have now been identified as promising or potential targets for antiviral intervention. HIV infection is initiated by interaction of the viral envelope glycoprotein complex with receptors on the cellular surface. The functional glycoprotein is composed of three heterodimers consisting of the transmembrane (gp41) and surface (gp120) subunits, which are noncovalently linked. The gp41 units are anchored in the virus membrane and contain an internal, a transmembrane, and an external region. Gp120 is situated outside of the viral membrane. Contact of the envelope complex with the CD4 receptor molecule leads to conformational changes that expose new binding regions of the glycoproteins. Gp120 interacts via the exposed V3 loop with one of two alternative coreceptors, the chemokine receptor molecules

CXCR4 and CCR5. Certain sequences in V3 define the preferences for either one of the coreceptors (Cocchi *et al.,* 1996; Speck *et al.,* 1997). Interaction of gp120 with CD4 and the coreceptor results in a conformational change of gp41 that exposes a hydrophobic peptide sequence of 12 amino acids close to the N-terminus of gp41 normally buried in the envelope complex, the fusion domain. The hydrophobic fusion domain inserts into the membrane of the cell. Present C-terminally from the fusion domain are the N-terminal and C-terminal heptad repeat sequences 1 (HR1) and 2 (HR2), which exhibit chemical affinity due to molecular complementarity. The affinity of HR1 for HR2 favors close contact between the two peptide regions in an antiparallel fashion. This leads to the formation of a hairpin structure in gp41, which draws the viral surface membrane closer to the lipid membrane of the cell. This facilitates fusion of the viral with the cellular membrane and allows subsequent entry of the nucleocapsid into the cytoplasm (LaBranche *et al.,* 2001).

2. Inhibition of Virus Entry

Several inhibitors of HIV replication that interfere with the process of virus entry have been tested in the laboratory. Although some of them are currently undergoing clinical testing, none has been licensed for therapy at the time of writing (LaBranche *et al.,* 2001). Some entry inhibitors interact with the viral glycoproteins, others target CD4 or the coreceptors. According to the level at which the drugs interfere, substances have been labeled as attachment inhibitors, which interfere with CD4 binding, chemokine receptor inhibitors, and fusion inhibitors. Drugs most advanced in regard to clinical testing are in the group of fusion inhibitors comprising substances such as T-20 (Jiang *et al.,* 1993; Kilby *et al.,* 1998; Wild *et al.,* 1994) and T-1249 (LaBranche *et al.,* 2001). These drugs are oligopeptides corresponding to specific HR1 or HR2 sequences that compete with the original peptide stretch for binding to the complementary HR domain. Binding of the oligopeptide thus displaces the original HR sequence of gp41 and prevents formation of the hairpin structure. Another fusion inhibitor acting by a related mechanism is the oligopeptide 5-helix (Root *et al.,* 2001).

Drugs that target receptor binding act on either CD4 or a coreceptor. For instance, SCH-C and SCH-D interact with CCR5; and AMD3100 and ALX40-4C bind to CXCR4 (Doranz *et al.,* 1997; Schols *et al.,* 1997). The IgG2 fusion protein PRO 542, which represents a tetravalent derivative of CD4, targets the envelope glycoprotein (Allaway *et al.,* 1995; D'Souza *et al.,* 2000; LaBranche *et al.,* 2001).

D. Integrase

Another interesting target for anti-HIV therapy is the viral integrase. This enzyme is a recombinase unique to retroviruses and mediates insertion of the proviral DNA into the host cell chromosome by a mechanism comprising several coordinated steps and involving *in vivo* additional proteins. The enzyme and the process catalyzed have recently been reviewed in great detail in several complementary articles edited by Skalka in this series (Skalka, 1999). Several inhibitors of the HIV integrase that are active *in vitro* have been developed (reviewed in Pommier *et al.,* 2000). Drugs that target integrase most selectively are the low molecular weight inhibitor S-1360 and several representatives of the class of diketo acids (Hazuda *et al.,* 2000; Pommier *et al.,* 2000). The latter substances bind to the integrase donor substrate DNA complex and inhibit strand transfer by competing with the target DNA (Espeseth *et al.,* 2000).

III. Mechanisms of Antiviral Drug Resistance

A. General Aspects

In HIV infection the term resistance has been used to describe several distinct conditions. Clinical resistance refers to a rebound in virus load levels after primary successful treatment. In cell culture, resistance is the ability of the virus to replicate or to express a marker gene in the presence of a drug. Moreover, resistance may refer to the activity of an isolated mutated enzyme in a biochemical reaction in the presence of an enzyme inhibitor. Finally, the word resistance is used in the context of genetic changes that, empirically, are associated with resistance in phenotypic or biochemical analyses.

The molecular mechanisms that underly resistance have usually been determined by biochemical analyses of isolated enzymes. On the molecular level, acquired drug resistance is a genetically encoded alteration of the protein sequence of an enzyme that confers on the pathogen the ability to replicate in the presence of a particular enzyme inhibitor. Drug resistance is usually expressed as the ratio of the drug concentration that inhibits replication of the mutated virus by 50% (50% inhibitory concentration, IC_{50}) or by 90% (IC_{90}) to the corresponding drug concentration for the wild-type strain.

In HIV, as in other retroviruses, spontaneous genetic mutations are relatively frequent due to the poor accuracy of the RT. In the presence of adequate selection pressure, resistant mutants occur quickly. If drug levels are low, there is no selection pressure. Theoretically, high drug concentrations may completely suppress virus replication and associated generation of mutants. Thus, the drug concentration optimal for accumulation of mutated virus variants permits low-level or intermittent virus replication. Some of the drug-resistant virus variants replicate more slowly than their progenitors. However, secondary mutations may occur on the basis of the low-performing enzyme that restore the functional activity and increase viral replicative fitness. For instance, mutations that confer greater enzymatic processivity may compensate for reduced binding caused by mutations in the active site (Borman *et al.,* 1996; Pazhanisamy *et al.,* 1996; Schock *et al.,* 1996).

Discrepancies have sometimes been observed between the biochemical properties of isolated mutated enzymes and recombinant viruses displaying the identical mutations. In particular, reverse transcriptase mutants that confer 100-fold resistance to AZT, if tested in cell culture remained almost as sensitive to AZT triphosphate as wild-type RT under typical *in vitro* DNA polymerase conditions (Lacey *et al.,* 1992; Wainberg *et al.,* 1990). The discrepancy observed with AZT may be due to assay conditions that did not adequately represent the situation in the cell. *In vitro* conditions that more precisely mimic the physiological situation in the cytoplasma recently led to biochemical assay results that correlate better with those observed with virus-infected cells. In particular, resistance was observed in enzymatic tests under physiological ATP and pyrophosphate concentrations indicating that dNTPs pyrophosphate, or both are involved in the AZT resistance mechanism (Arion *et al.,* 1998; Meyer *et al.,* 1999; Lennerstrand *et al.,* 2001). Significant differences have also been noted between biochemical data with purified mutated protease and cell culture results with variant recombinant viruses (Klabe *et al.,* 1998; Partaledis *et al.,* 1995). The reason for this divergence is not understood (Erickson *et al.,* 1999).

B. Resistance to Reverse Transcriptase Inhibitors

The mechanism of resistance to nucleosidic and nucleotidic reverse transcriptase inhibtors (NTRIs) is complex. Several single-nucleotide mutations or combinations thereof mediate resistance to one, two, or multiple NRTIs. Mutations relevant to resistance are predominantly observed at particular sites in the p66 polypeptide. They occur either at or adjacent to the substrate-binding site, the catalytic site, or the nucleic acid cleft at a distance from the active site of the polymerase and affect nucleotide binding, positioning of the primer/template, the nucleotide incorporation rate, or a combination of these reactions.

As an example, it has been proposed that mutations of or adjacent to the dNTP-binding site, such as Lys65Arg, Gln151Met, and Met184Ile/Val, preferentially decrease the affinity or correct positioning of the drug versus the primer terminus and template (Huang *et al.,* 1998). Moreover, mutations in the nucleic acid-binding cleft, several bases away from the active site, such as the Leu74Val mutation in the fingers domain, may lead to subtle alterations of the positioning of the primer/template. These increase the specificity of substrate binding and reduce the chance for an incoming nucleotide drug to be connected to the 3′ terminal primer nucleotide (Boyer *et al.,* 1994; Jonckheere *et al.,* 2000). Finally, amino acid changes close to the active site affect the kinetics of the enzymatic processes. For instance, mutations Asp67Asn and Lys70Arg increase the rate of pyrophosphorylysis, the reverse reaction of the polymerase reaction. Increased pyrophosphorylysis means that newly incorporated nucleotides, including potentially chain-terminating nucleotide analogues, are excised at a higher rate. This has been demonstrated for AZT. Selective removal of chain-terminating AZT triphosphate by the mutant RT increased the chance for elongation of the nucleic acid strand (Arion *et al.,* 1998). Alternatively, certain mutations locked AZT triphosphate at the substrate-binding site longer than other dNTPs, allowing removal of AZT by the reverse reaction and subsequent replacement by ATP (Lennerstrand *et al.,* 2001; Meyer *et al.,* 1999).

Other mutations alter the processivity of the enzyme. These mutations are usually located along regions of the nucleic acid cleft that are in contact with the template/primer complex. For instance, the Lys65Arg mutation has been associated with increased processivity of the RT (Arion *et al.,* 1996). Likewise, it was demonstrated that mutations Thr215Tyr and Lys219Gln increase the processivity through reduced RT dissociation from the nucleic acid strand (Arion *et al.,* 1998; Caliendo *et al.,* 1996). Increased processivity may compensate for decreased affinity of the enzyme for the dNTP substrate (Lys65Arg) or increased reverse reaction mediated by primary mutations (Arion *et al.,* 1998). The compensatory mutations confer on the mutated virus replication kinetics similar to those of wild-type HIV.

Resistance to the nonnucleoside RT inhibitors (NNRTIs) is conferred by mutations that affect binding of the drug. Mutations are located within and surrounding the NNRTI-binding pocket and may hinder drug binding sterically or disrupt functional interactions of drug and protein such as electrostatic or van der Waals forces (Smerdon *et al.,* 1994). Since the inhibitors chiefly interact with two β-sheet structures comprising amino acid residues 100–110 and 180–190 (Kohlstaedt *et al.,*

1992), resistance mutations primarily affect these regions (Hirsch *et al.*, 2000) and cross-resistance between different NNRTIs is frequently observed. Single-amino-acid mutations, such as Lys103Asn or Tyr181Cys, may confer resistance to multiple drugs. In previous studies, NNRTI monotherapy resulted in the emergence of drug-resistant virus variants in only a few weeks (Richman *et al.*, 1994; Wei *et al.*, 1995).

C. Resistance to Protease Inhibitors

More than 20 of the 99 amino acids of the protease can mutate under the selective pressure of protease inhibitors (PIs) to yield a replication-competent, yet drug-resistant enzyme (Boden and Markowitz, 1998). The pattern of resistance mutations that results from treatment with particular protease inhibitors is characteristic for the drug employed. The accumulation of mutations both in virus culture and in treated individuals resembles an ordered process. As a rule of thumb, the degree of resistance correlates with the number of mutations (Tisdale *et al.*, 1995). Frequently, a mutation in the substrate-binding site occurs first, followed by additional amino acid changes away from the active site (Eastman *et al.*, 1998; Molla *et al.*, 1996). In most instances, at least two and up to four mutations are required to yield measurable resistance (Tisdale *et al.*, 1995). Variants containing triple or quadruple mutations not only exhibit high-level resistance to a particular inhibitor, but usually demonstrate cross-resistance to more than one drug (Tisdale *et al.*, 1995). For instance, four mutations at residues 46, 63, 82, and 84 are sufficient for significant resistance to indinavir, saquinavir, amprenavir, and other protease inhibitors (Condra *et al.*, 1995). High-level drug resistance to more than five inhibitors is usually associated with constellations of 8–12 amino acid changes (Olsen *et al.*, 1999).

Amino acid mutations associated with resistance to single or multiple drugs are widely scattered over the polypeptide sequence. Favored active-site mutations are at positions 30, 48, 50, 82, and 84. Amino acid changes in the molecular loops of the flap region at positions not directly in contact with the substrate are predominantly at residues 45–47 and 54. However, mutations are not restricted to these regions.

Primary mutations located in the substrate-binding region reduce the affinity between the mutant protease and the drug. This is reflected by a larger dissociation rate of the enzyme/drug intermediate (Maschera *et al.*, 1996; Partaledis *et al.*, 1995). As an example, a reduced stability of the enzyme/drug intermediate that affects binding of the inhibitor indinavir can be caused by subtle alterations at the substrate-binding cleft that disrupt van der Waals interactions (Ile84Val), the introduction of an unfavorable hydrophilic moiety (Val82Thr) (Chen *et al.,* 1995), or both. Alterations at non-active-site amino acid positions similarly affect binding of the substance (Olsen *et al.,* 1999) and/or alter the stability of the dimer (Xie *et al.,* 1999) or may cause reshaping of the active site through what has been called long-range structural perturbations (Erickson *et al.,* 1999).

Secondary mutations may affect the catalytic activity of the enzyme. For instance, mutations Met46Ile and Leu63Pro increase catalysis of oligopeptide substrates *in vitro* by up to a factor of 3.6 compared to the nonmutated original protease (Schock *et al.,* 1996). Higher efficiency in substrate processing may compensate for reduced binding caused by mutations in the active site. Therefore, mutations in the protease enzyme that affect drug binding are frequently accompanied by secondary mutations that offset the binding disadvantage for the relevant substrate and increase survival of the virus variant in the presence of the drug (Pazhanisamy *et al.,* 1996; Schock *et al.,* 1996). Other secondary mutations that contribute to resistance to protease inhibitors occur at the enzyme cleavage sites. Relevant mutations have thus far only been observed at the cleavage sites between nucleocapsid (p7) and p1 (Ala431Val) and between p1 and p6 (Leu449Phe) of the p15 precursor protein (Doyon *et al.,* 1996; Zhang *et al.,* 1997). Mutations at the protease cleavage sites are exclusively compensatory, that is, they increase the processing of the p15 precursor protein by mutated enzymes and the replication capabilities of variant viruses. Cleavage-site mutations do not confer resistance on protease inhibitors themselves (Doyon *et al.,* 1996).

D. Resistance to Fusion and Integrase Inhibitors

In vitro experiments with fusion and integrase inhibitors have resulted in the generation of drug-resistant virus variants. The resistance mutations associated with the fusion inhibitor T-20 occur near the N-terminus of HR1, in the binding region of the oligopeptide. The mutations observed make up the otherwise highly conserved Gly–Ile–Val motif (amino acids 36–38) and result in the peptide sequences Asp–Ile–Met or Ser–Ile–Met, which confer reduced susceptibility to T-20 due to reduced binding affinity for the drug (Rimsky *et al.,* 1998).

Mutations in the active site of the HIV integrase, including Thr66Ile, Leu74Met, Ser153Tyr, Met154Ile, and combinations thereof, are associated with reduced susceptibility to representatives of the diketo acid class of integrase inhibitors (Hazuda *et al.,* 2000 and S-1360).

IV. Technologies for Measuring Drug Sensitivity

A. *Overview*

Considerable technological advances in the past decade have made HIV drug sensitivity testing available for clinical practice. According to the underlying test principle involved, the methods for determining drug sensitivity can be differentiated into those that use phenotypic and those that use genotypic systems. In phenotypic systems, infection and replication of the virus in the presence of antiviral drugs is measured directly and the IC_{50} is determined. The IC_{50} value obtained with the test sample is compared with the IC_{50} value of a drug-sensitive reference strain of HIV and the ratio of the drug concentrations required to inhibit replication reflects the degree of drug resistance of the test specimen. Results are relatively easy to interpret. However, thresholds levels that correlate well with clinically relevant resistance are not well established. Genotypic drug sensitivity analysis is based on the identification of mutations previously shown to be associated with phenotypic drug resistance. Genotypic assays are highly reproducible and relatively easy and rapid to perform. However, the analysis of results from genotypic assays requires detailed knowledge of resistance-associated mutation patterns. Interactions among different resistance mutations are complex and often difficult to interpret.

A major problem of both genotypic and phenotypic resistance assays is the low capability of detecting minor fractions of drug-resistant virus variants. Futhermore, phenotypic and genotypic assay systems require isolation and/or amplification of virus or parts of the viral genome. This procedure can result in selection of virus strains that do not reflect the majority species *in vivo*. The practicability and usefulness of both phenotypic and genotypic HIV sensitivity testing in monitoring treatment of infected patients was investigated in several retrospective and prospective studies. Recommendations for the application of HIV sensitivity testing in the clinical management of HIV infection have recently been published (EuroGuidelines Group, 2001; Hirsch *et al.,* 1998, 2000).

B. *Phenotypic Sensitivity Testing Using Primary HIV Isolates*

Phenotypic assays for the measurement of drug resistance have been in use since treatment of HIV-infected patients by RT inhibitors became available. The original approach for HIV resistance testing was to examine the replication of viruses isolated from treated individuals *in vitro* in the presence of the drug. As a first step in this process, virus is obtained by cocultivation of peripheral blood mononuclear cells (PBMCs) from the test subject with prestimulated PBMCs from seronegative donors or with lymphoblastoid cells, such as the cell line MT2. The titer of infectious virus in the culture supernatant is determined. PBMCs or a susceptible cell line are infected with a defined virus inoculum in the presence of serial dilutions of the antiviral compounds. Virus replication is analyzed by quantification of RT activity or the amount of Gag/p24 in each of the cell cultures or by microscopic analysis of syncytium formation. The IC_{50} obtained refers to the drug concentration that reduces RT activity, p24 concentration, or syncytium formation by 50% (Boucher *et al.*, 1990; Japour *et al.*, 1993; Johnson *et al.*, 1991; Land *et al.*, 1990; Larder *et al.*, 1989, 1990; Mayers *et al.*, 1992; Shafer *et al.*, 1993). Quantification of virus infectivity was simplified by using enzymatic cell viability assays based on the tetrazolium dye XTT (Jellinger *et al.*, 1997) or by the use of cell lines that contain reporter genes under transcriptional control of the Tat-responsive HIV LTR (Aguilar Cordova *et al.*, 1994; Gervaix *et al.*, 1997; Hachiya *et al.*, 2001; Pirounaki *et al.*, 2000). This approach to HIV sensitivity testing remains time-consuming and labor-intensive and involves the generation and handling of high-titer HIV stocks.

Isolation of HIV by cocultivation has several significant drawbacks. It may cause selection of viral subpopulations that do not reflect the majority species *in vivo* (Kusumi *et al.*, 1992). Furthermore, the long culture time required to isolate HIV and to obtain sufficient amounts of virus may result in selection against drug-resistant virus strains because drug resistance is frequently associated with reduced viral fitness (Croteau *et al.*, 1997; Devereux *et al.*, 2001; Harrigan *et al.*, 1998; Kosalaraska *et al.*, 1999; Maeda *et al.*, 1998; Mammano *et al.*, 1998; Martinez Picado *et al.*, 1999; Zennou *et al.*, 1998).

C. Phenotypic Sensitivity Testing with Recombinant HIV

Second-generation phenotypic resistance assays are based on recombinant DNA technology and were designed to circumvent the HIV isolation procedure by cocultivation and long-term culture. To generate recombinant HIV, viral sequences are amplified by reverse transcription PCR (RT-PCR) from virus present in the plasma of infected individuals. To achieve optimal sensitivity, a two-step nested PCR protocol is usually employed. The amplified sequences are then inserted into full-length molecular HIV plasmids with appropriate deletions of the corresponding RT and PR sequences. Transfection of the plasmids in eukaryotic cells leads to the production of infectious chimeric virus particles for further drug sensitivity testing.

This approach was used by a variety of laboratories with several specific technical variations (Boucher *et al.,* 1996; Hertogs *et al.,* 1998; Kellam and Larder, 1994; Shi and Mellors, 1997; Walter *et al.,* 1999). For instance, patient-derived RT and PR sequences were amplified by nested RT-PCR from PBMCs or from plasma and cotransfected in a CD4+ T cell line together with a proviral HIV plasmid containing deletions in the PR- and/or RT-coding sequences (Boucher *et al.,* 1996; Hertogs *et al.,* 1998; Kellam and Larder, 1994). Cotransfection resulted in homologous recombination of the HIV genome and, after about 10 days, production of sufficient amounts of recombinant chimeric virus for further sensitivity testing. Cytopathic effects due to replication of the recombinant virus in the presence of serial dilutions of the antiviral compound were quantified either by scoring the number of syncytia in virus plaque reduction assays or by using an MTT-based cell viability test. Using this method, drug susceptibility profiles of recombinant chimeric HIV were produced for the antiviral drugs in clinical use (Hertogs *et al.,* 1998). However, the need for virus propagation early in the assay procedure bears a theoretical risk of selection of virus variants with superior replication potential.

Instead of relying on homologous recombination for the generation of chimeric proviral plasmids, an alternative approach to generating recombinant virus is to ligate the amplified RT and PR sequences into the proviral plasmid by standard cloning procedures (Shi and Mellors, 1997; Walter *et al.,* 1999). For this approach, competent *Escherichia coli* were transformed after the ligation reaction and propagated in liquid culture and the plasmids were purified. Recombinant chimeric viruses that contain patient-derived PR and RT sequences were generated upon transfection of eukaryotic cells. Transfection procedures using 293T cells yielded sufficiently high quantities of virus after 2 days. Since 293T cells are CD4−, superinfection or virus replication does not occur (Walter *et al.,* 1999). Subsequent drug susceptibility testing has been performed either by virus plaque reduction assay (Shi and Mellors, 1997) or by the use of a cell line containing an indicator gene coding for a secretable alkaline phosphatase (SEAP) under control of the HIV LTR promoter (Walter *et al.,* 1999). Infection of these cells and subsequent production of the viral transactivator protein Tat induced expression of SEAP. The activity of the enzyme was quantified in the supernatant and reflected the number of infected cells.

Results obtained by drug sensitivity assays using recombinant HIV correlated favorably with tests using the parental virus isolates and with genotypic data when virus strains with defined mutations were examined (Bacheler *et al.,* 2001; Boucher *et al.,* 1996; Hertogs *et al.,* 1998; Walter *et al.,* 1999). Although selection on grounds of viral replication capacities does not play a role when polymerase sequences are amplified by RT-PCR and sufficient quantities of virus are obtained without *in vitro* replication, other technical factors may influence the spectrum of virus recombinants retrieved from the plasma. For instance, the selection of primers for the PCR reaction affects the spectrum of virus quasi-species amplified. Furthermore nucleotide mismatches and other copying errors may be caused by the reverse transcriptase and polymerase during RT-PCR. The chances of introducing new mutations can be kept low by using optimal reverse transcriptase conditions and DNA polymerases with proofreading capabilities and by limiting the number of PCR amplification cycles to the minimum.

Sensitivity testing with recombinant chimeric viruses has several advantages compared to that with primary HIV isolates (Table I). For instance, because patient-derived PR and RT sequences are tested in the context of a standard molecular HIV clone, selection events that may occur during the virus isolation procedure are minimized or avoided. This is particularly the case if secondary rounds of HIV infection are

TABLE I
RELATIVE ADVANTAGES AND DISADVANTAGES OF DIFFERENT HIV DRUG SENSITIVITY TESTS

Test	Advantages	Disadvantages
Culture of virus isolates	Suitable for testing of sensitivity to any antiviral agent	Labor-intensive, time-consuming, handling of infectious HIV, selection against viruses of reduced fitness
Recombinant virus assay	Less time-consuming than culture of HIV isolates	Labor-intensive, handling of infectious HIV
Recombinant HIV vector	Reduced assay time, no handling of infectious HIV required	Labor-intensive
Enzymatic analysis	No handling of infectious HIV required	Specific test required for each enzyme tested, physiologic conditions have to be carefully simulated *in vitro*
Genotyping	Short assay time, detection of minority mutants, commercial kits available	Mutations that confer resistance must be known, quantification of resistance not possible, interpretation complex

not required for the generation of sufficient amounts of virus. In addition, coreceptor preferences relevant with primary isolates do not influence testing with recombinant chimeric viruses. Finally, if virus propagation is not required because sufficient amounts are produced upon transfection, resistance testing is further speeded up.

To handle large amounts of patient specimens, the resistance assay based on production of recombinant chimeric HIV by homologous recombination developed by Hertogs and colleagues has been automated (Antivirogram assay, Virco, Inc.). Patient samples with a viral load greater than 1000 copies/ml can routinely be analyzed by this assay system (Hertogs *et al.,* 1998, 2000b). A variety of HIV drug resistance studies have been performed using this phenotypic assay (Adje *et al.,* 2001; Briones *et al.,* 2001; Caride *et al.,* 2000; Casado *et al.,* 2000; Conway *et al.,* 2001; Cozzi Lepri *et al.,* 2000; Harrigan *et al.,* 1999; Hertogs *et al.,* 2000a; Larder *et al.,* 1999, 2000; Miller *et al.,* 1998a, b, 2000a, b; Montaner *et al.,* 2001; Wegner *et al.,* 2000; Weidle *et al.,* 2001). Recently, a survey of over 6000 samples tested with this assay has been published (Hertogs *et al.,* 2000b).

D. Replication-Incompetent HIV Particles for Phenotypic Sensitivity Testing

An alternative phenotypic drug sensitivity test was developed using methods of the emerging lentiviral vector technology (for review see Amado and Chen, 1999; Buchschacher and Wong-Staal, 2000; Naldini and Verma, 2000). To construct an HIV vector from its pathogenic progenitor, at least one viral open reading frame that encodes for a protein essential for virus replication is removed from the genome of an infectious molecular HIV clone. All *cis*-acting sequences required for virus replication and the RNA-packaging sequence Ψ remain on the rudimentary proviral genome. A marker gene expression cassette is inserted into the vector genome. A second plasmid, the packaging plasmid, codes for the deleted HIV proteins, but does not contain the RNA-packaging signal. The vector genome plasmid is transfected into eukaryotic "producer" cells together with the packaging construct that delivers the additional genes essential for the production of virions in *trans*. Upon transfection, virus vector particles are generated that contain as genome only the rudimentary viral RNA and the marker gene. The vector particles are infectious. However, because one or more essential regions are lacking in the vector genome, complete cycles of virus replication are not possible and infection by vector particles is restricted to a single round.

The most convenient approach to rendering HIV replication-incompetent is deleting the envelope gene from the viral genome and providing the envelope protein in *trans* (Buchschacher and Wong-Staal, 2000). To avoid recombination events resulting in replication-competent virus, the vesicular stomatitis virus (VSV) G or the murine leukemia virus (MuLV) envelope glycoprotein is used for *trans*-complementation. Both envelope proteins can efficiently pseudo-type HIV capsids resulting in infectious particles (Akkina *et al.,* 1996; Naldini *et al.,* 1996a, b; Page *et al.,* 1990; Reiser *et al.,* 1996). HIV vector particles pseudo-typed with VSV G or MuLV Env infect a wide variety of cell lines and infection is not restricted to cells bearing the HIV receptor and coreceptor. Target cell infection can be quantified by measuring the expression of the marker gene present in the vector genome. Alternatively, if the vector expresses Tat, infection of the target cell by the replication-incompetent HIV vector particle can be detected by using cell lines containing an indicator gene under control of the HIV LTR promoter as described above.

The use of HIV vectors to determine the sensitivity of HIV to antiviral drugs was first suggested by Page *et al.* (1990). The feasibility was subsequently demonstrated for inhibitors of RT and PR by several groups (Hecht *et al.,* 1998; Jarmy *et al.,* 2001; Petropoulos *et al.,* 2000; Piketty *et al.,* 1999). In addition, the test principle can be used for the analysis of inhibitors of the viral integrase (Jarmy *et al.,* 2001). Piketty and colleagues deleted the protease-, reverse transcriptase-, and envelope-coding genomic regions from the HIV plasmid pNL4-3. Patient-derived PR and RT sequences were amplified and recombined by homologous recombination with the modified pNL4-3 plasmid. The VSV G envelope protein was provided in *trans* to produce infectious vector particles. Drug susceptibility assays were performed on indicator cells containing a Tat-responsive marker gene.

To render quantification of vector infection independent of Tat expression, a marker gene expression cassette containing the firefly luciferase, the secretable alkaline phosphatase, or a similar marker gene under control of a heterologous eukaryotic promotor can be inserted into the vector genome. In the assay system of Petropoulos and colleagues the patient-derived RT and PR sequences were ligated into an env-deleted HIV vector plasmid. The ligated plasmids were transformed in bacteria and recombinant vector plasmids prepared from bulk cultures of the transformed bacteria. Vector particles were produced by cotransfection with a packaging plasmid containing the amphotropic MuLV env. Drug sensitivity was measured by quantification of luciferase activity in the target cells (Petropoulos *et al.,* 2000).

An alternative HIV vector developed in our laboratory for HIV resistance testing contains specific modifications to minimize the risk of genetic recombination and outgrowth of replicating infectious viruses (Jarmy *et al.,* 2001). In this vector, the accessory and regulatory HIV genes vif, vpu, vpr, tat, and nef were deleted in addition to the env gene. Deletion of the Tat transactivator required substitution of the viral 5′ U3 promoter and thus the constitutively active CMV promoter was introduced (Kim *et al.,* 1998). In addition, the 3′ U3 region was deleted to obtain a self-inactivating vector construct (Dull *et al.,* 1998; Zufferey *et al.,* 1998), which lost the ability to express any vector protein in the infected target cell except for the marker protein. With this vector system, the risk of recombination to replication-competent HIV was reduced to practically nil and maximum biosafety was achieved.

For determination of sensitivity to antiviral drugs with the HIV vectors, RT and PR inhibitors have to be added at specific time points during the assay procedure. RT (and similarly integrase) inhibitors were added upon infection of the target cells. In contrast, since the HIV protease completes maturation of the virion, PR inhibitors were added to the producer cells in which, upon transfection with the plasmids, vector particles are made (Petropoulos *et al.,* 2000). These tests can be performed within 10 working days (Jarmy *et al.,* 2001; Petropoulos *et al.,* 2000) and do not require the production of infectious and replication-competent virus. Evaluation of tests based on replication-incompetent HIV particles demonstrated that sensitivity testing based on HIV vectors is highly reproducible (Jarmy *et al.,* 2001; Petropoulos *et al.,* 2000). In addition, results obtained with this assay system compared favorably with those obtained by genotypic analyses (Dunne *et al.,* 2001; Jarmy *et al.,* 2001; Petropoulos *et al.,* 2000) and other phenotypic sensitivity assays such as those that use replication-competent recombinant virus (Miller *et al.,* 2001; Qari *et al.,* 2000).

Similar to the approach utilizing recombinant replicating HIV, the procedure has been automated for resistance testing in a high-throughput format. The PhenoSense assay from ViroLogic, Inc., and the Phenoscript test from Viralliance Corporation represent examples of commercial test systems using the vector technology. The sensitivity of the vector-based drug sensitivity test systems maximally reaches 400–500 RNA copies/ml plasma (Petropoulos *et al.,* 2000). The results of a variety of HIV patient studies indicate the practicability of vector-based HIV sensitivity testing (Brown *et al.,* 2000; Condra *et al.,* 2000; Deeks *et al.,* 1999, 2001; Havlir *et al.,* 2000; Hecht *et al.,* 1998; Johnson *et al.,* 2001; Little *et al.,* 1999; Martinez Picado *et al.,* 2000; Parkin *et al.,* 1999, 2000; Piketty *et al.,* 1999, 2000; Weinstock *et al.,* 2000; Ziermann *et al.,* 2000).

E. Enzymatic Assays for the Analysis of Drug Resistance

Several non-cell-culture-based enzymatic assays for the determination of PR and RT activity have been developed and represent another alternative to phenotypic tests with infectious HIV particles. For instance, a drug sensitivity test based on a cell-free enzymatic system has been developed and used for phenotypic resistance analysis to nevirapine and lamivudine (Garcia Lerma *et al.,* 1999; Vazquez Rosales *et al.,* 1999). With this test system, the activity of the viral RT in plasma-derived virions was quantified in a reaction mixture containing a known nonretroviral heteropolymeric RNA template. The amount of reverse-transcribed cDNA was measured by quantitative PCR. The results obtained with this method correlated well with the genotype.

Analogous tests for PR inhibitors were also developed. In these tests, protease genes were derived from test specimens by RT-PCR, cloned in plasmids, and expressed in bacteria. Specific protease substrates were used and cleavage by the bacterially expressed HIV protease analyzed in the absence or presence of antiviral drugs. Cleavage of the substrate was measured by reverse-phase high-pressure liquid chromatography (Melnick *et al.,* 1998). Phenotypic resistance data obtained by this assay system correlated well with those obtained with cell-based phenotypic and genotypic assays (Maschera *et al.,* 1995; Melnick *et al.,* 1998). However, as discussed in Section IIIA sensitivity data obtained with cell-free enzymatic assays did not always correlate with results gained with cell-based test systems. This limits the general use of enzymatic tests for clinical sensitivity testing at this time.

F. Genotyping

Genotypic resistance testing is based on the detection of mutations that are known to be associated with reduced drug sensitivity of the virus. Genotypic assays are technically less complex than phenotypic assays, rapid to perform, and highly reproducible. Similar to phenotypic assays, they require amplification of the relevant HIV genome sequences by RT-PCR prior to sequence analysis. The length of the genome segment amplified depends on the specific assay system, but contains at least the protease and the relevant parts of the RT gene. In general, genotypic resistance assays require a viral load of >500–1000 copies/ml. Sensitivity is increased if several shorter overlapping fragments spanning the whole gene of interest are amplified instead of one long-range PCR.

After amplification of the relevant genome segments, two different techniques are used for the detection of resistance-associated mutations. Most in-house assays and two commercial kits, the ViroSeq (Applied Biosystems, Foster City, CA) and the TruGene assay (Visible Genetics, Toronto, Canada), make use of the dideoxynucleotide sequencing technology. Alternatively, hybridization methods are employed for the detection of specific mutations. A commercial test using this technology is the LineProbeAssay (LiPA HIV-1, Innogenetics) (Stuyver *et al.*, 1997). This assay detects only selected key RT mutations and in a newer version also PR mutations. It has a high sensitivity for detecting proportions as low as 10% of resistant strains in virus mixtures. However, polymorphisms or mutations near the specific site of hybridization may occasionally lead to false-negative results (Koch *et al.*, 1999; Puchhammer-Stockl *et al.*, 1999). This problem is aggravated in non-B subtypes (Kijak *et al.*, 2001).

A microarray-based assay to detect resistance-associated mutations has been developed by Affymetrix (HIV PRT GeneChip, Affymetrix, Santa Clara, CA). After amplification of PR and RT gene sequences with primers containing T3 and T7 promotor sequences, fluorescein-labeled cRNA is produced. The cRNA is then fragmented and subsequently hybridized to chips containing more than 16,000 unique oligonucleotide probes complementary to the HIV PR and RT genes. Hybridized cRNAs are detected by a laser scanner. The HIV sequence of the sample is calculated from the patten of hybridization signals by specialized computer software (Kozal *et al.*, 1996). This method is potentially useful for high-throughput genotyping, but problems with the accurate detection of length polymorphisms and of non-B-subtype sequences have been described (Vahey *et al.*, 1999). It is expected that newer GeneChip versions will be improved in this respect.

Since genotype determination with commercial kits or with automated sequencing equipment is straightforward, many laboratories perform it. As with all other laboratory tests, comparative studies of different genotyping techniques as well as external quality control measures are essential in order to ensure accurate and reliable results for clinical use. Hybridization methods and dideoxynucleotide sequencing have been compared in several publications. A study of 29 clinical specimens, 6 virus isolates, and 13 $\text{HIV}_{\text{NL4-3}}$ clones found an overall nucleotide concordance of 99.1% between dideoxynucleotide sequencing and GeneChip (Gunthard *et al.*, 1998). Artificial mixing experiments with wild-type and resistant strains analyzed by the GeneChip technology showed a bias toward wild-type bases. These results were essentially confirmed in another study of 49 samples from 22 highly treatment-experienced patients (Hanna *et al.*, 2000). Though the overall discordance rate between GeneChip and dideoxynucleotide sequencing was only 0.8%, it was significantly higher for resistance-associated codons (3.9%). The higher rate of discordances was mainly caused by genetic mixtures within or adjacent to discordant codons. Dideoxynucleotide sequencing more commonly found a known resistance-associated amino acid than GeneChip. A third study comparing GeneChip, the LineProbe assay, and a dideoxynucleotide sequencing method found that all three systems were capable of identifying clinically relevant mutations associated with contemporary drug therapies (Wilson *et al.*, 2000). Discordances up to 8% in pairwise comparisons were mainly seen for secondary mutations and polymorphisms of unrecognized clinical significance.

A number of multicenter evaluations have been performed in order to investigate the quality and sensitivity of standard sequencing methods. In a study involving 13 experienced laboratories with various sequencing protocols, agreement of the nucleotide assignments was demonstrated in more than 99% of the sequenced RT nucleotide positions in five HIV isolates (Demeter *et al.*, 1998). Discrepant results were obtained for only 0.29% of all nucleotides. These discrepancies did not appear to result from transcription errors. However, the results of two other sequencing quality control studies (ENVA-1, ENVA-2) involving a larger number of laboratories have been less encouraging and have revealed problems especially in the detection of virus mixtures (Schuurman *et al.*, 1999a, b). The ENVA-1 panel consisted of nine samples containing different mixtures of wild-type and mutant HIV-1 RT plasmids. Mutations in different ratios to the wild-type HXB2 were present at positions 41, 215, and 184. The presence of mutations was reported qualitatively and quantitatively. The genotyping results demonstrated extensive interlaboratory variation among the participating 23 laboratories. Two laboratories reported mutations in wild-type samples, whereas three laboratories failed to detect mutations in completely mutant samples. Mutations at positions 41, 215, and 184 present at relative concentrations of 25% were correctly identified by 13, 10, and 16 laboratories, respectively. The ENVA-2 panel consisted of five plasma samples containing mixtures of nonmutated laboratory HXB2 virus and of two other HIV-1 strains, one with defined mutations in the RT gene and the other with defined mutations in the PR gene. Although wild-type codons were correctly identified by almost all of the participating 33 laboratories, only about 70% of the participants detected all resistance mutations in the samples containing only the mutant variants. For the 1:1 mixtures of mutant and wild-type virus, the percentage of correct results was even lower. These studies clearly underscore the need for external quality control schemes. Another recent study, which compared the sequencing results from plasma samples of 46 heavily treated HIV-1 patients in two laboratories, has demonstrated that genotyping may be highly reproducible (Shafer *et al.,* 2001). The rates of complete sequence concordance between both laboratories were 99.1% for the PR and 99.0% for the RT gene. Of the remaining 1% of discordant results, about 90% were partial.

G. Measuring Sensitivity to Inhibitors of Virus Entry and the Integrase

Testing of sensitivity to fusion inhibitors such as T-20 has been performed experimentally using classical culture of wild-type and mutated HIV on cell lines and determination of the IC_{50} by the Reed and Muench method and by genetic analysis (Rimsky *et al.,* 1998). Because T-20 and related drugs are in advanced stages of clinical evaluation, the development of specific methods for routine testing of individuals treated with entry inhibitors is urgently needed (O'Brien, 2001). Phenotypic sensitivity assays based on recombinant replication-competent HIV or replication-incompetent vectors are currently being developed.

Sensitivity analysis of virus isolates to integrase inhibitors will become an issue with future clinical availability of these drugs. Theory and preliminary data indicate that as far as basic principles are concerned, genetic and phenotypic tests such as those used for resistance analysis with protease and reverse transcriptase inhibitors are suitable for integrase inhibitor sensitivity testing as well (Hazuda *et al.,* 2000; Jarmy *et al.,* 2001).

V. Clinical Implications

A. Interpretation of HIV Sensitivity Assays

Although the technology for generating resistance data is quite advanced, the interpretation of the test results remains a complex issue with both genotypic and phenotypic resistance assays. The approach most commonly used for genotype interpretation is "rules-based," that is, it relies on the identification of key mutations associated with resistance to each drug. Overviews for relevant mutations can be found in several publications and on several websites (Hirsch *et al.,* 2000; Schinazi *et al.,* 1999; Shafer and Deresinski, 2000). Expert advice is then used to construct algorithms for interpretation of the mutation data. These algorithms can also be used as the basis for genotype interpretation software. However, this approach poses a considerable challenge with respect to the prediction of cross-resistance and mutational interactions, especially when the mutation pattern is complex. Because knowledge about resistance-associated mutations is increasing rapidly, frequent updating of rules-based algorithms and software is necessary in order to use genotypic test results effectively.

Databases with paired genotypes and phenotypes form the basis for different strategies to interpret genotypes. The VirtualPhenotype, developed by Virco, is an example of this approach (Larder *et al.,* 2000b). Using codons identified as being critical for the resistance properties of HIV-1 samples, a database search is conducted for genotypes of patient samples in order to identify similar genotypes for which the phenotype is known. Subsequently, an average phenotype is calculated for all available matches in the database. This average phenotype represents an estimation of the phenotype of the patient sample. In results presented by Virco, the actual and the virtual phenotype were highly correlated (Larder *et al.,* 2000b) and the VirtualPhenotype was similar to the actual phenotype in predicting treatment response (Graham *et al.,* 2001). Although these data are promising, validation of this approach in prospective studies is necessary in order to assess its clinical usefulness. A potential limitation of the VirtualPhenotype is the analysis of resistance to new drugs and of unusual genotypes because there will be fewer database matches in these situations.

Geno2pheno is another prediction model of resistance based on a database, with genotype–phenotype pairs derived from more than 470 clinical isolates (Beerenwinkel *et al.,* 2001). This method uses mathematical modeling based on the machine-learning technique of decision trees. Error estimates were below 15% for most of the drugs. For stavudine, zalcitabine, and didanosine error estimates were $>25\%$, but accurate prediction of resistance is difficult for these three drugs with any method that is currently available. Other potential limitations are similar to those for the VirtualPhenotype. Geno2pheno is freely available on the Internet.

Although the results of phenotypic assays reported as IC_{50} values and fold-resistance data appear straightforward, their interpretation is challenging. Currently, the most important question is how to obtain clinically relevant data. In the Antivirogram from Virco and the Phenosense test from ViroLogic cutoff values for classifying HIV strains as resistant were initially based on the interassay variation of wild-type virus strains. Recently, this approach has been modified because the natural variation of susceptibility of wild-type strains to a certain drug, the biological cutoff, was found to be greater than the interassay variation observed with the original control strains. Virco has therefore defined new drug susceptibility ranges based on the assay data of HIV-1 strains from 1000 treatment-naive patients (Harrigan *et al.,* 2001). In addition, the technical cutoffs did not take into account the clinical response to the treatment. Meanwhile, drug sensitivity cutoff levels based on clinical response data have become available. This information is now being used for the interpretation of results obtained by the Phenosense and Phenoscript (Viralliance) HIV drug sensitivity tests (Lanier *et al.,* 2001). Both adaptations of the cutoff values may in the future further improve the clinical relevance of the drug sensitivity test systems.

Another unresolved issue is how to correlate resistance and pharmacological data in an adequate manner. One approach is to calculate the so-called inhibitory quotient (IQ), which represents the ratio of the achievable or measured trough concentration (C_{min}) and the IC_{50} of a certain drug. Initial studies have shown that the IQ for lopinavir and indinavir can predict treatment response and that the IQ appears to be superior to the genotype or phenotype in this respect (Hsu *et al.,* 2000; Kempf *et al.,* 2001). Based on the IQ it may be possible in certain situations to adjust drug doses in order to overcome partial resistance. It appears likely that combining resistance data and drug levels will become increasingly important.

B. Clinical Use of HIV Drug Sensitivity Assays

Evidence for the clinical significance of resistance testing has come from both retrospective and prospective studies. Several retrospective studies have shown an association between baseline resistance to antiviral drugs and clinical outcome (Deeks *et al.,* 1999; Harrigan *et al.,* 1999; Lorenzi *et al.,* 1999; Miller *et al.,* 2000a; Perez-Elias *et al.,* 2000; Piketty *et al.,* 1999; Tebas *et al.,* 1999; Van Vaerenbergh *et al.,* 2000; Zolopa *et al.,* 1999). A uniform statistical reanalysis of eight retrospective and two prospective studies found a consistent association between drug resistance and virologic failure after 24 weeks (DeGruttola *et al.,* 2000).

The question of whether resistance testing is useful for improving the clinical outcome has been assessed in prospective trials. Results from several randomized studies on the use of both phenotypic and genotypic resistance testing as a guide for selecting salvage therapy regimens have been reported (Baxter *et al.,* 2000; C. Cohen *et al.,* 2000; De Luca *et al.,* 2001; Durant *et al.,* 1999; Meynard *et al.,* 2000; Tural *et al.,* 2001; Haubrich *et al.,* 2001). The results of these studies can only partially be compared because of different designs with respect to several aspects. These include the treatment experience of the patient populations, the observation period, the methods used for resistance testing and data interpretation, the provision of expert advice, and the detection limit of viral load assays. Nevertheless, a greater reduction of HIV-1 RNA from baseline of about 0.5 $\log_{10}$ and a higher proportion of patients achieving a viral load below the detection limit or both has been reported in most of the studies for the group of patients treated on the basis of drug resistance data. A weakness of the trials reported so far is the short follow-up period of only 24 weeks or less in most of the studies. Therefore, the duration of the beneficial effect of sensitivity testing on the virologic response is unknown.

In addition to sensitivity testing, expert interpretation of drug resistance results and advice on optimal drug selection may improve clinical outcome. For instance, in the Havana study, in which genotypic resistance analysis was investigated, the effect of expert advice was studied as an independent factor (Tural *et al.,* 2001). It was observed that expert advice alone was associated with a significantly better outcome compared to standard care. In particular, the positive effect of expert advice on clinical outcome was comparable to the effect of genotyping without expert advice. A combination of both expert advice and genotyping was associated with the best outcome. Unfortunately, in other prospective studies, expert advice was provided only to the resistance assay arms, but not to the control arms.

It is still controversial which group of patients benefits the most from genotyping. Although the Havana study demonstrated that genotyping was most beneficial with patients who had failed three or more previous treatment regimens, the ARGENTA and the NARVAL studies revealed a durable effect of genotyping only for patients with less treatment experience (De Luca *et al.,* 2001; Meynard *et al.,* 2000).

The effect of phenotypic tests on patient management was studied in the VIRA3001, the NARVAL, and the CCTG 575 trials (C. Cohen *et al.,* 2000; Meynard *et al.,* 2000; Haubrich *et al.,* 2001). Interpretation of the drug sensitivity data obtained with the patient specimens was based on technical cutoffs, but not on either biological or clinical cutoff values. Although patients in the phenotypic arm of the VIRA3001 study had a significantly better response after 16 weeks compared to the standard care arm, there was no benefit by phenotyping in the NARVAL and CCTG 575 studies. The contrary results may primarily be due to different interpretation schemes in regard to the potential clinical relevance of the *in vitro* drug sensitivity data. Therefore, the realization of the necessity to establish clinically relevant drug sensitivity and resistance cutoff data may be regarded as the main benefit of these studies.

More prospective trials with clinically relevant interpretation methods of the phenotypic assays are necessary to clarify whether and in which clinical situations phenotypic drug sensitivity testing will be beneficial for the HIV-infected individual.

Despite these difficulties and open questions associated with the interpretation of genotypes and phenotypes, drug sensitivity testing has become part of the routine management of HIV-infected individuals in specific clinical situations. This is reflected in guidelines published by, for example, the U.S. Department of Health and Human Services (2001), the International AIDS Society—USA (Hirsch *et al.,* 2000), and the EuroGuidelines Group for HIV resistance (EuroGuidelines Group, 2001). At present, recommendations are being made for resistance testing in the following clinical situations: suspected recent transmission of resistant virus, initiation of treatment during primary HIV infection, therapy change in treated patients due to virologic failure, and pregnancy with detectable viral load levels (EuroGuidelines Group, 2001) (Hirsch *et al.,* 2000). The spectrum of clinical conditions for which resistance testing will turn out to be benefical will possibly change as more data from appropriate clinical trials become available.

VI. Conclusions

The advent of highly active antiviral combination therapy has converted infection with HIV into a treatable though not curable chronic disease. The necessity for long-term treatment and the ability of HIV to rapidly mutate leads to a substantial risk for the development of virus variants resistant to the drugs and subsequent treatment failure. Theory and experience with resistance testing in bacterial diseases indicate that analysis of drug resistance and determination of the sensitivity to alternative drugs offers the chance for improved clinical management of HIV-infected persons. However, classical methods for the measurement of antiviral drug sensitivity are particularly time-consuming and labor-intensive if applied to HIV infection. Therefore, several alternative methods have been developed to make analysis of drug resistance feasible for clinical practice. Although these methods are far from being as convenient and low cost as analogous methods used in bacterial diseases, the tests are now able to determine drug sensitivity of sufficient numbers of specimens and in a reasonable time frame, in particular if the processes are automated. Clinical evaluation of the benefit of resistance testing by these methods has recently begun. The methods developed and the experience that is being gained with sensitivity testing in HIV infection will be beneficial for the management of other chronic viral diseases when appropriate treatment options become available.

Acknowledgments

The authors thank Sieghart Sopper for critical review of the manuscript. This work was made possible through a research grant from the H. W. & J. Hector Foundation.

References

Adje, C., Cheingsong, R., Roels, T. H., Maurice, C., Djomand, G., Verbiest, W., Hertogs, K., Larder, B., Monga, B., Peeters, M., *et al.* (2001). High prevalence of genotypic and phenotypic HIV-1 drug-resistant strains among patients receiving antiretroviral therapy in Abidjan, Cote d'Ivoire. *J. Acquir. Immune Defic. Syndr.* **26,** 501–506.

Aguilar Cordova, E., Chinen, J., Donehower, L., Lewis, D. E., and Belmont, J. W. (1994). A sensitive reporter cell line for HIV-1 tat activity, HIV-1 inhibitors, and T cell activation effects. *AIDS Res. Hum. Retrovir.* **10,** 295–301.

Akkina, R. K., Walton, R. M., Chen, M. L., Li, Q. X., Planelles, V., and Chen, I. S. (1996). High-efficiency gene transfer into CD34+ cells with a human immunodeficiency virus type 1-based retroviral vector pseudotyped with vesicular stomatitis virus envelope glycoprotein G. *J. Virol.* **70,** 2581–2585.

Allaway, G. P., Davis-Bruno, K. L., Beaudry, G. A., Garcia, E. B., Wong, E. L., Ryder, A. M., Hasel, K. W., Gauduin, M. C., Koup, R. A., McDougal, J. S., *et al.* (1995). Expression and characterization of CD4–IgG2, a novel heterotetramer that neutralizes primary HIV type 1 isolates. *AIDS Res. Hum. Retrovir.* **11,** 533–539.

Amado, R. G., and Chen, I. S. (1999). Lentiviral vectors the promise of gene therapy within reach. *Science* **285,** 674–676.

Arion, D., Borkow, G., Gu, Z., Wainberg, M. A., and Parniak, M. A. (1996). The K65R mutation confers increased DNA polymerase processivity to HIV-1 reverse transcriptase. *J. Biol. Chem.* **271,** 19860–19864.

Arion, D., Kaushik, N., McCormick, S., Borkow, G., and Parniak, M. A. (1998). Phenotypic mechanism of HIV-1 resistance to 3′-azido-3′-deoxythymidine (AZT): Increased polymerization processivity and enhanced sensitivity to pyrophosphate of the mutant viral reverse transcriptase. *Biochemistry* **37,** 15908–15917.

Babine, R., and Bender, S. (1997). Molecular recognition of protein–ligand complexes: Application to drug design. *Chem. Rev.* **97,** 1359–1472.

Bacheler, L., Jeffrey, S., Hanna, G., D'Aquila, R., Wallace, L., Logue, K., Cordova, B., Hertogs, K., Larder, B., Buckery, R., *et al.* (2001). Genotypic correlates of phenotypic resistance to efavirenz in virus isolates from patients failing nonnucleoside reverse transcriptase inhibitor therapy. *J. Virol.* **75,** 4999–5008.

Balfour, H. H., Jr. (1999). Antiviral drugs. *N. Engl. J. Med.* **340,** 1255–1268.

Baxter, J. D., Mayers, D. L., Wentworth, D. N., Neaton, J. D., Hoover, M. L., Winters, M. A., Mannheimer, S. B., Thompson, M. A., Abrams, D. I., and Brizz, B. J., *et al.* (2000). A randomized study of antiretroviral management based on plasma genotypic antiretroviral resistance testing in patients failing therapy. CPCRA 046 Study Team for the Terry Beirn Community Programs for Clinical Research on AIDS. *AIDS* **14,** F83–F93.

Bebenek, K., Abbotts, J., Wilson, S. H., and Kunkel, T. A. (1993). Error-prone polymerization by HIV-1 reverse transcriptase. Contribution of template-primer misalignment, miscoding, and termination probability to mutational hot spots. *J. Biol. Chem.* **268,** 10324–10334.

Becerra, S. P., Kumar, A., Lewis, M. S., Widen, S. G., Abbotts, J., Karawya, E. M., Hughes, S. H., Shiloach, J., and Wilson, S. H. (1991). Protein–protein interactions of HIV-1 reverse transcriptase: Implication of central and C-terminal regions in subunit binding. *Biochemistry* **30,** 11707–11719.

Beerenwinkel, N., Schmidt, B., Walter, H., Kaiser, R., Lengauer, T., Hoffmann, D., Korn, K., and Selbig, J. (2001). Geno2pheno: Interpreting genotypic HIV drug resistance tests. *IEEE Intelligent Syst. Biol.* **16,** 35–41.

Boden, D., and Markowitz, M. (1998). Resistance to human immunodeficiency virus type 1 protease inhibitors. *Antimicrob. Agents Chemother.* **42,** 2775–2783.

Borman, A. M., Paulous, S., and Clavel, F. (1996). Resistance of human immunodeficiency virus type 1 to protease inhibitors: Selection of resistance mutations in the presence and absence of the drug. *J. Gen. Virol.* **77,** 419–426.

Boucher, C. A., Tersmette, M., Lange, J. M., Kellam, P., de Goede, R. E., Mulder, J. W., Darby, G., Goudsmit, J., and Larder, B. A. (1990). Zidovudine sensitivity of human immunodeficiency viruses from high-risk, symptom-free individuals during therapy. *Lancet* **336,** 585–590.

Boucher, C. A., Keulen, W., van Bommel, T., Nijhuis, M., de Jong, D., de Jong, M. D., Schipper, P., and Back, N. K. (1996). Human immunodeficiency virus type 1 drug susceptibility determination by using recombinant viruses generated from patient sera tested in a cell-killing assay. *Antimicrob. Agents Chemother.* **40,** 2404–2409.

Boyer, P. L., Tantillo, C., Jacobo-Molina, A., Nanni, R. G., Ding, J., Arnold, E., and Hughes, S. H. (1994). Sensitivity of wild-type human immunodeficiency virus type 1 reverse transcriptase to dideoxynucleotides depends on template length; the sensitivity of drug-resistant mutants does not. *Proc. Natl. Acad. Sci. USA* **91,** 4882–4886.

Briones, C., Perez Olmeda, M., Rodriguez, C., del Romero, J., Hertogs, K., and Soriano, V. (2001). Primary genotypic and phenotypic HIV-1 drug resistance in recent seroconverters in Madrid. *J. Acquir. Immune Defic. Syndr.* **26,** 145–150.

Brown, A. J., Precious, H. M., Whitcomb, J. M., Wong, J. K., Quigg, M., Huang, W., Daar, E. S., D'Aquila, R. T., Keiser, P. H., Connick, E., *et al.* (2000). Reduced susceptibility of human immunodeficiency virus type 1 (HIV-1) from patients with primary HIV infection to nonnucleoside reverse transcriptase inhibitors is associated with variation at novel amino acid sites. *J. Virol.* **74,** 10269–10273.

Buchschacher, G. J., and Wong-Staal, F. (2000). Development of lentiviral vectors for gene therapy for human diseases. *Blood* **95,** 2499–2504.

Caliendo, A. M., Savara, A., An, D., DeVore, K., Kaplan, J. C., and D'Aquila, R. T. (1996). Effects of zidovudine-selected human immunodeficiency virus type 1 reverse transcriptase amino acid substitutions on processive DNA synthesis and viral replication. *J. Virol.* **70,** 2146–2153.

Caride, E., Brindeiro, R., Hertogs, K., Larder, B., Dehertogh, P., Machado, E., de Sa, C. A., Eyer Silva, W. A., Sion, F. S., Passioni, L. F., *et al.* (2000). Drug-resistant reverse transcriptase genotyping and phenotyping of B and non-B subtypes (F and A) of human immunodeficiency virus type I found in Brazilian patients failing HAART. *Virology* **275,** 107–115.

Carpenter, C. C., Cooper, D. A., Fischl, M. A., Gatell, J. M., Gazzard, B. G., Hammer, S. M., Hirsch, M. S., Jacobsen, D. M., Katzenstein, D. A., Montaner, J. S., *et al.* (2000). Antiretroviral therapy in adults: Updated recommendations of the International AIDS Society—USA Panel. *JAMA* **283,** 381–390.

Casado, J. L., Hertogs, K., Ruiz, L., Dronda, F., Van Cauwenberge, A., Arno, A., Garcia Arata, I., Bloor, S., Bonjoch, A., Blazquez, J., *et al.* (2000). Non-nucleoside reverse transcriptase inhibitor resistance among patients failing a nevirapine plus protease inhibitor-containing regimen. *AIDS* **14,** F1–F7.

Chen, Z., Li, Y., Schock, H. B., Hall, D., Chen, E., and Kuo, L. C. (1995). Three-dimensional structure of a mutant HIV-1 protease displaying cross-resistance to all protease inhibitors in clinical trials. *J. Biol. Chem.* **270,** 21433–21436.

Cocchi, F., DeVico, A., Garzino-Demo, A., Cara, A., Gallo, R., and Lusso, P. (1996). The V3 domain fo the HIV-1 gp120 envelope glycoprotein is critical for chemokine-mediated blockade of infection. *Nature Med.* **2,** 1244–1247.

Cohen, C., Kessler, H., Hunt, S., *et al.* (2000). Phenotypic resistance testing significantly improves response to therapy: Final analysis of a randomized trial (VIRA3001). *Antivir. Ther.* **5**(Suppl. 3), 67 [Abstract 84].

Cohen, O., and Fauci, A. (2001). Pathogenesis and medical aspects of HIV-1 infection. *In* "Fields Virology" (D. Knipe and P. Howley, Eds.), 4th ed., pp. 2043–2094. Lippincott Williams & Wilkins, Philadelphia.

Condra, J. H., Schleif, W. A., Blahy, O. M., Gabryelski, L. J., Graham, D. J., Quintero, J. C., Rhodes, A., Robbins, H. L., Roth, E., Shivaprakash, M., *et al.* (1995). *In vivo* emergence of HIV-1 variants resistant to multiple protease inhibitors. *Nature* **374,** 569–571.

Condra, J. H., Petropoulos, C. J., Ziermann, R., Schleif, W. A., Shivaprakash, M., and Emini, E. A. (2000). Drug resistance and predicted virologic responses to human immunodeficiency virus type 1 protease inhibitor therapy. *J. Infect. Dis.* **182,** 758–765.

Conway, B., Wainberg, M., Hall, D., Harris, M., Reiss, P., Cooper, D., Vella, S., Curry, R., Robinson, P., Lange, J., *et al.* (2001). Development of drug resistance in patients receiving combinations of zidovudine, didanosine and nevirapine. *AIDS* **15,** 1269–1274.

Cozzi Lepri, A., Sabin, C. A., Staszewski, S., Hertogs, K., Muller, A., Rabenau, H., Phillips, A. N., and Miller, V. (2000). Resistance profiles in patients with viral rebound on potent antiretroviral therapy. *J. Infect. Dis.* **181,** 1143–1147.

Croteau, G., Doyon, L., Thibeault, D., McKercher, G., Pilote, L., and Lamarre, D. (1997). Impaired fitness of human immunodeficiency virus type 1 variants with high-level resistance to protease inhibitors. *J. Virol.* **71,** 1089–1096.

Crumpacker, C. (2001). Antiviral therapy. *In* "Fields Virology" (D. Knipe and P. Howley, Eds.), 4th ed., pp. 393–433. Lippincott Williams & Wilkins, Philadelphia.

Deeks, S. G., Hellmann, N. S., Grant, R. M., Parkin, N. T., Petropoulos, C. J., Becker, M., Symonds, W., Chesney, M., and Volberding, P. A. (1999). Novel four-drug salvage treatment regimens after failure of a human immunodeficiency virus type 1 protease inhibitor-containing regimen: Antiviral activity and correlation of baseline phenotypic drug susceptibility with virologic outcome. *J. Infect. Dis.* **179,** 1375–1381.

Deeks, S. G., Wrin, T., Liegler, T., Hoh, R., Hayden, M., Barbour, J. D., Hellmann, N. S., Petropoulos, C. J., McCune, J. M., Hellerstein, M. K., *et al.* (2001). Virologic and immunologic consequences of discontinuing combination antiretroviral-drug therapy in HIV-infected patients with detectable viremia. *N. Engl. J. Med.* **344,** 472–480.

DeGruttola, V., Dix, L., D'Aquila, R., Holder, D., Phillips, A., Ait-Khaled, M., Baxter, J., Clevenbergh, P., Hammer, S., Harrigan, R., *et al.* (2000). The relation between baseline HIV drug resistance and response to antiretroviral therapy: Re-analysis of retrospective and prospective studies using a standardized data analysis plan. *Antivir. Ther.* **5,** 41–48.

De Luca, A., Antinori, A., Cingolani, A., *et al.* (2001). A prospective, randomized study on the usefulness of genotypic resistance testing and the assesment of patient-reported adherence in unselected patients failing potent HIV therapy (ARGENTA): Final 6-month results. *In* "Program and Abstract of the 8th Conference on Retroviruses and Opportunistic Infections," Chicago, Illinois [Abstract 433].

Demeter, L. M., D'Aquila, R., Weislow, O., Lorenzo, E., Erice, A., Fitzgibbon, J., Shafer, R., Richman, D., Howard, T. M., Zhao, Y., *et al.* (1998). Interlaboratory concordance of DNA sequence analysis to detect reverse transcriptase mutations in HIV-1 proviral DNA. ACTG Sequencing Working Group. AIDS Clinical Trials Group. *J. Virol. Meth.* **75,** 93–104.

Devereux, H. L., Emery, V., Johnson, M., and Loveday, C. (2001). Replicative fitness *in vivo* of HIV-1 variants with multiple drug resistance-associated mutations. *J. Med. Virol.* **65,** 218–224.

di Marzo-Veronese, F., Copeland, T., Rahman, R., Oroszlán, S., Gallo, R., and Sarngadharan, M. (1986). Characterization of highly immunogenic p66/p51 as the reverse transcriptase of HTLV-III/LAV. *Science* **231,** 1289–1291.

Ding, J., Das, K., Moereels, H., Koymans, L., Andries, K., Janssen, P. A., Hughes, S. H., and Arnold, E. (1995a). Structure of HIV-1 RT/TIBO R 86183 complex reveals similarity in the binding of diverse nonnucleoside inhibitors. *Nature Struct. Biol.* **2,** 407–415.

Ding, J., Das, K., Tantillo, C., Zhang, W., Clark, A. D., Jr., Jessen, S., Lu, X., Hsiou, Y., Jacobo-Molina, A., Andries, K., *et al.* (1995b). Structure of HIV-1 reverse transcriptase in a complex with the non-nucleoside inhibitor alpha-APA R 95845 at 2.8 Å resolution. *Structure* **3,** 365–379.

Doranz, B. J., Grovit-Ferbas, K., Sharron, M. P., Mao, S. H., Goetz, M. B., Daar, E. S., Doms, R. W., and O'Brien, W. A. (1997). A small-molecule inhibitor directed against the chemokine receptor CXCR4 prevents its use as an HIV-1 coreceptor. *J. Exp. Med.* **186,** 1395–1400.

Doyon, L., Croteau, G., Thibeault, D., Poulin, F., Pilote, L., and Lamarre, D. (1996). Second locus involved in human immunodeficiency virus type 1 resistance to protease inhibitors. *J. Virol.* **70,** 3763–3769.

D'Souza, M. P., Cairns, J. S., and Plaeger, S. F. (2000). Current evidence and future directions for targeting HIV entry: Therapeutic and prophylactic strategies. *JAMA* **284,** 215–222.

Dull, T., Zufferey, R., Kelly, M., Mandel, R. J., Nguyen, M., Trono, D., and Naldini, L. (1998). A third-generation lentivirus vector with a conditional packaging system. *J. Virol.* **72,** 8463–8471.

Dunne, A., Mitschel, F., Coberly, S., Hellmann, N., Hoy, J., Mijch, A., Petropoulos, C., Mills, J., and Crowe, S. (2001). Comparison of genotyping and phenotyping methods for determining susceptibility of HIV-1 to antiretroviral drugs. *AIDS* **15,** 1471–1475.

Durant, J., Clevenbergh, P., Halfon, P., Delgiudice, P., Porsin, S., Simonet, P., Montagne, N., Boucher, C. A., Schapiro, J. M., and Dellamonica, P. (1999). Drug-resistance genotyping in HIV-1 therapy: The VIRADAPT randomised controlled trial. *Lancet* **353,** 2195–2199.

Eastman, P. S., Mittler, J., Kelso, R., Gee, C., Boyer, E., Kolberg, J., Urdea, M., Leonard, J. M., Norbeck, D. W., Mo, H., *et al.* (1998). Genotypic changes in human immunodeficiency virus type 1 associated with loss of suppression of plasma viral RNA levels in subjects treated with ritonavir (Norvir) monotherapy. *J. Virol.* **72,** 5154–5164.

Erickson, J. W., Gulnik, S. V., and Markowitz, M. (1999). Protease inhibitors: Resistance, cross-resistance, fitness and the choice of initial and salvage therapies. *AIDS* **13**(Suppl. A), S189–S204.

Esnouf, R., Ren, J., Ross, C., Jones, Y., Stammers, D., and Stuart, D. (1995). Mechanism of inhibition of HIV-1 reverse transcriptase by non-nucleoside inhibitors. *Nature Struct. Biol.* **2,** 303–308.

Espeseth, A. S., Felock, P., Wolfe, A., Witmer, M., Grobler, J., Anthony, N., Egbertson, M., Melamed, J. Y., Young, S., Hamill, T., *et al.* (2000). HIV-1 integrase inhibitors that compete with the target DNA substrate define a unique strand transfer conformation for integrase. *Proc. Natl. Acad. Sci. USA* **97,** 11244–11249.

EuroGuidelines Group. 2001. Clinical and laboratory guidelines for the use of HIV-1 drug resistance testing as part of treatment management: Recommendations for the European setting. The EuroGuidelines Group for HIV resistance. *AIDS* **15,** 309–320.

Furman, P. A., Fyfe, J. A., St Clair, M. H., Weinhold, K., Rideout, J. L., Freeman, G. A., Lehrman, S. N., Bolognesi, D. P., Broder, S., Mitsuya, H., *et al.* (1986). Phosphorylation of 3′-azido-3′-deoxythymidine and selective interaction of the 5′-triphosphate with human immunodeficiency virus reverse transcriptase. *Proc. Natl. Acad. Sci. USA* **83,** 8333–8337.

Garcia Lerma, J., Schinazi, R. F., Juodawlkis, A. S., Soriano, V., Lin, Y., Tatti, K., Rimland, D., Folks, T. M., and Heneine, W. (1999). A rapid non-culture-based assay for clinical monitoring of phenotypic resistance of human immunodeficiency virus type 1 to lamivudine (3TC). *Antimicrob. Agents Chemother.* **43,** 264–270.

Gervaix, A., West, D., Leoni, L. M., Richman, D. D., Wong Staal, F., and Corbeil, J. (1997). A new reporter cell line to monitor HIV infection and drug susceptibility *in vitro*. *Proc. Natl. Acad. Sci. USA* **94,** 4653–4658.

Goff, S. (2001). Retroviridae: The retroviruses and their replication. *In* "Fields Virology" (D. Knipe and P. Howley, Eds.), 4th ed., pp. 1871–1939. Lippincott Williams & Wilkins, Philadelphia.

Gopalakrishnan, V., Peliska, J. A., and Benkovic, S. J. (1992). Human immunodeficiency virus type 1 reverse transcriptase: Spatial and temporal relationship between the polymerase and RNase H activities. *Proc. Natl. Acad. Sci. USA* **89,** 10763–10767.

Graham, N., Peeters, M., Verbiest, W., Harrigan, R., and Larder, B. (2001). The Virtual Phenotype is an independent predictor of clinical response. *In* "Program and Abstracts of the 8th Conference on Retroviruses and Opportunistic Infections," Chicago, Illinois [Abstract 524].

Gunthard, H. F., Wong, J. K., Ignacio, C. C., Havlir, D. V., and Richman, D. D. (1998). Comparative performance of high-density oligonucleotide sequencing and dideoxynucleotide sequencing of HIV type 1 pol from clinical samples. *AIDS Res. Hum. Retrovir.* **14,** 869–876.

Hachiya, A., Aizawa Matsuoka, S., Tanaka, M., Takahashi, Y., Ida, S., Gatanaga, H., Hirabayashi, Y., Kojima, A., Tatsumi, M., and Oka, S. (2001). Rapid and simple phenotypic assay for drug susceptibility of human immunodeficiency virus type 1 using CCR5-expressing HeLa/CD4(+) cell clone 1-10 (MAGIC-5). *Antimicrob. Agents Chemother.* **45,** 495–501.

Hanna, G. J., Johnson, V. A., Kuritzkes, D. R., Richman, D. D., Martinez-Picado, J., Sutton, L., Hazelwood, J. D., and D'Aquila, R. T. (2000). Comparison of sequencing by hybridization and cycle sequencing for genotyping of human immunodeficiency virus type 1 reverse transcriptase. *J. Clin. Microbiol.* **38,** 2715–2721.

Harrigan, P. R., Bloor, S., and Larder, B. A. (1998). Relative replicative fitness of zidovudine-resistant human immunodeficiency virus type 1 isolates *in vitro*. *J. Virol.* **72,** 3773–3778.

Harrigan, P. R., Hertogs, K., Verbiest, W., Pauwels, R., Larder, B., Kemp, S., Bloor, S., Yip, B., Hogg, R., Alexander, C., *et al.* (1999). Baseline HIV drug resistance profile predicts response to ritonavir–saquinavir protease inhibitor therapy in a community setting. *AIDS* **13,** 1863–1871.

Harrigan, P. R., Montaner, J. S., Wegner, S. A., Verbiest, W., Miller, V., Wood, R., and Larder, B. A. (2001). World-wide variation in HIV-1 phenotypic susceptibility in untreated individuals: Biologically relevant values for resistance testing. *AIDS* **15,** 1671–1677.

Harris, D., Lee, R., Misra, H. S., Pandey, P. K., and Pandey, V. N. (1998). The p51 subunit of human immunodeficiency virus type 1 reverse transcriptase is essential in loading the p66 subunit on the template primer. *Biochemistry* **37,** 5903–5908.

Haubrich, R., Keiser, P., Kemper, C., *et al.* (2001). *In* "First IAS Conference on HIV Pathogenesis and Treatment," Buenos Aires, Argentina [Abstract 127].

Havlir, D. V., Hellmann, N. S., Petropoulos, C. J., Whitcomb, J. M., Collier, A. C., Hirsch, M. S., Tebas, P., Sommadossi, J. P., and Richman, D. D. (2000). Drug susceptibility in HIV infection after viral rebound in patients receiving indinavir-containing regimens. *JAMA* **283,** 229–234.

Hazuda, D. J., Felock, P., Witmer, M., Wolfe, A., Stillmock, K., Grobler, J. A., Espeseth, A., Gabryelski, L., Schleif, W., Blau, C., *et al.* (2000). Inhibitors of strand transfer that prevent integration and inhibit HIV- 1 replication in cells. *Science* **287,** 646–650.

Hecht, F. M., Grant, R. M., Petropoulos, C. J., Dillon, B., Chesney, M. A., Tian, H., Hellmann, N. S., Bandrapalli, N. I., Digilio, L., Branson, B., *et al.* (1998). Sexual transmission of an HIV-1 variant resistant to multiple reverse-transcriptase and protease inhibitors. *N. Engl. J. Med.* **339,** 307–311 (1998).

Hertogs, K., de Bethune, M. P., Miller, V., Ivens, T., Schel, P., Van Cauwenberge, A., Van Den Eynde, C., Van Gerwen, V., Azijn, H., Van Houtte, M., *et al.* (1998). A rapid method for simultaneous detection of phenotypic resistance to inhibitors of protease and reverse transcriptase in recombinant human immunodeficiency virus type 1 isolates from patients treated with antiretroviral drugs. *Antimicrob. Agents Chemother.* **42,** 269–276.

Hertogs, K., Bloor, S., De Vroey, V., van Den Eynde, C., Dehertogh, P., van Cauwenberge, A., Sturmer, M., Alcorn, T., Wegner, S., van Houtte, M., *et al.* (2000a). A novel human immunodeficiency virus type 1 reverse transcriptase mutational pattern confers phenotypic lamivudine resistance in the absence of mutation 184V. *Antimicrob. Agents Chemother.* **44,** 568–573.

Hertogs, K., Bloor, S., Kemp, S. D., Van den Eynde, C., Alcorn, T. M., Pauwels, R., Van Houtte, M., Staszewski, S., Miller, V., and Larder, B. A. (2000b). Phenotypic and genotypic analysis of clinical HIV-1 isolates reveals extensive protease inhibitor cross-resistance: A survey of over 6000 samples. *AIDS* **14,** 1203–1210.

Hirsch, M., Kaplan, J., and D'Aquila, R. (1996). Antiviral agents. *In* "Fields' Virology" (B. Fields, D. Knipe, P. Howley, *et al.*, Eds.), 3rd ed., pp. 431–466. Lippincott-Raven, Philadelphia.

Hirsch, M. S., Conway, B., D'Aquila, R. T., Johnson, V. A., Brun Vezinet, F., Clotet, B., Demeter, L. M., Hammer, S. M., Jacobsen, D. M., Kuritzkes, D. R., *et al.* (1998). Antiretroviral drug resistance testing in adults with HIV infection: Implications for clinical management. International AIDS Society—*USA Panel. JAMA* **279,** 1984–1991.

Hirsch, M. S., Brun Vezinet, F., D'Aquila, R. T., Hammer, S. M., Johnson, V. A., Kuritzkes, D. R., Loveday, C., Mellors, J. W., Clotet, B., Conway, B., *et al.* (2000). Antiretroviral drug resistance testing in adult HIV-1 infection: Recommendations of an International AIDS Society—USA Panel. *JAMA* **283,** 2417–2426.

Hizi, A., Tal, R., Shaharabany, M., and Loya, S. (1991). Catalytic properties of the reverse transcriptases of human immunodeficiency viruses type 1 and type 2. *J. Biol. Chem.* **266,** 6230–6239.

Hsieh, J. C., Zinnen, S., and Modrich, P. (1993). Kinetic mechanism of the DNA-dependent DNA polymerase activity of human immunodeficiency virus reverse transcriptase. *J. Biol. Chem.* **268,** 24607–24613.

Hsu, A., Granneman, G., Kempf, D., *et al.* (2000). The C_{trough} inhibitory quotient predicts virologic response to ABT-378/ritonavir (ABT-378/r) therapy in treatment-experienced patients. *AIDS* **14**(Suppl. 4), S12 [Abstract PL9.4].

Huang, H., Chopra, R., Verdine, G. L., and Harrison, S. C. (1998). Structure of a covalently trapped catalytic complex of HIV-1 reverse transcriptase: Implications for drug resistance. *Science* **282,** 1669–1675.

Jacobo-Molina, A., Ding, J., Nanni, R. G., Clark, A. D., Jr., Lu, X., Tantillo, C., Williams, R. L., Kamer, G., Ferris, A. L., Clark, P., *et al.* (1993). Crystal structure of human immunodeficiency virus type 1 reverse transcriptase complexed with double-stranded DNA at 3.0 Å resolution shows bent DNA. *Proc. Natl. Acad. Sci. USA* **90,** 6320–6324.

Jacques, P. S., Wohrl, B. M., Howard, K. J., and Le Grice, S. F. (1994). Modulation of HIV-1 reverse transcriptase function in "selectively deleted" p66/p51 heterodimers. *J. Biol. Chem.* **269,** 1388–1393.

Japour, A. J., Mayers, D. L., Johnson, V. A., Kuritzkes, D. R., Beckett, L. A., Arduino, J. M., Lane, J., Black, R. J., Reichelderfer, P. S., D'Aquila, R. T., *et al.* (1993). Standardized peripheral blood mononuclear cell culture assay for determination of drug susceptibilities of clinical human immunodeficiency virus type 1 isolates. The RV-43 Study Group, the AIDS Clinical Trials Group Virology Committee Resistance Working Group. *Antimicrob. Agents Chemother.* **37,** 1095–1101.

Jarmy, G., Heinkelein, M., Weissbrich, B., Jassoy, C., and Rethwilm, A. (2001). Phenotypic analysis of the sensitivity of HIV-1 to inhibitors of the reverse transcriptase, protease, and integrase using a self-inactivating virus vector system. *J. Med. Virol.* **64,** 223–231.

Jellinger, R. M., Shafer, R. W., and Merigan, T. C. (1997). A novel approach to assessing the drug susceptibility and replication of human immunodefficiency virus type 1 isolates. *J. Infect. Dis.* **175,** 561–566.

Jiang, S., Lin, K., Strick, N., and Neurath, A. R. (1993). HIV-1 inhibition by a peptide. *Nature* **365,** 113.

Johnson, V. A., Merrill, D. P., Videler, J. A., Chou, T. C., Byington, R. E., Eron, J. J., D'Aquila, R. T., and Hirsch, M. S. (1991). Two-drug combinations of zidovudine, didanosine, and recombinant interferon-alpha A inhibit replication of zidovudine-resistant human immunodeficiency virus type 1 synergistically *in vitro*. *J. Infect. Dis.* **164,** 646–655.

Johnson, V. A., Petropoulos, C. J., Woods, C. R., Hazelwood, J. D., Parkin, N. T., Hamilton, C. D., and Fiscus, S. A. (2001). Vertical transmission of multidrug-resistant human immunodeficiency virus type 1 (HIV-1) and continued evolution of drug resistance in an HIV-1-infected infant. *J. Infect. Dis.* **183,** 1688–1693.

Jonckheere, H., Anne, J., and De Clercq, E. (2000). The HIV-1 reverse transcription (RT) process as target for RT inhibitors. *Med. Res. Rev.* **20,** 129–154.

Kamer, G., and Argos, P. (1984). Primary structural comparison of RNA-dependent polymerases from plant, animal and bacterial viruses. *Nucleic Acids Res.* **12,** 7269–7282.

Kellam, P., and Larder, B. A. (1994). Recombinant virus assay: A rapid, phenotypic assay for assessment of drug susceptibility of human immunodeficiency virus type 1 isolates. *Antimicrob. Agents Chemother.* **38,** 23–30.

Kempf, D., Hsu, A., Jiang, P., *et al.* (2001). Response to ritonavir (RTV) intensification in indinavir (IDV) recipients is highly correlated with virtual inhibitory quotient. *In* "Program and Abstracts of the 8th Conference on Retroviruses and Opportunistic Infections," Chicago, Illinois [Abstract 523].

Kijak, G. H., Rubio, A. E., Quarleri, J. F., and Salomon, H. (2001). HIV Type 1 genetic diversity is a major obstacle for antiretroviral drug resistance hybridization-based assays. *AIDS Res. Hum. Retrovir.* **17,** 1415–1421.

Kilby, J. M., Hopkins, S., Venetta, T. M., DiMassimo, B., Cloud, G. A., Lee, J. Y., Alldredge, L., Hunter, E., Lambert, D., Bolognesi, D., *et al.* (1998). Potent suppression of HIV-1 replication in humans by T-20, a peptide inhibitor of gp41-mediated virus entry. *Nature Med.* **4,** 1302–1307.

Kim, V. N., Mitrophanous, K., Kingsman, S. M., and Kingsman, A. J. (1998). Minimal requirement for a lentivirus vector based on human immunodeficiency virus type 1. *J. Virol.* **72,** 811–816.

Klabe, R. M., Bacheler, L. T., Ala, P. J., Erickson-Viitanen, S., and Meek, J. L. (1998). Resistance to HIV protease inhibitors: A comparison of enzyme inhibition and antiviral potency. *Biochemistry* **37,** 8735–8742.

Klarmann, G. J., Schauber, C. A., and Preston, B. D. (1993). Template-directed pausing of DNA synthesis by HIV-1 reverse transcriptase during polymerization of HIV-1 sequences *in vitro*. *J. Biol. Chem.* **268,** 9793–9802.

Koch, N., Yahi, N., Colson, P., Fantini, J., and Tamalet, C. (1999). Genetic polymorphism near HIV-1 reverse transcriptase resistance-associated codons is a major obstacle for the line probe assay as an alternative method to sequence analysis. *J. Virol. Meth.* **80,** 25–31.

Kohlstaedt, L. A., Wang, J., Friedman, J. M., Rice, P. A., and Steitz, T. A. (1992). Crystal structure at 3.5 Å resolution of HIV-1 reverse transcriptase complexed with an inhibitor. *Science* **256,** 1783–1790.

Kosalaraska, P., Kavlick, M., V., M., Le, R., and Mitsuya, H. (1999). Comparative fitness of multi-dideoxinucleoside-resistant HIV-1 in an *in vitro* competitive HIV-1 replication assay. *J. Virol.* **73,** 5356–5365.

Kozal, M. J., Shah, N., Shen, N., Yang, R., Fucini, R., Merigan, T. C., Richman, D. D., Morris, D., Hubbell, E., Chee, M., *et al.* (1996). Extensive polymorphisms observed in HIV-1 clade B protease gene using high-density oligonucleotide arrays. *Nature Med.* **2,** 753–759.

Kusumi, K., Conway, B., Cunningham, S., Berson, A., Evans, C., Iversen, A. K., Colvin, D., Gallo, M. V., Coutre, S., Shpaer, E. G., *et al.* (1992). Human immunodeficiency virus type 1 envelope gene structure and diversity *in vivo* and after cocultivation *in vitro*. *J. Virol.* **66,** 875–885.

LaBranche, C. C., Galasso, G., Moore, J. P., Bolognesi, D. P., Hirsch, M. S., and Hammer, S. M. (2001). HIV fusion and its inhibition. *Antivir. Res.* **50,** 95–115.

Lacey, S. F., Reardon, J. E., Furfine, E. S., Kunkel, T. A., Bebenek, K., Eckert, K. A., Kemp, S. D., and Larder, B. A. (1992). Biochemical studies on the reverse transcriptase and RNase H activities from human immunodeficiency virus strains resistant to 3′-azido-3′-deoxythymidine. *J. Biol. Chem.* **267,** 15789–15794.

Land, S., Terloar, G., McPhee, D., Birch, C., Doherty, R., Cooper, D., and Gust, I. (1990). Decreased *in vitro* susceptibility to zidovudine of HIV isolates obtained from patients with AIDS. *J. Infect. Dis.* **161,** 326–329.

Lanier, E. R., Hellmann, N., Scott, J., *et al.* (2001). Determination of a clinically relevant phenotypic "cut-off" for abacavir using the PhenoSense assay. *In* "Program and Abstracts of the 8th Conference on Retroviruses and Opportunistic Infections," Chicago, Illinois [Abstract 254].

Lapatto, R., Blundell, T., Hemmings, A., Overington, J., Wilderspin, A., Wood, S., Merson, J. R., Whittle, P. J., Danley, D. E., Geoghegan, K. F., *et al.* (1989). X-ray analysis of HIV-1 proteinase at 2.7 Å resolution confirms structural homology among retroviral enzymes. *Nature* **342,** 299–302.

Larder, B. A., Purifoy, D. J., Powell, K. L., and Darby, G. (1987). Site-specific mutagenesis of AIDS virus reverse transcriptase. *Nature* **327,** 716–717.

Larder, B. A., Darby, G., and Richman, D. D. (1989). HIV with reduced sensitivity to zidovudine (AZT) isolated during prolonged therapy. *Science* **243,** 1731–1734.

Larder, B. A., Chesebro, B., and Richman, D. D. (1990). Susceptibilities of zidovudine-susceptible and -resistant human immunodeficiency virus isolates to antiviral agents determined by using a quantitative plaque reduction assay. *Antimicrob. Agents Chemother.* **34,** 436–441.

Larder, B. A., Bloor, S., Kemp, S. D., Hertogs, K., Desmet, R. L., Miller, V., Sturmer, M., Staszewski, S., Ren, J., Stammers, D. K., *et al.* (1999). A family of insertion mutations between codons 67 and 70 of human immunodeficiency virus type 1 reverse transcriptase confer multinucleoside analog resistance. *Antimicrob. Agents Chemother.* **43,** 1961–1967.

Larder, B. A., Hertogs, K., Bloor, S., van den Eynde, C. H., DeCian, W., Wang, Y., Freimuth, W. W., and Tarpley, G. (2000a). Tipranavir inhibits broadly protease inhibitor-resistant HIV-1 clinical samples. *AIDS* **14,** 1943–1948.

Larder, B. A., Kemp, S. D., and Hertogs, K. (2000b). Quantitative prediction of HIV-1 phenotypic drug resistance from genotpes: The virtual phenotype (VirtualPhenotype). *Antivir. Ther.* **5**(Suppl. 3), 49 [Abstract 63].

Lennerstrand, J., Hertogs, K., Stammers, D. K., and Larder, B. A. (2001). Correlation between viral resistance to zidovudine and resistance at the reverse transcriptase level for a panel of human immunodeficiency virus type 1 mutants. *J. Virol.* **75,** 7202–7205.

Levy, J. (1994). "HIV and the Pathogenesis of AIDS." ASM Press, Washington. D.C.

Little, S. J., Daar, E. S., D'Aquila, R. T., Keiser, P. H., Connick, E., Whitcomb, J. M., Hellmann, N. S., Petropoulos, C. J., Sutton, L., Pitt, J. A., *et al.* (1999). Reduced antiretroviral drug susceptibility among patients with primary HIV infection. *JAMA* **282,** 1142–1149.

Lorenzi, P., Opravil, M., Hirschel, B., Chave, J. P., Furrer, H. J., Sax, H., Perneger, T. V., Perrin, L., Kaiser, L., and Yerly, S. (1999). Impact of drug resistance mutations on virologic response to salvage therapy. Swiss HIV Cohort Study. *AIDS* **13,** F17–F21.

Maeda, Y., Venzon, D., and Mitsuya, H. (1998). Altered drug sensitivity, fitness, and evolution of HIV-1 with pol gene mutations conferring multi-dideoxynucleoside resistance. *J. Infect. Dis.* **177,** 1207–1213.

Majumdar, C., Abbotts, J., Broder, S., and Wilson, S. H. (1988). Studies on the mechanism of human immunodeficiency virus reverse transcriptase. Steady-state kinetics, processivity, and polynucleotide inhibition. *J. Biol. Chem.* **263,** 15657–15665.

Mammano, F., Petit, C., and Clavel, F. (1998). Resistance-associated loss of viral fitness in human immunodeficiency virus type 1: Phenotypic analysis of protease and gag coevolution in protease inhibitor-treated patients. *J. Virol.* **72,** 7632–7637.

Martinez Picado, J., Savara, A. V., Sutton, L., and D'Aquila, R. T. (1999). Replicative fitness of protease inhibitor-resistant mutants of human immunodeficiency virus type 1. *J. Virol.* **73,** 3744–3752.

Martinez Picado, J., DePasquale, M. P., Kartsonis, N., Hanna, G. J., Wong, J., Finzi, D., Rosenberg, E., Gunthard, H. F., Sutton, L., Savara, A., *et al.* (2000). Antiretroviral resistance during successful therapy of HIV type 1 infection. *Proc. Natl. Acad. Sci. USA* **97,** 10948–10953.

Maschera, B., Furfine, E., and Blair, E. D. (1995). Analysis of resistance to human immunodeficiency virus type 1 protease inhibitors by using matched bacterial expression and proviral infection vectors. *J. Virol.* **69,** 5431–5436.

Maschera, B., Darby, G., Palu, G., Wright, L. L., Tisdale, M., Myers, R., Blair, E. D., and Furfine, E. S. (1996). Human immunodeficiency virus. Mutations in the viral protease that confer resistance to saquinavir increase the dissociation rate constant of the protease-saquinavir complex. *J. Biol. Chem.* **271,** 33231–33235.

Mayers, D. L., McCutchan, F. E., Sanders Buell, E. E., Merritt, L. I., Dilworth, S., Fowler, A. K., Marks, C. A., Ruiz, N. M., Richman, D. D., Roberts, C. R., *et al.* (1992). Characterization of HIV isolates arising after prolonged zidovudine therapy. *J. Acquir. Immune Defic. Syndr.* **5,** 749–759.

Mellors, J., Rinaldo, C., Gupta, P., White, R., Todd, J., and Kingsley, L. (1996). Prognosis in HIV-1 infection predicted by the quantity of virus in plasma. *Science* **272,** 1167–1170.

Melnick, L., Yang, S. S., Rossi, R., Zepp, C., and Heefner, D. (1998). An *Escherichia coli* expression assay and screen for human immunodeficiency virus protease variants with decreased susceptibility to indinavir. *Antimicrob. Agents Chemother.* **42,** 3256–3265.

Meyer, P. R., Matsuura, S. E., Mian, A. M., So, A. G., and Scott, W. A. (1999). A mechanism of AZT resistance: An increase in nucleotide-dependent primer unblocking by mutant HIV-1 reverse transcriptase. *Mol. Cells* **4,** 35–43.

Meynard, J. L., Vray, M., Morand-Joubert, L., *et al.* (2000). Impact of treatment guided by phenotypic or genotypic resistance tests on the response to antiretroviral therapy: A randomized trial (NARVAL, ANRS 088). *Antivir. Ther.* **5**(Suppl. 3), 67–68 [Abstract 85].

Miller, M., Schneider, J., Sathyanarayana, B. K., Toth, M. V., Marshall, G. R., Clawson, L., Selk, L., Kent, S. B., and Wlodawer, A. (1989). Structure of complex of synthetic HIV-1 protease with a substrate-based inhibitor at 2.3 Å resolution. *Science* **246**(4934), 1149–52.

Miller, V., de Bethune, M. P., Kober, A., Sturmer, M., Hertogs, K., Pauwels, R., Stoffels, P., and Staszewski, S. (1998a). Patterns of resistance and cross-resistance to human immunodeficiency virus type 1 reverse transcriptase inhibitors in patients treated with the nonnucleoside reverse transcriptase inhibitor loviride. *Antimicrob. Agents Chemother.* **42,** 3123–3129.

Miller, V., Sturmer, M., Staszewski, S., Groschel, B., Hertogs, K., de Bethune, M. P., Pauwels, R., Harrigan, P. R., Bloor, S., Kemp, S. D., *et al.* (1998b). The M184V mutation in HIV-1 reverse transcriptase (RT) conferring lamivudine resistance does not result in broad cross-resistance to nucleoside analogue RT inhibitors. *AIDS* **12,** 705–712.

Miller, V., Cozzi-Lepri, A., Hertogs, K., Gute, P., Larder, B., Bloor, S., Klauke, S., Rabenau, H., Phillips, A., and Staszewski, S. (2000a). HIV drug susceptibility and treatment response to mega-HAART regimen in patients from the Frankfurt HIV cohort. *Antivir. Ther.* **5,** 49–55.

Miller, V., Sabin, C., Hertogs, K., Bloor, S., Martinez Picado, J., D'Aquila, R., Larder, B., Lutz, T., Gute, P., Weidmann, E., *et al.* (2000b). Virological and immunological effects of treatment interruptions in HIV-1 infected patients with treatment failure. *AIDS* **14,** 2857–2867.

Miller, V., Schuurman, R., and Clavel, F. (2001). *Comparison of HIV drug susceptibility (phenotype) results reported by three major laboratories.* Presented at the 5th International Workshop on HIV Drug Resistance and Treatment Strategies [Abstract 169].

Molla, A., Korneyeva, M., Gao, Q., Vasavanonda, S., Schipper, P. J., Mo, H. M., Markowitz, M., Chernyavskiy, T., Niu, P., Lyons, N., *et al.* (1996). Ordered accumulation of mutations in HIV protease confers resistance to ritonavir. *Nature Med.* **2,** 760–766.

Montaner, J., Harrigan, P., Jahnke, N., Raboud, J., Castillo, E., Hogg, R., Yip, B., Harris, M., Montessori, V., and O'Shaughnessy, M. (2001). Multiple drug rescue therapy for HIV-infected individuals with prior virologic failure to multiple regimens. *AIDS* **15,** 61–69.

Naldini, L., and Verma, I. M. (2000). Lentiviral vectors. *Adv. Virus Res.* **2000,** 55599–55609.

Naldini, L., Blomer, U., Gage, F. H., Trono, D., and Verma, I. M. (1996a). Efficient transfer, integration, and sustained long-term expression of the transgene in adult rat brains injected with a lentiviral vector. *Proc. Natl. Acad. Sci. USA* **93,** 11382–11388.

Naldini, L., Blomer, U., Gallay, P., Ory, D., Mulligan, R., Gage, F. H., Verma, I. M., and Trono, D. (1996b). *In vivo* gene delivery and stable transduction of nondividing cells by a lentiviral vector. *Science* **272,** 263–267.

Navia, M. A., Fitzgerald, P. M., McKeever, B. M., Leu, C. T., Heimbach, J. C., Herber, W. K., Sigal, I. S., Darke, P. L., and Springer, J. P. (1989). Three-dimensional structure of aspartyl protease from human immunodeficiency virus HIV-1. *Nature* **337,** 615–620.

O'Brien, W. (2001). *HIV entry. From molecular insights to specific inhibitors.* Presented at First IAS Conference on HIV Pathogenesis and Treatment, Buenos Aires.

Olsen, D. B., Stahlhut, M. W., Rutkowski, C. A., Schock, H. B., vanOlden, A. L., and Kuo, L. C. (1999). Non-active site changes elicit broad-based cross-resistance of the HIV-1 protease to inhibitors. *J. Biol. Chem.* **274,** 23699–23701.

Page, K. A., Landau, N. R., and Littman, D. R. (1990). Construction and use of a human immunodeficiency virus vector for analysis of virus infectivity. *J. Virol.* **64,** 5270–5276.

Parkin, N. T., Lie, Y. S., Hellmann, N., Markowitz, M., Bonhoeffer, S., Ho, D. D., and Petropoulos, C. J. (1999). Phenotypic changes in drug susceptibility associated with failure of human immunodeficiency virus type 1 (HIV-1) triple combination therapy. *J. Infect. Dis.* **180,** 865–870.

Parkin, N. T., Deeks, S. G., Wrin, M. T., Yap, J., Grant, R. M., Lee, K. H., Heeren, D., Hellmanna, N. S., and Petropoulos, C. J. (2000). Loss of antiretroviral drug susceptibility at low viral load during early virological failure in treatment-experienced patients. *AIDS* **14,** 2877–2887.

Partaledis, J. A., Yamaguchi, K., Tisdale, M., Blair, E. E., Falcione, C., Maschera, B., Myers, R. E., Pazhanisamy, S., Futer, O., Cullinan, A. B., *et al.* (1995). *In vitro* selection and characterization of human immunodeficiency virus type 1 (HIV-1) isolates with reduced sensitivity to hydroxyethylamino sulfonamide inhibitors of HIV-1 aspartyl protease. *J. Virol.* **69,** 5228–5235.

Patel, P. H., Jacobo-Molina, A., Ding, J., Tantillo, C., Clark, A. D., Jr., Raag, R., Nanni, R. G., Hughes, S. H., and Arnold, E. (1995). Insights into DNA polymerization mechanisms from structure and function analysis of HIV-1 reverse transcriptase. *Biochemistry* **34,** 5351–5363.

Pazhanisamy, S., Stuver, C. M., Cullinan, A. B., Margolin, N., Rao, B. G., and Livingston, D. J. (1996). Kinetic characterization of human immunodeficiency virus type-1 protease-resistant variants. *J. Biol. Chem.* **271,** 17979–17985.

Perez-Elias, M. J., Lanier, R., Munoz, V., Garcia-Arata, I., Casado, J. L., Marti-Belda, P., Moreno, A., Dronda, F., Antela, A., Marco, S., *et al.* (2000). Phenotypic testing predicts virological response in successive protease inhibitor-based regimens. *AIDS* **14,** F95–F101.

Petropoulos, C. J., Parkin, N. T., Limoli, K. L., Lie, Y. S., Wrin, T., Huang, W., Tian, H., Smith, D., Winslow, G. A., Capon, D. J., *et al.* (2000). A novel phenotypic drug susceptibility assay for human immunodeficiency virus type 1. *Antimicrob. Agents Chemother.* **44,** 920–928.

Pettit, S. C., Simsic, J., Loeb, D. D., Everitt, L., Hutchison, C. A., 3rd, and Swanstrom, R. (1991). Analysis of retroviral protease cleavage sites reveals two types of cleavage sites and the structural requirements of the P1 amino acid. *J. Biol. Chem.* **266,** 14539–14547.

Piketty, C., Race, E., Castiel, P., Belec, L., Peytavin, G., Si-Mohamed, A., Gonzalez-Canali, G., Weiss, L., Clavel, F., and Kazatchkine, M. D. (1999). Efficacy of a five-drug combination including ritonavir, saquinavir and efavirenz in patients who failed on a conventional triple-drug regimen: Phenotypic resistance to protease inhibitors predicts outcome of therapy. *AIDS* **13,** F71–F77.

Piketty, C., Race, E., Castiel, P., Belec, L., Peytavin, G., Si Mohamed, A., Gonzalez Canali, G., Weiss, L., Clavel, F., and Kazatchkine, M. D. (2000). Phenotypic resistance to protease inhibitors in patients who fail on highly active antiretroviral therapy predicts the outcome at 48 weeks of a five-drug combination including ritonavir, saquinavir and efavirenz. *AIDS* **14,** 626–628.

Pirounaki, M., Heyden, N. A., Arens, M., and Ratner, L. (2000). Rapid phenotypic drug susceptibility assay for HIV-1 with a CCR5 expressing indicator cell line. *J. Virol. Meth.* **85,** 151–161.

Pommier, Y., Marchand, C., and Neamati, N. (2000). Retroviral integrase inhibitors year 2000: Update and perspectives. *Antivir. Res.* **47,** 139–148.

Puchhammer-Stockl, E., Schmied, B., Mandl, C. W., Vetter, N., and Heinz, F. X. (1999). Comparison of line probe assay (LIPA) and sequence analysis for detection of HIV-1 drug resistance. *J. Med. Virol.* **57,** 283–289.

Qari, S., Respess, R., and Weinstock, H. (2000). A comparative analysis of Virco Antivirogram and ViroLogic PhenoSense phenotypic assays for drug suscebility of HIV-1. *Antivir. Ther.* **5**(Suppl. 3), 49.

Reiser, J., Harmison, G., Kluepfel Stahl, S., Brady, R. O., Karlsson, S., and Schubert, M. (1996). Transduction of nondividing cells using pseudotyped defective high-titer HIV type 1 particles. *Proc. Natl. Acad. Sci. USA* **93,** 15266–15271.

Ren, J., Esnouf, R., Garman, E., Somers, D., Ross, C., Kirby, I., Keeling, J., Darby, G., Jones, Y., Stuart, D., *et al.* (1995). High resolution structures of HIV-1 RT from four RT-inhibitor complexes. *Nature Struct. Biol.* **2,** 293–302.

Richman, D. D., Havlir, D., Corbeil, J., Looney, D., Ignacio, C., Spector, S. A., Sullivan, J., Cheeseman, S., Barringer, K., Pauletti, D., *et al.* (1994). Nevirapine resistance mutations of human immunodeficiency virus type 1 selected during therapy. *J. Virol.* **68,** 1660–1666.

Rimsky, L. T., Shugars, D. C., and Matthews, T. J. (1998). Determinants of human immunodeficiency virus type 1 resistance to gp41-derived inhibitory peptides. *J. Virol.* **72,** 986–993.

Ripka, A. S., and Rich, D. H. (1998). Peptidomimetic design. *Curr. Opin. Chem. Biol.* **2,** 441–452.

Rodgers, D. W., Gamblin, S. J., Harris, B. A., Ray, S., Culp, J. S., Hellmig, B., Woolf, D. J., Debouck, C., and Harrison, S. C. (1995). The structure of unliganded reverse transcriptase from the human immunodeficiency virus type 1. *Proc. Natl. Acad. Sci. USA* **92,** 1222–1226.

Root, M. J., Kay, M. S., and Kim, P. S. (2001). Protein design of an HIV-1 entry inhibitor. *Science* **291,** 884–888.

Schinazi, R. F., Larder, B. A., and Mellors, J. W. (1999). Resistance table: Mutations in retroviral genes associated with drug resistance: 1999–2000 update. *Int. Antivir. News* **7,** 46–69.

Schock, H. B., Garsky, V. M., and Kuo, L. C. (1996). Mutational anatomy of an HIV-1 protease variant conferring cross-resistance to protease inhibitors in clinical trials. Compensatory modulations of binding and activity. *J. Biol. Chem.* **271,** 31957–31963.

Schols, D., Este, J. A., Henson, G., and De Clercq, E. (1997). Bicyclams, a class of potent anti-HIV agents, are targeted at the HIV coreceptor fusin/CXCR-4. *Antivir. Res.* **35,** 147–156.

Schuurman, R., Brambilla, D. J., de Groot, T., and Boucher, C. (1999a). Second worldwide evaluation of HIV-1 drug resistance genotyping quality, using the ENVA-2 panel. *In* "Program and Abstracts of the 39th Interscience Conference on Antimicrobial Agents and Chemotherapy," San Francisco [Abstract 1168].

Schuurman, R., Demeter, L., Reichelderfer, P., Tijnagel, J., de Groot, T., and Boucher, C. (1999b). Worldwide evaluation of DNA sequencing approaches for identification of drug resistance mutations in the human immunodeficiency virus type 1 reverse transcriptase. *J. Clin. Microbiol.* **37,** 2291–2296.

Seelmeier, S., Schmidt, H., Turk, V., and von der Helm, K. (1988). Human immunodeficiency virus has an aspartic-type protease that can be inhibited by pepstatin A. *Proc. Natl. Acad. Sci. USA* **85,** 6612–6616.

Shafer, R. W., and Deresinski, S. C. (2000). Human immunodeficiency virus on the web: A guided tour. *Clin. Infect. Dis.* **31,** 568–577.

Shafer, R. W., Kozal, M. J., Katzenstein, D. A., Lipil, W. H., Johnstone, I. F., and Merigan, T. C. (1993). Zidovudine susceptibility testing of human immunodeficiency virus type 1 (HIV) clinical isolates. *J. Virol. Meth.* **41,** 297–310.

Shafer, R. W., Hertogs, K., Zolopa, A. R., Warford, A., Bloor, S., Betts, B. J., Merigan, T. C., Harrigan, R., and Larder, B. A. (2001). High degree of interlaboratory reproducibility of human immunodeficiency virus type 1 protease and reverse transcriptase sequencing of plasma samples from heavily treated patients. *J. Clin. Microbiol.* **39,** 1522–1529.

Shi, C., and Mellors, J. W. (1997). A recombinant retroviral system for rapid *in vivo* analysis of human immunodeficiency virus type 1 susceptibility to reverse transcriptase inhibitors. *Antimicrob. Agents Chemother.* **41,** 2781–2785.

Skalka A., Ed. (1999). "Retroviral Integration." Advances in Virus Research, Vol. 52. Academic Press, San Diego, California.

Smerdon, S. J., Jager, J., Wang, J., Kohlstaedt, L. A., Chirino, A. J., Friedman, J. M., Rice, P. A., and Steitz, T. A. (1994). Structure of the binding site for nonnucleoside inhibitors of the reverse transcriptase of human immunodeficiency virus type 1. *Proc. Natl. Acad. Sci. USA* **91,** 3911–3915.

Speck, R. F., Wehrly, K., Platt, E. J., Atchison, R. E., Charo, I. F., Kabat, D., Chesebro, B., and Goldsmith, M. A. (1997). Selective employment of chemokine receptors as human immunodeficiency virus type 1 coreceptors determined by individual amino acids within the envelope V3 loop. *J. Virol.* **71,** 7136–7139.

St Clair, M. H., Richards, C. A., Spector, T., Weinhold, K. J., Miller, W. H., Langlois, A. J., and Furman, P. A. (1987). 3′-Azido-3′-deoxythymidine triphosphate as an inhibitor and substrate of purified human immunodeficiency virus reverse transcriptase. *Antimicrob. Agents Chemother.* **31,** 1972–1977.

Stuyver, L., Wyseur, A., Rombout, A., Louwagie, J., Scarcez, T., Verhofstede, C., Rimland, D., Schinazi, R. F., and Rossau, R. (1997). Line probe assay for rapid detection of drug-selected mutations in the human immunodeficiency virus type 1 reverse transcriptase gene. *Antimicrob. Agents Chemother.* **41,** 284–291.

Swanstrom, R., and Eron, J. (2000). Human immunodeficiency virus type-1 protease inhibitors: Therapeutic successes and failures, suppression and resistance. *Pharmacol. Ther.* **86,** 145–170.

Swanstrom, R., and Wills, J. (1997). Synthesis, assembly and processing of viral poteins.

In "Retroviruses" (J. Coffin, S. Huges, and H. Varmus, Eds.), pp. 263–334. Cold Spring Harbor Press, Cold Spring Harbor, New York.

Tantillo, C., Ding, J., Jacobo-Molina, A., Nanni, R. G., Boyer, P. L., Hughes, S. H., Pauwels, R., Andries, K., Janssen, P. A., and Arnold, E. (1994). Locations of anti-AIDS drug binding sites and resistance mutations in the three-dimensional structure of HIV-1 reverse transcriptase. Implications for mechanisms of drug inhibition and resistance. *J. Mol. Biol.* **243,** 369–387.

Tasara, T., Amacker, M., and Hubscher, U. (1999). Intramolecular chimeras of the p51 subunit between HIV-1 and FIV reverse transcriptases suggest a stabilizing function for the p66 subunit in the heterodimeric enzyme. *Biochemistry* **38,** 1633–1642.

Tebas, P., Patick, A. K., Kane, E. M., Klebert, M. K., Simpson, J. H., Erice, A., Powderly, W. G., and Henry, K. (1999). Virologic responses to a ritonavir–saquinavir-containing regimen in patients who had previously failed nelfinavir. *AIDS* **13,** F23–F28.

Tisdale, M., Myers, R. E., Maschera, B., Parry, N. R., Oliver, N. M., and Blair, E. D. (1995). Cross-resistance analysis of human immunodeficiency virus type 1 variants individually selected for resistance to five different protease inhibitors. *Antimicrob. Agents Chemother.* **39,** 1704–1710.

Tural, C., Ruiz, L., Holtzer, C., *et al.* (2001). Utility of HIV genotyping and clinical expert advice—The Havana Trial. *In* "Program and Abstracts of the 8th Conference on Retroviruses and Opportunistic Infections," Chicago, Illinois [Abstract 434].

U.S. Deptartment of Health and Human Services. 2001. Guidelines for the use of antiretroviral agents in HIV-infected adults and adolescents. Available at http://hivatis.org/trtgdlns.html [August 13, 2001].

Vahey, M., Nau, M. E., Barrick, S., Cooley, J. D., Sawyer, R., Sleeker, A. A., Vickerman, P., Bloor, S., Larder, B., Michael, N. L., *et al.* (1999). Performance of the Affymetrix GeneChip HIV PRT 440 platform for antiretroviral drug resistance genotyping of human immunodeficiency virus type 1 clades and viral isolates with length polymorphisms. *J. Clin. Microbiol.* **37,** 2533–2537.

Van Vaerenbergh, K., Van Laethem, K., Van Wijngaerden, E., Schmit, J. C., Schneider, F., Ruiz, L., Clotet, B., Verhofstede, C., Van Wanzeele, F., Muyldermans, G., *et al.* (2000). Baseline HIV type 1 genotypic resistance to a newly added nucleoside analog is predictive of virologic failure of the new therapy. *AIDS Res. Hum. Retrovir.* **16,** 529–237.

Vazquez Rosales, G., Garcia Lerma, J. G., Yamamoto, S., Switzer, W. M., Havlir, D., Folks, T. M., Richman, D. D., and Heneine, W. (1999). Rapid screening of phenotypic resistance to nevirapine by direct analysis of HIV type 1 reverse transcriptase activity in plasma. *AIDS Res. Hum. Retrovir.* **15,** 1191–1200.

Wainberg, M. A., Tremblay, M., Rooke, R., Blain, N., Soudeyns, H., Parniak, M. A., Yao, X. J., Li, X. G., Fanning, M., Montaner, J. S., *et al.* (1990). Characterization of reverse transcriptase activity and susceptibility to other nucleosides of AZT-resistant variants of HIV-1. Results from the Canadian AZT Multicentre Study. *Ann. N.Y. Acad. Sci.* **616,** 346–355.

Walter, H., Schmidt, B., Korn, K., Vandamme, A. M., Harrer, T., and Uberla, K. (1999). Rapid, phenotypic HIV-1 drug sensitivity assay for protease and reverse transcriptase inhibitors. *J. Clin. Virol.* **13,** 71–80.

Wang, J., Smerdon, S. J., Jager, J., Kohlstaedt, L. A., Rice, P. A., Friedman, J. M., and Steitz, T. A. (1994). Structural basis of asymmetry in the human immunodeficiency virus type 1 reverse transcriptase heterodimer. *Proc. Natl. Acad. Sci. USA* **91,** 7242–7246.

Wegner, S. A., Brodine, S. K., Mascola, J. R., Tasker, S. A., Shaffer, R. A., Starkey, M. J., Barile, A., Martin, G. J., Aronson, N., Emmons, W. W., *et al.* (2000). Prevalence of genotypic and phenotypic resistance to anti-retroviral drugs in a cohort of therapy-naive HIV-1 infected US military personnel. *AIDS* **14,** 1009–1015.

Wei, X., Ghosh, S., Taylor, M., Johnson, V., Emini, E., Deutsch, P., Lifson, J., Bonhoeffer, S., Nowak, M., Hahn, B., *et al.* (1995). Viral dynamics in human immunodeficiency virus type 1 infection. *Nature* **373,** 117–122.

Weidle, P. J., Kityo, C. M., Mugyenyi, P., Downing, R., Kebba, A., Pieniazek, D., Respess, R., Hertogs, K., De Vroey, V., Dehertogh, P., *et al.* (2001). Resistance to antiretroviral therapy among patients in Uganda. *J. Acquir. Immune. Defic. Syndr.* **26,** 495–500.

Weinstock, H., Respess, R., Heneine, W., Petropoulos, C. J., Hellmann, N. S., Luo, C. C., Pau, C. P., Woods, T., Gwinn, M., and Kaplan, J. (2000). Prevalence of mutations associated with reduced antiretroviral drug susceptibility among human immunodeficiency virus type 1 seroconverters in the United States, 1993–1998. *J. Infect. Dis.* **182,** 330–333.

Wild, C., Shugars, D., Greenwell, T., McDanal, C., and Matthews, T. (1994). Peptides corresponding to a predictive α-helical domain of human immunodeficiency virus type 1 gp41 are potent inhibitors of virus infection. *Proc. Natl. Acad. Sci. USA* **91,** 9770–9774.

Wilk, T., and Fuller, S. D. (1999). Towards the structure of the human immunodeficiency virus: Divide and conquer. *Curr. Opin. Struct. Biol.* **9,** 231–243.

Wilson, J. W., Bean, P., Robins, T., Graziano, F., and Persing, D. H. (2000). Comparative evaluation of three human immunodeficiency virus genotyping systems: The HIV-GenotypR method, the HIV PRT GeneChip assay, and the HIV-1 RT line probe assay. *J. Clin. Microbiol.* **38,** 3022–3028.

Wlodawer, A., and Erickson, J. W. (1993). Structure-based inhibitors of HIV-1 protease. *Annu. Rev. Biochem.* **62,** 543–585.

Wlodawer, A., Miller, M., Jaskolski, M., Sathyanarayana, B. K., Baldwin, E., Weber, I. T., Selk, L. M., Clawson, L., Schneider, J., and Kent, S. B. (1989). Conserved folding in retroviral proteases: Crystal structure of a synthetic HIV-1 protease. *Science* **245,** 616–621.

Xie, D., Gulnik, S., Gustchina, E., Yu, B., Shao, W., Qoronfleh, W., Nathan, A., and Erickson, J. W. (1999). Drug resistance mutations can effect dimer stability of HIV-1 protease at neutral pH. *Protein Sci.* **8,** 1702–1707.

Zennou, V., Mammano, F., Paulous, S., Mathez, D., and Clavel, F. (1998). Loss of viral fitness associated with multiple Gag and Gag–Pol processing defects in human immunodeficiency virus type 1 variants selected for resistance to protease inhibitors *in vivo*. *J. Virol.* **72,** 3300–3306.

Zhang, Y. M., Imamichi, H., Imamichi, T., Lane, H. C., Falloon, J., Vasudevachari, M. B., and Salzman, N. P. (1997). Drug resistance during indinavir therapy is caused by mutations in the protease gene and in its Gag substrate cleavage sites. *J. Virol.* **71,** 6662–6670.

Ziermann, R., Limoli, K., Das, K., Arnold, E., Petropoulos, C. J., and Parkin, N. T. (2000). A mutation in human immunodeficiency virus type 1 protease, N88S, that causes *in vitro* hypersensitivity to amprenavir. *J. Virol.* **74,** 4414–4419.

Zolopa, A. R., Shafer, R. W., Warford, A., Montoya, J. G., Hsu, P., Katzenstein, D., Merigan, T. C., and Efron, B. (1999). HIV-1 genotypic resistance patterns predict response to saquinavir–ritonavir therapy in patients in whom previous protease inhibitor therapy had failed. *Ann. Intern. Med.* **131,** 813–821.

Zufferey, R., Dull, T., Mandel, R. J., Bukovsky, A., Quiroz, D., Naldini, L., and Trono, D. (1998). Self-inactivating lentivirus vector for safe and efficient *in vivo* gene delivery. *J. Virol.* **72,** 9873–9880.

ADVANCES IN VIRUS RESEARCH, VOL. 58

PERSPECTIVES ON POLYDNAVIRUS ORIGINS AND EVOLUTION

Matthew Turnbull and Bruce Webb

Department of Entomology
University of Kentucky
Lexington, Kentucky 40546-0091

I. Introduction
A. Life Cycle and Virus Transmission
B. Genome Organization
II. Classification of Polydnaviruses as Viruses
A. Rationale for Comparison
B. Consideration of Polydnavirus Origins
III. Ichnoviruses
A. Brief History and Ichnovirus Phylogeny
B. Hosts
C. Genome Organization
D. Gene Families
E. Ichnovirus Functions
IV. Bracoviruses
A. System Characteristics and Origin
B. Bracovirus Life Cycle
C. Genome Organization
D. Gene Families
E. Bracovirus Functions
V. Origin of the Polydnaviridae
A. Evolutionary Model I: A Viral Ancestor
B. Evolutionary Model II: A Wasp-Encoded Episome
C. Possible Future Avenues of Research
VI. Conclusion
References

I. Introduction

A. *Life Cycle and Virus Transmission*

Polydnaviruses exist in obligate mutualisms with some parasitic Hymenoptera. A given polydnavirus (PDV) species replicates only in a single wasp species and all members of that wasp species carry the associated virus. Two genera of PDVs are recognized, the bracoviruses (BVs) and the ichnoviruses (IVs), associated with braconid and ichneumonid wasps, respectively (Stoltz *et al.*, 1995). The PDV genera have similar

0065-3527/02 $35.00

genome organization and replication/transmission pathways, but have no known genetic similarity (Fleming and Summers, 1991; Gruber *et al.*, 1996; Stoltz *et al.*, 1986). Viral gene expression in insects parasitized by PDV-carrying wasps causes physiological alterations that are essential for parasitoid survival and development.

PDV life cycles have been described as having "two arms" (Stoltz, 1993). Virus replication and vertical transmission occur only in the wasp, whereas viral genes disrupting physiology in the parasitized host function only in that "arm" of the life cycle. Although the two PDV genera have similar life cycles and genomic organization, the viruses are morphologically and genetically distinct, suggesting that the genomic similarities result from selection pressures imposed by their unusual life cycles. In this review we first describe the PDV life cycle emphasizing the features shared between genera. We then consider the two genera individually and close by discussing the potential evolutionary forces that may have resulted in their acquiring shared characteristics.

PDVs replicate from proviral DNA in specialized cells of the wasp oviduct (i.e., the calyx cells). Virus replication is first detected in the late pupal stage with virus released from calyx cells by budding (ichnovirus) (Volkoff *et al.*, 1995) or cell lysis (bracovirus) (Stoltz *et al.*, 1976) and accumulating to high concentrations in the oviduct lumen. When the wasp parasitizes its insect host, usually a lepidopteran larva, a quantity of virus is delivered with the wasp egg. The virus enters lepidopteran cells, where a host-specific subset of viral genes is expressed without detectable virus replication (Theilmann and Summers, 1986). Viral gene expression inhibits the host immune responses to the parasite egg, thereby enabling wasp survival (Asgari *et al.*, 1996; Cui *et al.*, 2000; Lavine and Beckage, 1996; Prevost *et al.*, 1990; Strand, 1994). Virus transmission to subsequent generations is ensured by stable integration of proviral DNA in the wasp genome (Fleming and Summers, 1991; Gruber *et al.*, 1996; Savary *et al.*, 1999).

PDV-associated wasps are found in speciose wasp lineages, with estimates of over 30,000 species having PDV associations (Webb, 1998). These PDV-carrying lineages are also ancient, with the fossil record demonstrating their existence over at least 60 million years (Whitfield, 2000). Although the two wasp families are related, and are classified within the superfamily Ichneumonoidea, their common ancestors do not carry PDVs. This suggests that the origins of ichnoviruses and bracoviruses are distinct and that PDVs are paraphyletic. Thus, the ichnovirus–ichneumonid wasp and bracovirus–braconid wasp associations represent independent events linking wasps with their respective viruses. Additionally, although BV-associated braconids form a monophyletic lineage (Whitfield, 1997), ichneumonid phylogenies (Belshaw *et al.*, 1998) suggest that ichnovirus–ichneumonid mutualisms have evolved multiple times or, alternatively, that virus–wasp associations have arisen via horizontal transfer of virus between ichneumonid wasp species.

Horizontal transmission currently is limited to parasitization when virions are delivered with wasp eggs into the parasitized host (Stoltz and Vinson, 1979b). In the parasitized insect, virions enter host cells, where viral genes are expressed. However, PDV replication has never been detected in the parasitized host (Theilmann and Summers, 1986), nor has reinfection of parasitoids been documented. Therefore, infection of cells in the parasitized host is a dead end because it does not result in horizontal or vertical virus transmission. PDV transmission within a wasp species occurs only by transmission of integrated proviral segments from mother to offspring (Stoltz *et al.*, 1986; Fleming and Summers, 1991). In an evolutionary context, what is known of virus transmission and wasp phylogeny suggests that the ~30,000 wasp species carrying PDVs evolved from a small number (3–4) of associations between wasp and virus. The subsequent evolution of the extant number of species is an indication that this association is among the most successful mutualisms known.

B. Genome Organization

PDVs are unique among DNA viruses in having segmented double-stranded genomes. PDV segments are closed circular DNA molecules that are supercoiled in ichnoviruses, although the topology of bracovirus segments is less clear (Stoltz and Vinson, 1979b). PDV genomes are large and complex, with estimates ranging from 150 to over 275 kb, and have from 10 to 30 segments (Webb *et al.*, 2000). The genomes exist in two states, a linear proviral form integrated in the wasp genome and a circular episomal form present in both the wasp and infected cells of the parasitized host. Proviral segments seem to be dispersed in IVs (Fleming and Summers, 1986) but in tandem array in BVs (Savary *et al.*, 1997). Episomal virus replicates in the ovaries of pupal and adult wasps (Volkoff *et al.*, 1995) with large amounts of virus produced by relatively few replicative cells. This viral form is delivered during parasitization and causes host pathology in the absence of further replication.

PDVs are the biological mediators of parasitoid survival in the parasitized host and have evolved within the constraints of these two biological associations. Therefore, genome organization likely reflects selection pressure for both reliable genetic transmission and effective disruption of parasitized host physiological systems (Webb, 1998; Webb and Cui, 1998). In this light several characteristics of PDV genomes are notable. PDV genome segments are present in varying number and molarity following replication (Stoltz *et al.,* 1981), which results in unequal viral gene copy number and protein titers within the parasitized host (Webb and Cui, 1998). Segments contain large amounts of noncoding sequence, which may comprise highly repetitive regions (Webb *et al.,* unpublished). Proviral integration is stable through many wasp generations (Fleming and Summers, 1991), though low levels of excision occur in all tissues (Stoltz *et al.,* 1986) and some viral segments are able to integrate into cellular DNA after infection (Kim *et al.,* 1996; Mckelvey *et al.,* 1996; Volkoff *et al.,* 2001).

Viral gene expression alters parasitized host physiology, producing an environment suitable for larval wasp development. Viral proteins synthesized in the parasitized insect appear to be responsible for inhibiting host immune responses (Asgari *et al.,* 1996, 1997; Cui *et al.,* 1997, 2000; Lavine and Beckage, 1996; X. Li and Webb, 1994; Prevost *et al.,* 1990; Strand, 1994; Webb and Luckhart, 1994). Host resources are also redirected, supporting parasitoid development (Beckage and Kanost, 1993; Cusson *et al.,* 2000; Shelby and Webb, 1997; Soller and Lanzrein, 1996; Vinson *et al.,* 1979) with concomitant developmental alterations of the parasitized insect (Cusson *et al.,* 2000; Soller and Lanzrein, 1996). Ultimately, PDV infections allow parasitoid development and contribute to the eventual death of the parasitized host. PDV gene expression is required for successful parasitization, as natural or experimental inhibition of viral expression results in death of the wasp larva and normal development of the host.

II. Classification of Polydnaviruses as Viruses

A. Rationale for Comparison

Although PDVs are recognized as viruses having an incomplete life cycle by the International Committee on Viral Taxonomy (Webb *et al.,* 2000), their unusual life cycle periodically raises questions regarding their origin and identification as viruses. These are confounded by the failure to genetically relate PDVs to known viral lineages, although some morphological and symptomatic similarities do exist. In this review we consider the question of whether PDVs are viruses as an organizational theme and conclude with avenues of research that may contribute to elucidating their evolutionary origin.

B. *Consideration of Polydnavirus Origins*

There is not a clearly supported evolutionary pathway elucidating the evolutionary origin of polydnaviruses. Polydnaviruses are clearly "virus-like" in their morphology, replication, and ability to infect cells, but have other characteristics that are unlike those of conventional viruses. Based on the monophyly of braconid lineages, the absence of PDVs in common wasp ancestors, and their distinctive genetic, functional, and morphological characteristics, ichnoviruses and bracoviruses likely have separate origins (Stoltz, 1993; Whitfield, 1997). Therefore, one must assume that PDVs have evolved independently at least twice. Two hypotheses exist concerning the origin of PDVs: that PDVs are descendents of "free-living" viruses that are now fixed in their mutualism or that PDVs are derivatives of wasp episomal DNA (e.g., transposable elements) which acquired a protein coat.

If PDVs are derivatives of conventional viruses, parasitic wasps may have acquired viruses that normally infect the wasp host. Parasitoids mechanically vector pathogenic viruses of lepidopteran larvae (e.g., Miller, 1998) and exploit viruses that inhibit the parasitized host's immune responses through facultative associations (Bigot *et al.*, 1995; Lawrence and Akin, 1990; Stoltz and Makkay, 2000). The ovaries of ichneumonids have been hypothesized to be susceptible to a wide variety of asymptomatic virus infections, though pathological infection of parasitoids appears rare (Stoltz and Whitfield, 1992). It is possible to envision transition of a beneficial virus mechanically vectored by a wasp to an obligate mutualism. Such a transition may involve transfer of viral genes required for replication to the wasp genome, thereby fixing the genetic association. The ability of the virus to alter parasitized host physiology to improve survival of wasp larvae would then benefit both the wasp and virus, whereas viral phenotypes that did not enhance wasp fitness would be selected against.

There is less evidence supportive of the idea that PDVs are derived from wasp genetic secretions. If one assumes that PDVs are derived from transposable elements that acquired a protein coat, the transposable element must have acquired not only a protein coat, but also a function that would benefit parasitoid survival in the parasitized insect. Restriction of replication to the oviduct also would be required, although some transposable elements are known to function in a developmentally regulated manner (e.g., *gypsy;* Labrador and Corces, 1997). Functional proteins are delivered from the female wasp reproductive tract as venoms and the genes encoding these proteins may be a potential origin of PDV genes altering host physiology. If genes encoding venom proteins acquired the ability to excise and be encapsidated, then this could deliver wasp episomal DNA encoding a biologically active gene, allowing for prolonged expression of these biologically active genes. This scenario would allow for replication and packaging of nucleic acids in the calyx tissues, but does not account for mobilization and encapsidation of the ancestral venom genes.

Viruses are known to infect cells, replicate as episomes in specific types of differentiated cells and become encapsidated, inhibit host immunity, acquire genes from the cellular genome, and require "helper" gene functions for replication. However, the pathway for evolution of transposable elements into an infectious agent having PDV characteristics is not well supported by our present understanding of transposable element function and evolution. Similarly, the ability of a eukaryotic toxin gene to excise and become encapsidated is not well supported by the literature.

Regardless of their origin, PDVs represent an interesting and important evolutionary phenomenon. Consideration of the origin and evolution of the two PDV genera provides an interesting perspective from which to compare the two groups. Some viruses and mobile genetic elements are clearly related, and all viruses are thought to be evolutionarily derived from cellular genes. PDVs present an unusual biological system in which the viral genome may be becoming progressively more dependent on cellular functions for replication and simultaneously evolving virulence characteristics that are more selective, diversified, and pathological. Whether PDVs ultimately are shown to be virus derivatives or packaged wasp episomes has no impact on their biological relevance because they clearly represent unique and diverse biological entitities that are relevant to understanding the evolution and function of symbiotic genetic elements.

III. Ichnoviruses

A. *Brief History and Ichnovirus Phylogeny*

Ichnoviruses were initially described as virus-like particles in the oviduct of some ichneumonid wasps (Rotheram, 1967). The particles were found to contain DNA of varying sizes (Stoltz *et al.*, 1981) and to be

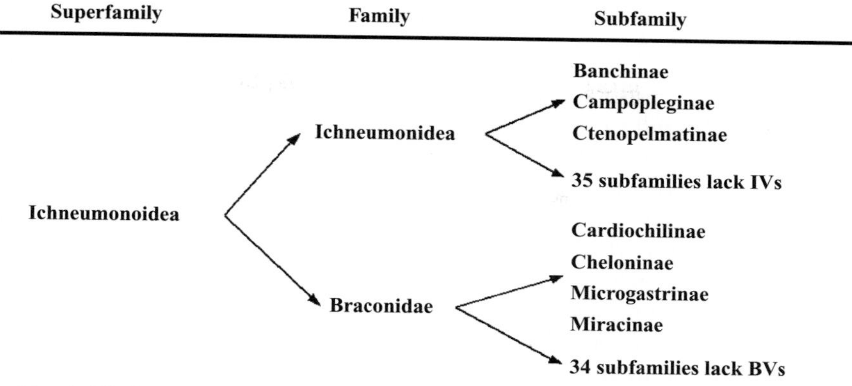

FIG. 1. Phylogenetic relationships within the Ichneumonoidea superfamily. Bracoviruses are limited to the microgastroid complex, made up of four subfamilies, and ichnovirus association is limited to three subfamilies of icheneumonid wasps.

present in the ovary of every female of the infected species (Stoltz *et al.*, 1986). Due to their distinctive genomes, viral morphology, and wasp associations, Brown (1986) recognized these wasp-associated viruses as polydnaviruses and separated those associated with ichneumonid wasps (ichnoviruses) from those associated with braconid wasps (bracoviruses). Ichnoviruses, or IV-like particles, are taxonomically described from approximately 20 species of Campopleginae, Ctenopelmatinae, and Banchinae wasps (Fig. 1), with surveys suggesting that all species in these genera carry a mutualistic IV (Stoltz *et al.*, 1981; Stoltz and Vinson, 1979b; Webb, 1998). Twenty-two ichneumonid subfamilies do not appear to have an associated ichnovirus (Webb, 1998). The most current ichneumonid phylogeny (Belshaw *et al.*, 1998) shows the IV-associated ichneumonids as a paraphyletic assemblage, suggesting either multiple IV–wasp associations or past horizontal transfer of viruses between species through unknown mechanisms. Because the most basal groups of ichneumonids lack IVs, it is likely that the IV–ichneumonid association arose independently of the BV–braconid association. Morphologies of the two genera are distinct, and there is no antigenic or nucleic acid similarity between the IVs and BVs (Cook and Stoltz, 1983; Stoltz and Whitfield, 1992), further supporting separation of the two groups. Based on the number of species in genera known to have ichnoviruses, greater than 10,000 species of ichneumonid wasps are expected to have mutualisms with an IV (Webb, 1998). Ichneumonid groups associated with an IV appear to be more speciose than related wasp genera lacking such (Webb, 1998), suggesting that this viral association enhances relative fitness, possibly by increasing host range.

B. Hosts

The studies of ichnoviruses have focused on those associated with *Hyposoter* spp. and *Campoletis sonorensis* wasps. *Hyposoter* spp. IVs have been utilized in many morphological and cytological studies (e.g., Stoltz and Vinson, 1979a, b; Volkoff *et al.,* 1995). *Campoletis sonorensis* IV (*Cs*IV) has been studied, with experimental emphases on genetic and pathological studies (e.g., Fleming and Summers, 1991; Krell *et al.,* 1982; Shelby and Webb, 1999; Webb, 1998). Virus-like particles (VLPs) that lack nucleic acid are described from the ovaries of one ichneumonid wasp, *Venturia canescens,* which should be associated with an IV based on its phylogenetic position. This suggests that the virus genome has been eliminated in this species and possibly in others.

IVs require two hosts for completion of their life cycle. Vertical transmission, excision of proviral DNA, and virus replication occur only in the replicative host, the parasitic wasp. Viral infection of the lepidopteran host causes pathology. Studies of *Cs*IV demonstrate that viral genes may be expressed in both hosts or only one, though no detrimental pathologies are apparent in the wasp. In a parasitized insect, or pathogenic host, negative effects are readily apparent and associated with viral expression.

1. Replicative Host

a. Replication. Although episomal viral DNA is detectable in peripheral tissues (Cui and Webb, 1997; Fleming and Summers, 1986; Stoltz *et al.,* 1986), virus replication is restricted to a single layer of cells at the junction of the common and lateral oviducts termed the calyx (Volkoff *et al.,* 1995). Prolific virus replication is observed in the hypertrophied calyx cell nuclei in late pupal and adult females (Norton and Vinson, 1983), but this replication does not cause pathology to the insect. Replication is closely linked with oviduct maturation and appears to be induced by increasing levels of the insect steroid hormone 20-hydroxyecdysone (Webb and Summers, 1992). The rate and cytology of replication differ between pupal and adult stages, with increased replication occurring following adult emergence (Volkoff *et al.,* 1995; Webb, 1998).

Little is known concerning the molecular mechanisms of ichnovirus replication. Genes corresponding to DNA replicases were not identified by sequence analyses of the *Cs*IV genome (Webb *et al.,* unpublished), suggesting that these functions are provided by the host cell. Replication clearly produces IV segments that are present in variable molar amounts (Stoltz *et al.,* 1981), presumably due to differences in segment replication rates because there is no evidence of variable proviral segment copy number. All studied proviral IV segments are integrated in a single copy at a single locus. Segments present in hypermolar quantities in the virion are also present in hypermolar ratios in the infected host. Increasing gene copy number within the infected host by increasing titer of infectious virus results in a concomitant increase in mRNA and protein levels (Cui *et al.,* 2000).

b. Segment Types. Ichnovirus segments are of at least two types, unique and nested segments. Unique segments are those which hybridize to a single episomal segment and a proviral integration site in viral and wasp genomic Southern blots, respectively. These segments also have little repetitive DNA in noncoding areas of the segment (Webb *et al.,* unpublished). Nested segments exist as single-copy proviral segments in the wasp genome, but hybridize to multiple episomal DNA segments at high stringency in viral genomic DNA Southern blots (Blissard *et al.,* 1986; Cui and Webb, 1997; Xu and Stoltz, 1993). Nested segments also have repetitive sequences encompassing both coding and noncoding sequences, suggesting that these segments have undergone repeated duplication. In *Cs*IV, segment nesting is restricted to segments that encode cys-motif genes, although not all segments encoding cys-motif genes are nested. Segment nesting results from intramolecular recombination at internal repeats (Fig. 2). The recognition sequences in some percentage of the "parental" segments recombine, allowing for removal of part of the segment and formation of one or more "daughter" segments (Cui and Webb, 1997). This process may be repeated such that three or four daughter segments are produced from a single parental segment with each daughter segment a partial duplicate of the parental segment. Interestingly, segment nesting appears to be nonrandom and seems to provide a mechanism for controlling gene copy number and level of gene expression in a group of viruses that cause pathology in the absence of virus replication. This process allows for increasing copy number of some cys-motif genes within the pathogenic host by providing for selective amplification of these genes (Webb and Cui, 1998). For example, the *Cs*IV *VHv1.4* gene is always retained on parental segment V (15.2 kb) and daughter segments C, K, L^2, and T, whereas the *VHv1.1* gene is present on segment V, but deleted from all daughter segments. This leads to ratios of the *VHv1.4:VHv1.1* genes of 1.67:1 within the parasitized host at the DNA level and 1.82:1 at the mRNA level. This difference also seems to reflect relative hemolymph concentrations of these two cys-motif proteins, although these have not been measured in absolute concentrations (B. A. Webb *et al.,* unpublished). The functional

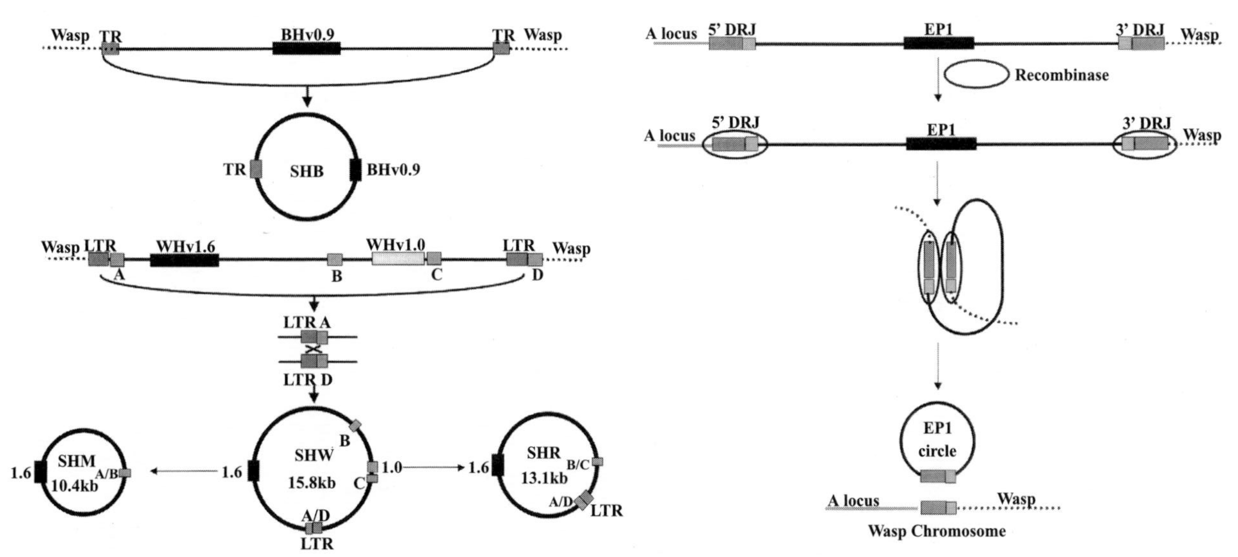

FIG. 2. Comparative mechanisms of integration, excision, and replication of ichnoviruses and bracoviruses. *Cs*IV segment B excision occurs through recombination of two 59-bp terminal repeats, where one repeat is lost during the excision event, resulting in an "empty" locus in the reformed wasp chromosome and superhelical segment B (Fleming and Summers, 1991). *Cs*IV segment W is an example of a nested segment, and can undergo further recombination following the initial excision. Recombination between the two LTRs results in segment excision from the wasp genome, where subsequent recombinations in part of the population of SH-W segments result in daughter segments such as SH-M and SH-R (modified from Cui and Webb, 1997). BV segment excision is exemplified in the excision and formation of a circular EPI locus from *Cotesia congregata* chromosome. Recombination between two terminal repeats with a Hin-like motif results in excision from the wasp genome, with reformation of the chromosome with an empty locus and the episomal viral DNA (modified from Savary *et al.,* 1997).

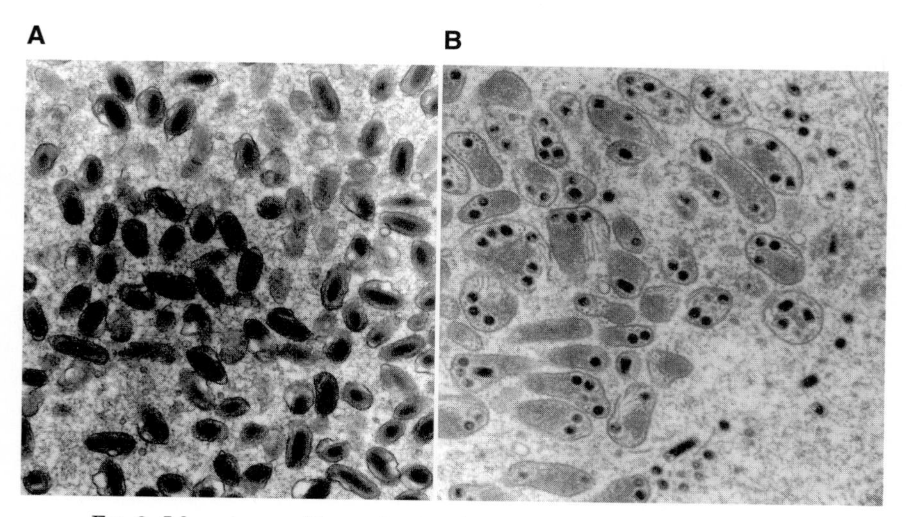

FIG. 3. Ichnovirus and bracovirus nucleocapsids (courtesy of D. B. Stoltz).

significance of selective gene amplification has not been demonstrated, but this pattern is observed on all nested segment families in which it has been studied.

c. Capsid Morphology and Composition. Ichnovirus nucleocapsids are lenticular and ~100 × 300 nm in size (Stoltz and Vinson, 1979b; Volkoff *et al.,* 1995; Webb, 1998). Size and capsid dimensions support encapsidation of the entire IV genome in a single capsid (Krell *et al.,* 1982). Morphologically, IVs are quite distinct from bracoviruses (Fig. 3). Nucleocapsids are surrounded by two envelopes, one obtained within the nuclear virogenic stroma, the second acquired during budding of the virion from the calyx cell (Volkoff *et al.,* 1995). In *Cs*IV the inner membrane extrudes into a "tail"-like bud, although an analogous structure was not detected in *Hf* IV (Stoltz and Vinson, 1979b). Additional membranes are also present intermittently during passage of the virus through the replicative cell, but are lost prior to exiting the cell (Volkoff *et al.,* 1995). In *Hd* IV, mature virions with single membranes have been detected in the lumen of the oviduct (Volkoff *et al.,* 1995), though whether these are produced by cell lysis or membrane loss in the oviduct lumen is not clear.

IV capsids are complex, made up of 15 or more polypeptides, some of which are glycosylated (Krell *et al.,* 1982; Cook and Stoltz, 1983).

Tertiary structure determinations of IV capsids have not been published, nor have the nonpeptide virion components been experimentally determined. Cross-reactions are observed when crude calyx fluid or purified virus isolated from a variety of IVs is probed with antisera raised against *He*IV (Cook and Stoltz, 1983). This cross-reactivity may be due to epitope similarity among wasp ovarian proteins, venom proteins, and ichnoviruses (Cook and Stoltz, 1983; Webb and Summers, 1990). The major cross-reacting proteins from other *Hyposoter* spp. IV were in the same weight range (40–55 kDa) as the two major structural *He*IV proteins (Cook and Stoltz, 1983), suggesting that IV virions retain related structural proteins.

Though the *Cs*IV capsid comprises ~25 peptides, only one structural protein gene (p12) is known to be encapsidated in *Cs*IV (Deng and Webb, 1999). The gene encoding the p44 virion structural protein is present in the wasp genome and is not encapsidated during replication (Deng *et al.,* 2000). Two additional genes exhibiting sequence similarity to the p12 and p44 virion structural proteins are present in wasp and viral genomes, respectively, suggesting that multiple virion protein gene families are distributed between wasp and viral genomes (B. A. Webb and Huang, unpublished). The size and complexity of PDV virions suggest that a minimum of 39.4 kb of DNA is required to encode virion structural proteins in the *Cs*IV genome (Krell *et al.,* 1982). In combination with the absence of other genes associated with viral replication in the *Cs*IV genome (Webb *et al.,* unpublished), these data support a progressive loss from the viral genome of those genes required only for virus replication. This failure to encapsidate genes required only for replication is similar to the pronounced loss of genes and concomitant reduction in genome size described from obligate symbiotic bacteria. This trend toward genome reduction is described as "reductive evolution" (Anderssen and Kurland, 1998) and presumably occurs in PDVs through horizontal transfer of genes encoding viral structural proteins from the encapsidated viral genome to the genome of the mutualistic wasp.

IV capsid morphology is unlike that of other common insect viruses (Miller, 1996), and no evolutionary or biochemical links to extant virus groups have been described. Some similarities were noted to two unclassified viruses observed in a fire ant (Avery *et al.,* 1977) and a whirligig beetle (Gouranton, 1972) based on nucleocapsid morphology (Stoltz and Faulkner, 1978). A virus-like particle having morphological similarities to IVs has also been described from the oviduct of the braconid wasp *Microctonus aethiopoides,* although the relatedness of this particle to IVs is not clear (Barratt *et al.,* 1999).

d. Virus Infection and Associated Factors. Following packaging, virions bud from calyx cells into the oviduct lumen and mix with ovarian proteins to form calyx fluid (Stoltz, 1993; Volkoff *et al.,* 1995). Virion budding is continual, proceeding for at least 10 days into adulthood in *H. didymator* (Volkoff *et al.,* 1995). During oviposition the female wasp injects virions, along with egg(s), venom, and ovarian proteins into the parasitized insect. The ovarian proteins (OPs) appear to suppress early immune responses in hosts of *C. sonorensis* (Luckhart and Webb, 1996; Webb and Luckhart, 1994) and *Venturia canescens* (Beck *et al.,* 2000). *C. sonorensis* OPs (*Cs*OPs) and venom proteins possess epitopes similar to some *Cs*IV-encoded proteins (Webb and Summers, 1990). Also, the egg surface may passively evade host immune responses based on surface characteristics (Kinuthia *et al.,* 1999). Similarly, a *V. canescens* protein mimicking a host protein induced during bacterial challenge may function to inhibit encapsulation (Berg *et al.,* 1988).

2. *Pathogenic Host*

Ichnoviruses are introduced into the body cavity (hemocoel) of lepidopteran larvae and have been observed in most tissues exposed to insect blood (hemolymph) including muscle, fat body, Malphigian tubules, and hemocytes (Stoltz and Vinson, 1979a). Viral replication is not observed in parasitized hosts, with virus titers remaining constant or only slightly reduced in permissive hosts (Theilmann and Summers, 1986); in nonpermissive hosts, virus protein expression declines concomitant with reduced viral DNA levels (Cui *et al.,* 2000). The outer virion membrane is lost as the virion penetrates the host basement membrane. The inner-envelope protrusion may be involved in mediating basement membrane penetration (Stoltz and Vinson, 1979a). The inner membrane appears to fuse to the cellular membrane with the nucleocapsid then internalized. Nucleocapsids then move to the nucleus, which they enter probably through nuclear pores, and release viral DNA (Webb, 1998). Viral gene expression is detected within two hours postparasitization (p.p.) and increases for about 24 hr. Viral nucleocapsids are no longer evident in the nucleus by 24 hr p.p. (Stoltz and Vinson, 1979a), and virus is undetectable by polymerase chain reaction (PCR) in cell-free hemolymph at this time (X. Li and B. A. Webb, unpublished).

C. *Genome Organization*

Ichnoviruses are double-stranded DNA viruses. The genome is present in two forms, an integrated proviral form and a supercoiled episomal form, both of which are detectable in every member of each

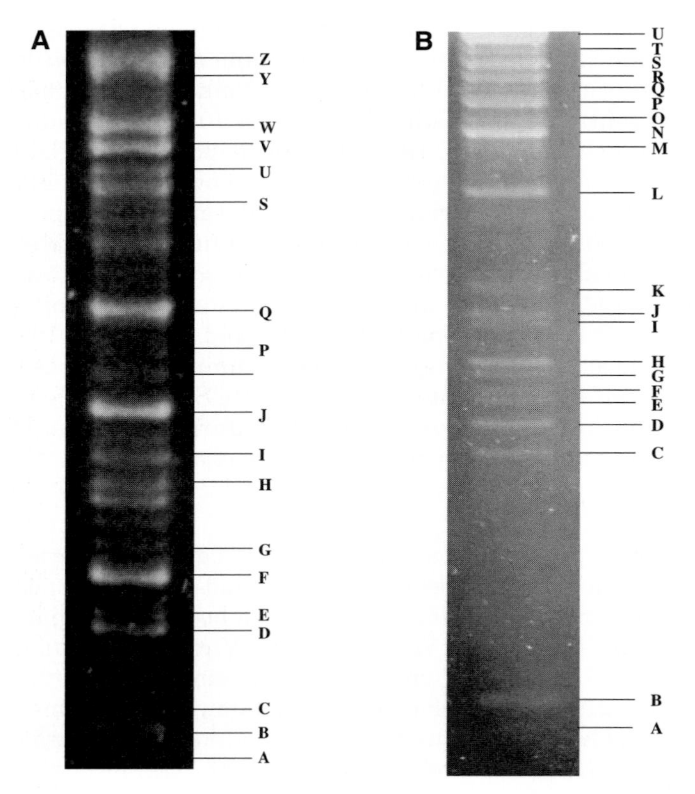

FIG. 4. *Cs*IV and *Cm*BV genomes. Nomenclature of segments is based on relative size, in alphabetical order starting with the smallest visible segment. *Cs*IV includes only superhelical DNA, and only unique and parental segments are labeled. *Cm*BV segments are labeled according to visible segments. Image courtesy of D. B. Stoltz.

virus-associated wasp species (Fleming and Summers, 1991; Stoltz *et al.*, 1986). Estimates of IV genome sizes range from ~150 kb to greater than 250 kb (Webb *et al.*, 2000). Greater than 20 segments are observed in all IVs (Krell *et al.*, 1982; Stoltz *et al.*, 1981). IV segments are named alphabetically in increasing order of size (e.g., segment A is ~6.2 kb, segment B is ~6.6 kb) (Fig. 4). *Cs*IV has a nonredundant genome size of 248 kb and 22 unique or parental segments (Webb *et al.*, unpublished).

1. *Molecular Characteristics*

IV genome segments exist in the proviral state in the wasp genome as discrete viral DNA segments flanked by nonviral wasp DNA, suggesting that the proviral segments are dispersed (Fleming and

Summers, 1991). However, the extent of segment dispersal is unknown because existing analyses have been limited to characterization of genomic clones of ~20 kb. It is clear that IV segments are not in tandem arrays as appears to typify BV segments (Savary *et al.,* 1997).

Integration sites for some segments have been identified in *Cs*IV, but the degree of conservation among various segments and between IV species is unknown. *Cs*IV segment B possesses a 59-bp imperfect repeat at each integrated terminus, but only a single repeat in episomal DNA (Fleming and Summers, 1991). Either imperfect repeat may be found in episomal segments, with the other repeat remaining in the wasp genome after segment excision (W. Rattanadechakul and B. A. Webb, manuscript in preparation). By contrast, *Cs*IV segment W, a nested segment, encodes a pair of identical ~1.2-kb terminal repeats containing ~350- and ~750-bp repeats (Cui and Webb, 1997). Cui and Webb suggest that the high degree of conservation in nested segment repeats enhances efficiency of excision and replication, thereby favoring the high relative molar concentrations of nested segments (Webb, 1998). Interestingly, a 350-bp repeat residing within the segment W recombination repeat is also the site of intramolecular recombination events that produce the segment W daughter segments. Taken together the data suggest that integration and recombination events have some features in common, but that some other features are specific for segment type (e.g., nested vs. unique repeats). Developing a better understanding of these conserved and distinctive features requires analyses of other segments, particularly from other IVs. Similarities between *Cs*IV terminal repeats and the structure of some mobile elements have been discussed. Like insertion sequences, the long terminal repeats (LTRs) of *Cs*IV segment W contain short direct terminal repeats, specifically a tetranucleotide palindrome GATC (Cui and Webb, 1997), similar to those implicated in insertion/excision sequences in mobile elements (Haren *et al.,* 1999). However, loss of one of the repeats during the excision of the IV segment is different from what typically occurs during excision of transposable elements, in which both repeats are maintained (Haren *et al.,* 1999). The molecular mechanism of IV excision and the precise role of the terminal repeats remain to be elucidated, but are highly relevant to elucidating putative relationships with other classes of mobile genetic elements.

2. *Ichnovirus Genomics*

Estimating IV genome size by simple addition of segment sizes from agarose gels is not accurate due to the presence of comigrating and nested segments. However, the sequence of the *Cs*IV genome has recently been completed and comprises 22 nonredundant segments (Webb *et al.*, unpublished). The nonredundant *Cs*IV genome size (which excludes nested "daughter" segments) is 248 kb with segments ranging in size from 6.2 kb (A) to 18.6 kb (Z). If partially redundant (daughter) segments totaling 67 kb are included, the *Cs*IV genome size is larger (~315 kb). The *Cs*IV genome is approximately 60% G/C and greater than 60% noncoding. Portions of essentially every *Cs*IV segment have homology to other segments, primarily in coding regions. This provides new insight into the results of Theilmann and Summers (1998), who observed cross-hybridization to almost every viral segment with a segment B-derived probe at 50% stringency. The majority of this cross-hybridization was due to genes encoding an ~540-bp repeat sequence present on nine viral segments. Other viral segments are also similar within coding sequences, although intrasegmental repeated sequences are evident. For example, *Cs*IV segment V (15.2 kb) is made up of five copies of a 3-kb region, each of which has 80–94% similarity (K. G. Lindstrom, L. Cui, and B. A. Webb, unpublished); *Cs*IV segment W (15.8 kb) is similarly organized with three similar copies of a 5-kb region (Cui and Webb, 1997). Every *Cs*IV segment encodes at least one gene, with the majority of the predicted genes transcribed in the same orientation within a segment. An exception is that the three genes on segment Q (12.1 kb) are encoded on opposite strands. Interestingly, segment Q is also the only segment known to encode members of more than one gene family (rep and vinnexin) (Webb *et al.*, unpublished).

D. Gene Families

Paradoxically, although the number of unrelated genes encoded within the *Cs*IV genome is small, the genome encodes a relatively large number of variants of several genes (Table I). The *Cs*IV genome encodes three major gene families totaling at least 35 genes (Webb *et al.*, unpublished). Initial open reading frame (ORF) analyses indicate that *Cs*IV encodes fewer than five genes having no similarity to other *Cs*IV genes, none of which is related to known viral structural or replication genes. Because *Cs*IV is the only sequenced IV genome, the following section focuses on this virus.

1. Cys-Motif Genes

The cys-motif gene family was initially identified based on the presence of an intercystine knotlike motif (Pallaghy *et al.*, 1994) present in several *Cs*IV peptide sequences (Dib-Hajj *et al.*, 1993). The motif consists of six conserved cysteine residues in an ~50-variable-residue

TABLE I

SELECTED PDV GENES. NOTE THE LARGE NUMBER OF DUPLICATED GENES IN EACH GENE FAMILY, AS WELL AS THE ENCODING SEGMENT OF THE GENES

Virus	Gene	Gene Family	Encoding Segment	Segment Type (IV)	Host	Citation
*Cs*IV	*VHv1.1*	cys-motif	V	nested	lepidopteran	Dib-Hajj, Webb, & Summers, 1993
	VHv1.4	cys-motif	V	↓	↓	Cui & Webb, 1996
	WHv1.0	cys-motif	W	↓	↓	Blissard, Smith, & Summers, 1987
	WHv1.6	cys-motif	W	↓	↓	Blissard, Smith, & Summers, 1987
	BHv0.9	rep	B	unique	lepidopteran	Theilmann & Summers, 1988
	SH2743	rep	O	↓	both	Theilmann & Summers, 1988
	I0.9	rep	I	↓	↓	Hilgarth and Webb, in prep.
	I1.1	rep	I	↓	↓	Hilgarth and Webb, in prep.
	I1.2	rep	I	↓	↓	Hilgarth and Webb, in prep.
	HC1185	rep	O	↓	↓	Theilmann & Summers, 1988
	VnxD-1	vinnexin	D	↓	both?	Turnbull, Lindstrom, & Webb, unpub.
	VnxG-1	vinnexin	G	↓	?	Turnbull, Lindstrom, & Webb, unpub.
	VnxQ-1	vinnexin	Q	↓	?	Turnbull, Lindstrom, & Webb, unpub.
	VnxQ-2	vinnexin	Q	↓	?	Turnbull, Lindstrom, & Webb, unpub.
*Hd*IV	*M24*	Gly-Pro rich	?*	nested?	*Sf9*-cell line	Volkoff *et al.,* 1999
	M27	Gly-Pro rich	↓	↓	↓	Volkoff *et al.,* 1999
	M40	Gly-Pro rich	↓	↓	↓	Volkoff *et al.,* 1999
	rep gene	rep	?	?	↓	Volkoff, per. comm.
	HdVnx1	vinnexin	?	?	↓	Volkoff, per. comm.
*Md*BV	*egf7.0*	EGF-like	O		lepidopteran	Strand, Witherell, and Trudeau, 1997
	egf7.5	EGF-like	O		↓	Strand, Witherell, and Trudeau, 1997
	egf0.4	EGF-like	O		↓	Trudeau, Witherell, and Strand, 2000
*Cc*BV	*EP1*	early	A		lepidopteran	Beckage, 1993

* The *Hd*IV genes *M24*, *M27*, *M40* are on different molecules of the *Hd*IV genome.

matrix (Blissard *et al.,* 1987; Cui and Webb, 1996; Dib-Hajj *et al.,* 1993). Ten members of the family are encoded by *Cs*IV on segments A, A′, F, L, U, V, and W (Webb *et al.,* unpublished). Some segments encoding cys-motif genes are of the nested segment type and these are the only nested segments in *Cs*IV. The smaller gene on segments V and W (*VHv1.1* and *WHv1.0*) is lost during segment nesting (Cui and Webb, 1997; Webb and Cui, 1998), leading to differential gene levels following replication. The different gene levels were shown to affect protein levels in *Cs*IV-infected hosts (Cui *et al.,* 2000). The segment W cys-motif genes each encode a single cys motif, whereas all the other genes encode two copies of the motif (Webb *et al.,* unpublished).

All cys-motif genes share a similar structure. Each gene possesses a large 5′ untranslated region (UTR) with an intron and a conserved intron immediately prior to the second cysteine of each cys motif. The cys-motif genes sharing the same segment are more similar to each other than to genes on other segments (M. W. Turnbull and B. A. Webb, unpublished), with the 5′ portion of each gene (5′ UTR, exon 1, intron 1, and exon 2, which encodes the bulk of the N-terminal cys motif) more conserved than the 3′ end. The intron residing within the cys motif is as conserved as or more highly than the adjacent exons at the nucleotide level (Cui and Webb, 1996; Dibb-Hajj *et al.,* 1993).

Expression of *VHv1.1* (Cui and Webb, 1996; X. Li and Webb, 1994) and *VHv1.4* (Cui *et al.,* 1997; Cui and Webb, 1996) mRNAs is observed at 2 hr postinfection (p.i.). All cys-motif genes encode signal peptides, suggesting that they are secreted. By western blotting VHv1.1 and VHv1.4 are detected in all tissues for the duration of infection, but the proteins are most abundant in plasma, hemocytes, and fat body (Cui *et al.,* 1997). Immunofluorescence studies suggest that detection of cys-motif proteins in fat body results from adherence of hemocytes to the tissue and uptake by fat body cells (Cui *et al.,* 1997). cys-motif proteins probably enter hemocytes through ATP-dependent endocytosis because uptake is inhibited at 4°C (Cui *et al.,* 1997). The fate of the proteins following uptake has not been studied, though large immunoreactive foci are observed in the cytoplasm of hemocytes, indicating that they may accumulate in lysosomes, as has been reported for *Cs*OPs (Luckhart and Webb, 1996).

Each cys-motif protein has at least one potential N-linked glycosylation site, with VHv1.4 having five sites. WHv1.6 is either weakly or nonglycosylated (Blissard *et al.,* 1989), whereas WHv1.0, VHv1.1, and VHv1.4 are strongly glycosylated (Soldevila *et al.,* 1997). EndoH and PNGase treatment of VHv1.1 (Soldevila *et al.,* 1997) and VHv1.4 (Cui *et al.,* 1997) demonstrates that some carbohydrate moieties are

PNGase-sensitive, but EndoH-resistant, indicating mannose trimming and complex glycosylation, which is atypical of insect cells. A recombinant VHv1.1 mutant in which the glycosylation site is ablated does not bind to hemocytes (A. Soldevila and B. A. Webb, unpublished). In the presence of sialic acid, but not mannose, a reduction of VHv1.4 uptake occurs (M. W. Turnbull and B. A. Webb, unpublished), suggesting that the CHO monomer competes for binding sites required for the VHv1.4 protein to enter cells. Combined, these pieces of data suggest that cys-motif glycoconjugates are important for cellular recognition and binding.

The involvement of cys-motif proteins in abrogating cellular immune responses is supported by several lines of evidence. Granulocytes bind and endocytose VHv1.1 and VHv1.4, whereas it appears that plasmatocytes bind VHv1.1, but not VHv1.4 (Cui *et al.,* 1997; X. Li and Webb, 1994). A recombinant VHv1.1 protein reduced encapsulation of *C. sonorensis* eggs 50% relative to saline-injected larvae (X. Li and Webb, 1994), whereas cells overlaid with recombinant VHv1.4 protein were morphologically similar to infected granulocytes (Cui *et al.,* 1997).

Recent studies have shown that cys-motif genes exhibit regional differences in evolution rate (Dupas *et al.,* submitted). Overall, selection pressures on cys-motif genes are strongly negative, with synonomous codon substitutions more frequent than nonsynonomous mutations. However, there is evidence of positive selection within the cys motif and within the region immediately posterior to the motif. Mutations within these regions preferentially alter residue charge. Charge-changing mutations are more likely to result in divergent proteins because of the importance of charge on tertiary protein structure. Thus, charge-changing mutations may produce functional cys-motif protein variants that have higher parasitization success or expanded host range (Dupas *et al.,* submitted). Similar mutation patterns are observed from and hypothesized to increase variation within the venom genes of predatory snails (*Conus* spp.; Conticello *et al.,* 2001).

2. Rep Genes

The *Cs*IV rep gene family was identified by Theilmann and Summers (1987). Cross-hybridization was observed between repetitive regions on segments B, H, M, and O[1] (all now known to be unique segments; Webb *et al.,* unpublished) and to other *Cs*IV genome segments. Sequence analysis of the cross-hybridizing region segments revealed an approximately 540-bp repetitive sequence present on the three segments. One copy of the repeat was identified on segment B, five on H, and three on O^1 with an average nucleotide similarity of 60–70% and regions of >90% similarity. Rep sequences on all three segments (B, H, and O^1) are now known to be part of expressed genes (Theilmann and Summers, 1988; Hilgarth and Webb, in press). A superhelical segment B (SH-B) rep probe hybridized to a 900-bp mRNA transcript from parasitized *H. virescens* (the *BHv0.9* gene), whereas an SH-H probe hybridized to an ~2.7-kb mRNA expressed in both parasitized *H. virescens* and in the *C. sonorensis* oviduct. A SH-O^1 rep probe hybridized to four mRNAs from both parasitized *H. virescens* and *C. sonorensis,* though one signal is likely cross-hybridization to H transcripts (Theilmann and Summers, 1988).

Sequencing the *Cs*IV genome identified 23 additional putative rep-containing genes on nine unique segments (R. Hilgarth and B. A. Webb, manuscript in preparation). Average similarity in the repeat portion of the rep proteins is 43% (R. Hilgarth and B. A. Webb, manuscript in preparation). A rep gene homologue has been isolated from *Hd*IV (A.-N. Volkoff *et al.,* in press). Pairwise amino acid similarity between the *Hd*IV rep protein and the *Cs*IV segment Z rep proteins is 42% (R. Hilgarth, A.-N. Volkoff, and B. A. Webb, unpublished). Two additional *Hd*IV rep ORFs are known from *Hd*IV segment E, although expression of these genes has not been proven (A.-N. Volkoff, personal communication).

The rep sequence was hypothesized to be ubiquitous and possibly to facilitate genome evolution by promoting recombination of *Cs*IV segments and genes (Theilmann and Summers, 1987). Because essentially all rep sequences are found in expressed genes, it seems less likely that this sequence has a structural role in segment excision or genome evolution. The functions of the rep proteins are unknown, and no homologous genes have been identified in public databases. The proteins do not encode signal peptides and are thus not predicted to be secreted from infected cells (Theilmann and Summers, 1987). Preliminary work using an anti-rep gene antibody supports intracellular protein localization (R. Hilgarth and B. A. Webb, unpublished). At least some rep genes are expressed in both *C. sonorensis* and parasitized *H. virescens,* though the levels of expression are generally lower in wasp tissues (R. Hilgarth and B. A. Webb, manuscript in preparation). The greater than 25 members makes the rep gene family the largest reported viral gene family.

3. *Vinnexin Genes*

A third *Cs*IV gene family with four members was recently identified by National Center for Biotechnology Information (NCBI) database comparisons (K. G. Lindstrom, M. W. Turnbull, and B. A. Webb unpublished); a homologue has also been isolated from *Hd*IV (A.-N. Volkoff *et al.*, unpublished). These viral genes show strong sequence similarity (E-values from 10^{-30} to 10^{-80}) in BlastX (Altschul *et al.*, 1997) analyses to the innexins (Phelan *et al.*, 1998a), and thus have been termed vinnexins. Innexin genes encode invertebrate gap junction structural proteins (Phelan *et al.*, 1998b). Innexins are functionally analogous to the vertebrate connexins, and have been identified in *Caenorhabditis elegans, Drosophila,* and grasshopper (Curtin *et al.*, 1999; Ganfornina *et al.*, 1999; Landesman *et al.*, 1999; Starich *et al.*, 1996). In vertebrates, hexamerization of connexins forms a hemichannel spanning the plasma membrane, which may be made up of one (homomeric) or multiple (heteromeric) connexin types (Jiang and Goodenough, 1996; Stauffer, 1995).

Gap junctions are formed between adjacent cells through hemichannel interaction between two cells to allow selective transfer of small molecules including ions, second messengers, and dyes. Based on structure–function analyses of connexins, it is likely that innexin composition governs gap junction formation (Foote *et al.*, 1998) and properties (Bevans *et al.*, 1998). Formation of functional gap junctions is mediated by the structure of the external loops, with activation of the gap junctions relying on the structure of the inner loop. The vinnexins possess the same predicted topology as innexins: cytosolic N- and C-termini, four transmembrane regions, and two extracellular loops each with a pair of conserved cysteine residues (M. W. Turnbull, A.-N. Volkoff, and B. A. Webb, unpublished). However, vinnexins possibly may form nonfunctional hemichannels as a means of disrupting immune responses of infected cells. The extracellular loops of the innexins must interact with extacellular loops from a paired cell to form a channel. The vinnexin extacellular loops are least similar to one another and to the innexins (M. W. Turnbull and B. A. Webb, unpublished), indicating a possible mechanism for disruption of gap junction formation.

The *Cs*IV vinnexin gene *CsVnx-Q2* is expressed in parasitized caterpillars and possibly female *C. sonorensis* (M. W. Turnbull and B. A. Webb, unpublished). Antisera raised against a unique region of CsVnx-Q2 localizes to the membrane of infected cells in immunofluorescence assays (M. W. Turnbull and B. A. Webb, unpublished). *Hd*IV vinnexin *HdVnx-1* is also expressed in *Hd*IV-infected cell lines (A.-N. Volkoff, unpublished), although its localization is not known.

Gap junctions form between hemocytes during encapsulation (Churchill *et al.*, 1993). Gap junctional communication also occurs between various immune tissues in mammals, including macrophages and neutrophils, and is observed by dye transfer between adjacent polymorphonuclear cells following lipopolysaccharide (LPS) induction (as reviewed in Saez *et al.,* 2000). We hypothesize that the vinnexins may represent viral orthologues of host innexins, and suspect that they alter gap junction formation and activity as a means of inhibiting appropriate intercellular communication during encapsulation. If this hypothesis is correct, it suggests that innexins may be more important in normal immune cell function than previously suspected.

4. Other Ichnovirus Genes and Gene Families

Volkoff *et al.* (2001) described two gene families in *Hd*IV in addition to the single vinnexin and rep genes mentioned above. The *M40*, *M27*, and *M24* genes are 2.3, 1.6, and 1.5 kb, respectively, and are characterized by strong regional similarities, a conserved intron in the 5′ UTR and Gly- and Pro-rich regions. The 5′ UTR of the three cDNAs has an average pairwise identity of approximately 90%. Each of the cDNAs has tandem repeats in the 5′ region, with *M27* and *M24* similar in composition and repeat number, whereas *M40* is more divergent. The M gene family members are encoded by three different *Hd*IV genome segments. Both the *M24* and *M27* genes are expressed in parasitized hosts. An antibody raised against the protein encoded by the *M24* gene, P40, binds to proteins of 40 and ~70 kDa, respectively. The larger protein may correspond to the predicted *M27*-encoded protein, P69. Interestingly, the antibody also cross-reacted with an ~80-kDa host protein in both parasitized and nonparasitized hosts, which may represent common host and virus antigens (Volkoff *et al.,* 1999).

Several other cDNA probes from *Hd*IV-infected Sf9 cells cross-hybridize with multiple bands when probed against RNA isolated from infected Sf9 cells, suggesting other multigene families exist in the *Hd*IV genome (Volkoff *et al.,* 1999). Probes specific for the C, D, and S group cDNAs hybridized to three RNAs each. The *Hd*IV data suggest common features among IV gene families, but further characterization of *Hd*IV gene families and genome structure is required before the similarities and differences in genome organization and gene families between *Hd*IV and *Cs*IV will be apparent.

A single expressed gene has been described from *Tr*IV-infected caterpillars (Beliveau *et al.,* 2000). This ~650-bp mRNA is detected in parasitized *Choristoneura fumiferana* probed with whole viral DNA. The predicted 9.3-kDa protein is similar to *Cs*IV VHv1.4 in the signal peptide region. The *TrV1* gene is located on segment G, and is expressed in fat body and epidermal cells. Rather than having an immunosuppressive function, the *Tr*IV gene has been hypothesized to have a role in alteration of host development.

E. *Ichnovirus Functions*

1. *Alteration of Immune Response*

a. Inhibition of Antiviral Response. Insect antiviral responses to IV infection are poorly characterized. The decline in IV transcription and DNA titer in nonpermissive lepidopteran hosts suggests that antiviral immune responses, including apoptosis, exist and can eliminate or suppress IV transcription in infected cells. Because *Cs*IV infects several major tissues, it is important to assay for tissue-specific effects (e.g., apoptosis) in semi- and nonpermissive hosts. Although apoptotic regulatory mechanisms are fairly well described in *Drosophila,* potential points of interaction of *Cs*IV and the lepidopteran apoptotic cascade have not been determined.

Coinfection of the baculovirus *Autographa californica* M nucleopolyhedrovirus (AcMNPV) with *Cs*IV allows replication of the former in a normally resistant host, *Helicoverpa zea* (Washburn *et al.,* 1995). Total and differential hemocyte populations change drastically following *Cs*IV infection (Davies *et al.,* 1987) and *Tr*IV (Doucet and Cusson, 1996). As discussed in the next section, these changes in hemocyte populations likely mediate the reduced antiviral response because hemocytes are primary mediators of the antibaculovirus response (Trudeau *et al.,* 2001).

b. Inhibition of Cellular Response. The IV effects on insect cellular immune responses is the most studied aspect of the IV–host interactions, with emphasis on the inhibition of cellular encapsulation, the main response to eukaryotic invaders. There are four mature hemocyte classes recognized in Lepidoptera: granulocytes, plasmatocytes, spherulocytes, and oenocytoids (Gardiner and Strand, 1999; Willott *et al.,* 1994). Granulocytes and plasmatocytes are the primary cells involved in recognition of and response to large foreign bodies and initiate encapsulation reactions (Pech and Strand, 1996). Granulocytes are macrophage-like cells capable of opsonization (Pech and Strand, 1996), and also secrete plasmatocyte spreading peptide (PSP-1), a *Pseudoplusia includens* cytokine that activates plasmatocyte responses to foreign objects (Clark *et al.,* 1998). Plasmatocytes respond to granulocyte-mediated opsonization of parasites by flattening and adhering to foreign objects, which results in the accumulation of multiple plasmatocyte layers around parasitoid eggs (Pech and Strand, 1996) and the eventual death of the parasite.

Disruption of encapsulation may reduce responses to all foreign objects or be more specific and inhibit only encapsulation of parasitoid eggs (Davies and Vinson, 1988). The virus-like particles of *V. canescens* have antigenic similarity to a host protein involved in antibacterial and encapsulation responses (Berg *et al.,* 1988), suggesting that encapsulation resonses in this system may be disrupted through molecular mimicry. Whether similar protein mimics exist in other IVs or whether host cellular responses are suppressed by virus homologues of host genes that regulate immune function is of considerable interest.

The *Cs*IV cys-motif proteins bind to both granulocytes and plasmatocytes (VHv1.1) (X. Li and Webb, 1994) or granulocytes alone (VHv1.4) (Cui *et al.,* 1997). VHv1.1 and VHv1.4 are expressed in all infected tissues, but the proteins are secreted and accumulate in the hemolymph, where they bind to hemocyte surface proteins (Cui *et al.,* 1997). Recombinant VHv1.1 protein induces some of the pathologies associated with *Cs*IV infection of hemocytes, including reduction of encapsulation of wasp eggs (X. Li and Webb, 1994).

IV-induced cytopathologies are varied, but often include disruption of the hemocyte cytoskeleton. F-Actin filaments fail to polymerize normally following parasitization or injection of purified virus. Presumably as a result, infected cells do not adhere efficiently to foreign objects and fail to attain their normal morphologies when activated by exposure to a glass slide. Similar phenomena are observed following injection of *Cs*OPs (Luckhart and Webb, 1996; Webb and Luckhart, 1994). Plasma from virus-infected animals also induces these characteristic morphological abnormalities in uninfected cells, suggesting that secreted CsIV proteins are involved. Because *Cs*IV and *Cs*OP symptoms are similar, it will be interesting to determine if there is sequence similarity among these proteins.

c. Inhibition of Humoral Response. Insect humoral immunity encompasses an array of immune factors including antimicrobial peptides, plasma-borne pathogen recognition molecules, and defensive melanization reactions (Gillespie *et al.,* 1997). IVs have an impact on several of these humoral immune responses. *Cs*IV inhibits synthesis of antibacterial peptides such as lysozyme and cecropin (Shelby *et al.,* 1998). Induction of antibacterial responses may itself reduce encapsulation (Berg *et al.,* 1988), although some parasitoid systems either have little impact on hemolymph antimicrobial activity or are not induced by parasitization (Nicolas *et al.,* 1996). IVs also inhibit hemolymph melanization. Defensive melanization involves induction of the phenoloxidase cascade in insect plasma. Two major biochemical pathways exist, both involving the conversion of the inactive zymogen prophenoloxidase to the active phenoloxidase (PO) following recognition of foreign objects through activation of a serine protease cascade (Sugumaran, 1996).

Recently, serine protease inhibitor activity was isolated from the calyx fluid of *V. canescens* (=*Vc*VLPs) (Beck *et al.*, 2000), though direct action on the serine proteases of the PO cascade was not shown. It is possible the *V. canescens* serine protease inhibitors block conversion of prophenoxidase to its active form as a mechanism for inhibiting defensive melanization. The second activation cascade involves synthesis of melanin by dopachrome isomerase activity on L-DOPA (Sugumaran, 1996; J. Li *et al.*, 1994). Reductions in PO activity and melanization have been observed in several IV systems, including *Tr*IV (Doucet and Cusson, 1996) and *Cs*IV (Stoltz and Cook, 1983). *Cs*IV infection reduces melanization by inhibiting synthesis of enzymes in the melanization pathway at the posttranscrptional level. The resultant changes in enzyme levels alter substrate availability and composition such that melanization is blocked (Hilgarth, 1995; Shelby *et al.*, 2000). In nonpermissive hosts of *C. sonorensis* melanization reactions are only temporarily depressed (Cui *et al.*, 2000) with recovery of melanization occurring as hemolymph *Cs*OPs and/or *Cs*IV protein titers decline.

2. *Developmental Regulation*

a. Hormonal Titers. Effects of IVs on host hormone titers and enzymes regulating hormone titers have been reported. Levels of juvenile hormone esterase (JHE), the primary enzyme involved in breakdown of juvenile hormone, are reduced significantly following *Cs*IV infection (Shelby and Webb, 1997). Injection of *Tranosema rostrale* calyx fluid (=*Tr*IV) inhibits JHE synthesis without altering JH titers, and also depresses 20-hydroxyecdysone titers (Cusson *et al.*, 2000). *Cs*IV infection of final-instar *H. virescens* larvae induces atrophy of the prothoracic glands (Dover *et al.*, 1988a) and a consequent reduction in ecdysteroid titers (Dover *et al.*, 1988b). These varied effects suggest that several mechanisms are used to delay (Doucet and Cusson, 1996) or arrest (Dover *et al.*, 1987) development of host larvae (Cusson *et al.*, 2000).

b. Hemolymph Composition. IV infection alters hemolymph protein levels drastically. Titers of some free amino acids increase (Vinson, 1990), whereas total levels of plasma protein drop (Shelby and Webb, 1994), possibly as a result of inhibition of protein synthesis at a post-transcriptional step. The effects are selective; arylphorin, a major hemolymph storage protein, is affected, but other major hemolymph proteins, including lipophorin and transferrin, are not (Shelby and Webb, 1994).

Carbohydrate and lipid metabolism are significantly altered by IV infection. Triglycerides and glycogen deposits in fat body of control insects are absent in *C. sonorensis*-parasitized larvae (Hilgarth, 1995).

Hemolymph titers of trehalose increase in parasitized insects (Vinson, 1990), suggesting that an overall impact of IV infection is mobilization of nutrient stores, which may increase their availability for developing parasite larvae.

IV. Bracoviruses

A. *System Characteristics and Origin*

1. *Brief History*

Bracoviruses were described when Vinson and Scott (1975) isolated DNA-containing particles from the oviducts of the braconid wasp *Cardiochiles nigriceps*. Earlier descriptions of virus-like particles from ichneumonids (Rotheram, 1967) suggested that the particles did not contain DNA, though we now know that the particles in the host studied in this work (*V. canescens*) are unusual in that they lack nucleic acid. Vinson and Scott (1975) showed that the *C. nigriceps* virus-like particles probably contained DNA. Stoltz *et al.* (1976) then isolated nucleic acid-containing particles from the oviducts of *C. nigriceps, Microplitis croceipes,* and *Chelonus texanus,* and described the nuclear origin of these particles and their morphological similarity to baculoviruses. Stoltz and Faulkner (1978) found baculovirus-like particles in the calyxes of *Cotesia melanoscela* and in the hemocytes of tussock moth larvae parasitized by this wasp. These early studies demonstrated that the bracovirus particles are formed in a "virogenic stroma" in wasp calyx cells in the absence of cellular division, and are packaged in capsids that are delivered to parasitized insects, where they infect lepidopteran host cells, but do not replicate. As a result of these descriptions these particles were classified as viruses, initially as a subgroup of baculoviruses (Stoltz and Faulkner, 1978).

2. *Associations*

a. Model / Studied Associations. BV biology has been studied in several BV–wasp systems. *Chelonus inanitus* BV (*Ci*BV), *Cotesia rubecula* BV (*Cr*BV), *Cotesia congregata* BV (*Cc*BV), and *Microplitis demolitor* BV (*Md*BV) have served as the primary models, with *Cotesia melanoscela* BV (*Cm*BV) identified as the type bracovirus (Webb *et al.*, 2000).

b. Phylogeny of Associations. The bracoviruses are found only in the "microgastroid" complex of braconid wasps (Stoltz and Whitfield,

1992), which comprises eight subfamilies. The Cardiochilinae, Cheloninae, Miracinae, and Microgastrinae are the four largest subfamilies, with all described members of these subfamilies having BV associations (Fig. 1) (Whitfield, 1997). Though possession of the virus has only been demonstrated in a few cases, all species of the eight subfamilies in the microgastroid complex are thought to have BVs (Stoltz and Whitfield, 1992; Whitfield, 1997). Morphological and molecular analyses of the microgastroid complex clearly support a common progenitor for the entire complex, suggesting the bracovirus lineage is also monophyletic (Whitfield, 1997, 2000).

Several characteristics of BVs also support monophyly including their DNA genome organization, envelope origin, and capsid structure (Stoltz *et al.,* 1976). Although all evidence indicates that the bracovirus lineages are monophyletic, the viruses themselves have clearly diverved significantly over their ~60 million-year association with braconid wasps.

c. Ecology of Associated Wasps. The microgastroid complex wasps are endoparasitoids, primarily of larval lepidopterans, although some are egg parasitoids (e.g., *Chelonus inanitus*). Interestingly, BVs are not associated with braconid ectoparasitoids or parasitoids of dipteran hosts (Stoltz and Vinson, 1979b).

B. Bracovirus Life Cycle

Like that of IVs, the bracovirus life cycle can be considered to have two "arms" with virus replication and transmission occuring in the wasp host and functional gene expression occuring in lepidopteran hosts (Stoltz, 1993). However, BVs are morphologically distinct, they are released from calyx cells by lysis, and there is no evidence of nucleic acid homology between the two PDV genera. Although no BV genomes have been fully sequenced, the available data suggest that their structural organization is distinctive, as is the distribution of viral genes among the viral segments. Bracovirus studies have focused on the virus function in the parasitized host and its role in altering host developmental and immunological systems.

1. Replicative Host

a. Replication. As in IVs, bracovirus genomes are integrated into the host wasp genome (Gruber *et al.,* 1996; Savary *et al.,* 1999), although proviral BV segments appear to be tandemly arrayed (Savary *et al.,* 1997) rather than dispersed. BVs are vertically transmitted as proviruses in the host wasp genome as demonstrated via maintenance of genetic markers through crosses (Stoltz *et al.,* 1986). Like IVs, excision of BV segments in the pupal and adult calyx cells is followed by massive replication in the calyx cells (Gruber *et al.,* 1996). BVs appear to exit cells by lysis, as budding has not been observed, and cellular debris is present in the calyx fluid of *Co. congregata,* although such debris is not always observed in braconid calyx fluid (Stoltz *et al.,* 1976). Braconid calyx cells appear to continually differentiate from stem cells at the calyx/ovary junction, as cells having incompletely developed virogenic stroma are observed (Stoltz and Vinson, 1979b).

Circular and integrated BV forms are present in males and females, though ratios of the circular to the integrated forms vary significantly depending on tissue (Savary *et al.,* 1997). However, episomal viral DNA was not detected in haploid males, indicating that BV segment excision from male genomes may require diploidy, an unusual state in braconid wasps. Excision of the *Cc*BV *EP1* gene demonstrated that recombination between viral terminal repeats results in formation of an episomal circle and the rejoined wasp chromosome, with loss of a single pair of the repeats (Savary *et al.,* 1997). The integration junctions consist of 282- and 177-bp palindromic direct repeats at the 5′ and the 3′ ends, respectively. The direct repeat junctions exhibit sequence similarity to the Hin family of recombinases, a characteristic shared with the *Ci*BV repeats (Gruber *et al.,* 1996). A single A/T base pair is lost during excision of the *EP1* gene, indicating specificity of the location of recombination, and excision results in an empty locus following rejoining of the wasp sequences (Savary *et al.,* 1997).

b. Bracovirus Capsids. Bracovirus nucleocapsids are assembled in specialized regions of calyx cell nuclei designated the virogenic stroma, which are characterized by an electron-dense center with partially assembled virions around the periphery (Stoltz and Vinson, 1979b). BV nucleocapsids exhibit more morphological heterogeneity than do IV nucleocapsids. Bracovirus nucleocapsids are uniform in diameter (30–40 nm), but variable in length (30–150 nm), although discrete size classes are discernible (Stoltz and Vinson, 1979b). Capsids may acquire a single envelope at the virogenic stroma or multiple nucleocapsids may be embedded within a protein matrix that is singly enveloped (Stoltz and Vinson, 1977). The envelope may be irregular and possess protrusions, particularly in *Cotesia* spp. BVs (Stoltz and Vinson, 1979b). A layer of electron-dense material may lie between the nucleocapsid and envelope, though the nature of this material has not been determined. The size and dimensions of BV capsids and genome segments suggest that individual genome segments may be encapsidated with variable capsid length reflecting packaging of genome segments of different sizes. Evidence for this is provided by the observation that different capsid sizes release nucleic acids with different molecular weights (Albrecht *et al.,* 1994).

Stoltz and Vinson (1979b) noted that BV capsids are morphologically similar to the *Oryctes* baculoviruses. Capsids of *Cotesia* spp. BVs have terminal envelope protrusion or "tails," particularly in cases where multiple nucleocapsids are enveloped. Tails range from 2 to 300 nm in length and are ~8.5 nm in diameter, and morphologically are similar to *Oryctes* baculoviruses capsids as well (Stoltz and Vinson, 1979b). *Cn*BV and *Ch. altitudinis* BV nucleocapsids also possess terminal tail-like structures, which average 60 nm in length (Stoltz and Vinson, 1979b; Varricchio *et al.,* 1999), suggesting that tails may be a ubiquitous BV feature.

*Cm*BV and *Cn*BV virions comprise 15–20 polypeptides (Stoltz and Vinson, 1979b). Antisera raised against *Ci*BV reacts with approximately six proteins from purified *Ci*BV (Soller and Lanzrein, 1996), though a more complex protein composition is possible if the immunoreactive proteins contain immunodominant epitopes. Bracovirus virions also have lipid and carbohydrate constituents (Webb *et al.,* 2000), though these have not been characterized in detail.

2. Pathogenic Host

a. Cell Entry. Bracovirus virions are observed in association with host basement membranes by 45 min p.p. (Stoltz and Vinson, 1979b). It has been suggested that the virion envelope protrusions are involved in either or both basement membrane penetration and cell entry (Stoltz and Vinson, 1979b), although phagocytosis is not evident. Nucleocapsids align with nuclear pores within 1 hr p.p., with the tail penetrating the nuclear envelope. Empty nucleocapsids are observed outside the nucleus by 2 hr p.p., suggesting that uncoating of viral DNA occurs through the nuclear pores (Stoltz and Vinson, 1979b). This process is similar to the entry of some baculoviruses into the nucleus because viral DNA appears to dissociate from the capsid without total capsid destruction.

BV nucleocapsids are observed primarily in the cytoplasm of fat body, muscle, and blood cells, but also are observed in other cell types at lower frequency (Stoltz and Vinson, 1977). In *P. includens, Md*BV DNA titers are highest in hemocytes, with an estimated 90% of viral DNA found in this tissue (Strand *et al.,* 1992). Intriguingly *Glyptapanteles indiensis* BV (*Gi*BV) infects some nonlepidopteran insect cell lines (Gundersen-Rindal *et al.,* 1999), suggesting that cell entry does not limit BV host range, but that other barriers to functional virus expression exist.

b. Viral DNA Persistence and Expression. In general bracovirus DNA can be detected in parasitized larvae from the time of infection to the time of parasite eclosion. However, viral DNA titer is known to decline in some systems, possibly due to host antiviral reponses, and transcript levels vary temporally as well. Strand *et al.* (1992) detected *Md*BV DNA in the host *P. includens* through at least 6 days p.p., whereas *Cc*BV DNA is present for up to 10 days p.p. in *M. sexta* (Lavine and Beckage, 1996). Viral titer does not increase after infection of lepidopteran cells, indicating that replication occurs only within the female wasp.

BV transcription varies widely, with viral genes often expressed for shorter durations than are IV genes. For example, four *Md*BV transcripts are detected within parasitized *P. includens* (Strand *et al.*, 1997; Trudeau *et al.*, 2000). The four genes are transcribed by 24 hr p.p., and peak levels are reached by 48 hr p.p. and then decline to become almost undetectable by 5 days p.p., implying conserved regulatory mechanisms. At least three *Cc*BV genes are transcribed in parasitized *M. sexta* by 30 min p.p., making up more than 10% of total hemolymph protein content by 24 hr p.p. (Harwood *et al.*, 1994). However, transcript levels drop quickly and by 8 days p.p. are no longer detectable, even though viral DNA remains detectable up to 10 days p.p. and *Co. congregata* larval development requires ~14 days (Lavine and Beckage, 1996). Interestingly, in *Md*BV-infected *Spodoptera frugiperda*, a nonpermissive host, viral genes are briefly transcribed and suppress the immune system during the first few days postinfection. By day 5 postparasitization, viral DNA is present, but not transcribed, and the host fully recovers to encapsulate parasitoid larvae (Trudeau and Strand, 1998).

One possible explanation for the drop of viral transcription is that virus-infected cells are removed by host immune functions (e.g., apoptosis). In braconid systems, persistent virus expression may not be essential because other wasp factors may maintain immune suppression. In particular, teratocytes, an unusual cell type derived from serosal membranes of parasitoid eggs, exist in braconids and some other parasitoids (Dahlman and Vinson, 1993). The role of teratocytes in later stages of braconid immunosuppression is not clear, although they do seem to have developmental effects. Teratocytes synthesize large amounts of proteins following egg hatch (Schepers *et al.*, 1998), with secreted teratocyte proteins partly responsible for inhibition of development in braconid–BV–lepidopteran systems by posttranscriptional inhibition of host protein synthesis (Webb *et al.*, 2001). The presence of the teratocytes complicates BV studies in parasitized animals after parasitoid egg hatch. However, the activity of teratocytes likely maintains host suitability in the latter stages of braconid parasitoid development as BV transcription declines (Webb *et al.,* 2001).

C. Genome Organization

Bracovirus genomes are typically made up of fewer, larger segments than IV genomes (Fig. 4) (Webb, 1998). For example, *Gi*BV has 11 segments ranging from ~10 to greater than 30 kb (Gundersen-Rindal and Dougherty, 2000) and *Cn*BV has 6 major segments ranging from 6 to over 23 kb (Varricchio *et al.,* 1999). *Cc*BV has an unusually large number of segments for a BV, at least 20, which range from less than 10 to greater than 20 kb in size (de Buron and Beckage, 1992). Overall genome size estimates for BVs are ~200 kb (Webb *et al.,* 2000). Interestingly, all isolated BV genes are encoded by one or two segments within each BV genome (Savary *et al.,* 1997; Strand *et al.,* 1997; Trudeau *et al.,* 2000), with other segments, representing the majority of the genome, not known to encode genes. Bracovirus genomes are AT-rich, consisting of approximately 35% G + C content (Stoltz and Vinson, 1979b). The extent of repetitive sequence within BV genomes is not known, though sequenced individual segments can be highly repetitive (Albrecht *et al.,* 1994; K. Kadesh and M. R. Strand, personal communication).

Polymorphisms are known from BV genomes, with polymorphic loci present in individual laboratory wasps of a *Cm*BV laboratory colony having a highly similar genetic background (Stoltz and Xu, 1990). Relaxed selection pressures due to the presence of duplicate genes may contribute to polymorphic loci in BV genomes. Stoltz and Xu (1990) suggested that inbreeding could lead to a higher level of polymorphisms due to continual interactions between homologous sequences. If this does occur, it may indicate that BV genomes not only tolerate genetic variation, but may have evolved to facilitate its generation.

D. Gene Families

Gene families are observed in BVs as well as IVs (Table I). BV genomes appear to have fewer gene families and gene family members than IVs, but this may reflect the localization of BV genes to a small subset of genome segments. Multigene families are described from most studied bracoviruses, with all members studied to date expressed in the parasitized host.

*Md*BV-parasitized *P. includens* hemocytes express at least six viral genes by 4 hr p.p. (Strand *et al.,* 1992). *Md*BV encodes at least a single multigene family of three members, all on segment O (Strand *et al.*, 1997; Trudeau *et al.*, 2000). The three genes encode a conserved 5′ cysteine rich-region which exhibits similarity to epidermal growth factor motifs (Strand *et al.*, 1997); the region also is similar to a hookworm serpin which has anticoagulant properties (Trudeau *et al.*, 2000). A fourth gene has been identified on *Md*BV segment O, but is not known to be part of a multigene family.

Three similar genes (*EP1*, *EP2*, *EP3*) are expressed by 30 min p.p. in *M. sexta* parasitized by *Co. congregata* (Beckage, 1993). By 24 hr p.p., the *EP* gene products make up over 10% of the total hemolymph proteins (Harwood *et al.*, 1994). The three early-expressed proteins are predicted to be highly glycosylated and similar in size (Beckage *et al.*, 1987; Harwood *et al.*, 1994). They also are encoded on the same *Cc*BV genome segment (Harwood *et al.*, 1994). An *EP1*-specific probe hybridized to two size classes of mRNA, indicating the possibility of additional family members (Harwood *et al.*, 1994).

CrV1 and *CrV2* are the only viral transcripts detected in *Cr*BV-infected animals (Asgari *et al.*, 1996). Both are expressed between 4 and 8 hr p.p. and reach peak levels at 6 hr p.p. *CrV2* transcript is found only in the fat body, whereas *CrV1* transcription is detected in fat body and hemocytes. *CrV1* transcripts are most abundant, though the functions of neither the *CrV1*- and nor *CrV2*-encoded proteins are known.

*Cn*BV potentially encodes a gene family. When *CnPDV1*, a gene transcribed in the lepidopteran host, was used to probe viral DNA, at least three cross-reacting bands were observed (Varricchio *et al.*, 1999), suggesting either segment nesting, a phenomenon which has not been reported in bracoviruses, or a multigene family.

Unlike IVs, in which each segment appears to encode one or more genes, BVs appear to encode all expressed genes on one or two segments (Table I). The functional and evolutionary significance of this phenomenon is unknown, but intriguing.

E. Bracovirus Functions

The majority of BV genes are thought to be involved in suppression of host immune responses. Although viral proteins are not noticeably similar structurally, their similar pathologies and the monophyly of BVs suggest that common pathways are manipulated by BVs, particularly those involved in host nonself recognition and regulation of hemocyte populations.

1. Immune System

a. Inhibition of Antiviral Response. Although insect antiviral immune responses are not well studied, some data suggest that BVs reduce the efficacy of lepidopteran antiviral immunity. *Manduca sexta* is a semipermissive host for the baculovirus *Autographa californica* M nucleopolyhedrovirus (AcMNPV), but is more permissive when coinfected with *Cc*BV (Washburn *et al.,* 2000). As in *Cs*IV-infected *H. zea* larvae, increased permissivity may be linked to alterations in hemocyte function. *H. zea* hemocytes aggregate around foci of AcMNPV infection and induce melanization, possibly as a means of reducing virus dispersal (Trudeau *et al.,* 2001). Changes in relative and absolute hemocyte numbers following BV infection (see below) likely influence antiviral responsiveness.

b. Inhibition of Cellular Response. Parasitization or injection of purified BV results in hemocyte clumping and morphological abnormalities. In *Cc*BV, these symptoms occur in the majority of hemocytes by 24 hr p.p., but are no longer evident by 8 days p.p. (Lavine and Beckage, 1996). In *Md*BV-infected *P. includens,* behavioral abnormalities are observed 2 hr p.p. (Strand and Noda, 1991), but last the duration of infection (approximately 2 weeks) (Strand *et al.,* 1992). In both systems viral DNA is present throughout the course of development.

The CrV1 protein from *Cor*BV induces actin filament destabilization by 6 hr postinfection (Asgari *et al.,* 1996); injection of cytochalasin D induces similar changes (Asgari *et al.,* 1997). Changes in hemocyte cytoskeletal proteins are correlated with reductions in hemocyte adhesion, spreading, and encapsulation. Virally induced actin cytoskeleton changes also correlate with changes in several surface markers on *Cor*BV-infected hemocytes. For example, both a mucin-like glycoprotein and phosphatidylserine, normally present on the surface of healthy hemocytes, are absent 24 hr after exposure to recombinant CrV1 protein (Asgari *et al.,* 1997). Mucins are a class of proteins which mediate cell–cell communication (Kramerov *et al.,* 1996) and cell adhesion during encapsulation (Kotani *et al.,* 1995) via lipophorin binding (Theopold and Schmidt, 1997). Mucin presence on the outer surfaces of healthy hemocytes is upregulated following multicellular immune challenge (Theopold and Schmidt, 1997). Mucin transport is mediated by actin filaments (Oliver and Specian, 1990), suggesting that cell surface mucins are directly impacted by virus-induced changes in the hemocyte cytoskeleton. External membrane levels of phosphatidylserine are reduced as well following application of recombinant CrV1 (Asgari *et al.,* 1997). In vertebrates, phosphatidylserine is involved in coagulation and utilizes the actin cytoskeleton for cytoplasmic transport (Kunzelmann-Marche *et al.,* 2000). Thus the alterations of these cell surface markers may be linked to inhibition of insect cellular immunity and mediated by actin cytoskeleton disruptions.

In some bracovirus systems, suppression of cellular immune responses to parasite eggs is accompanied by formation of hemocyte aggregations, which may indicate misdirection of cellular immune responses. Injection of the *Cc*BV virion causes hemocyte aggregation with the subsequent clearance of these aggregations accompanied by recovery of immune responses (Lovallo and Cox-Foster, 1999). FAD–glucose dehydrogenase (GLD) is activated only in immune-activated plasmatocytes, making it a useful immune response marker (Cox-Foster and Stehr, 1994). Lovallo and Cox-Foster (1999) observed activated GLD following injection of ultraviolet-inactivated *Cc*BV, indicating that viral transcription is not required to induce this immune response. Uninfected hemocytes overlaid with purified *Coc*BV were still able to adhere, but preferentially reacted with other hemocytes rather than to an introduced egg. Also, there was a reduction in overall adherence to glass (i.e., a foreign object) and formation of clumps of cells adhering to each other. Therefore, the virions were suggested to function as decoys that induced immune reactions to virus or virus-infected cells at higher frequency than to the wasp egg. Cells participating in such aggregations would be subject to GLD-induced effects, which include induction of apoptosis and cytolysis. Thus, GLD induction may contribute to changes in hemocyte populations and abrogation of an appropriately targeted encapsulation response to wasp eggs. The virus not destroyed during this "autoimmune response" is transcribed, presumably leading to sustained inhibition of immune responses via other mechanisms.

Relative hemocyte counts undergo radical shifts following bracovirus infection (reviewed in Stettler *et al.,* 1998). Total hemocyte counts temporarily increase in larvae parasitized by *M. demolitor,* although encapsulation of wasp eggs is inhibited (Strand and Noda, 1991). *Md*BV infection of *P. includens* preferentially targets granulocytes, the primary circulating nonself recognition cells, for apoptosis, leading to changes in relative hemocyte counts following *Md*BV infection (Strand and Pech, 1995). Whether other BVs target specific hemocyte subpopulations for death or alteration is unknown. Cellular debris and hemocyte surface blebbing are observed in *Cc*BV-infected *M. sexta,* indicating cytolysis (Lavine and Beckage, 1996), though apoptosis was not observed following either *Cr*BV infection or viral protein application (Asgari *et al.,* 1996); in neither case were relative hemocyte counts reported.

Prior to oviposition, the eggs of *Co. rubecula* are coated with proteins that inhibit encapsulation by *Pieris rapae* hemocytes. These wasp-encoded proteins adhere to the surface of the egg via a putative N-transmembrane region (Asgari *et al.,* 1998), but may be removed via sodium dodecyl sulfate (SDS) treatment (Asgari and Schmidt, 1994). Antisera raised against *Cr*BV cross-react with the egg surface proteins, but SDS treatment eliminates this signal. When this protein layer is removed by washing or altered by treatment of the egg surface with *Cr*BV antibody, eggs are encapsulated even in the presence of virions (Asgari and Schmidt, 1994). Inhibition of egg encapsulation can be rescued by application of recombinant Crp32, one of the wasp-encoded surface proteins (Asgari *et al.,* 1998). Interestingly, antisera raised against the Crp32 protein and *Cr*BV cross-react with two native *P. rapae* proteins (Asgari *et al.,* 1998). These data suggest that the proteins bound to the egg surface may "camouflage" the egg, allowing passive evasion of the immune response.

A novel type of protein encoded by *Co. kariyai* BV is observed in the hemolymph of parasitized *Pseudaletia separata* (Hayakawa *et al.,* 1994). The 470-kDa hexamer, similar to the lipid-carrier arylphorin, was present in high levels. The protein inhibits degranulation of hemocytes, presumably resulting in reduced release of cytotoxic molecules and granulocyte opsonins, although this has yet to be experimentally verified.

c. Inhibition of Humoral Response. *Cc*BV infection significantly reduces melanotic encapsulation in *M. sexta* larvae by 24 hr p.i., although ability to encapsulate returns to preinfection levels by 8 days p.i. (Lavine and Beckage, 1996). A dose-dependent effect is seen in *Md*BV infection of *P. includens,* which signficantly reduces PO activity; dose dependence is especially evident in synergistic interactions between venom and calyx fluid (Strand and Noda, 1991). Published accounts of *Cc*BV (Lavine and Beckage, 1996), on the other hand, do not report dose-dependent effects of the virus on the humoral response, though the studies did not look at venom–BV interactions.

2. Developmental Alterations

Lepidopteran larvae parasitized by wasps carrying BVs usually show aberrant growth and development. Initiation of metamorphosis may be induced prematurely (i.e., in an early instar) or delayed (i.e., a supernumery larval instar) (Vinson and Iwantsch, 1980; Lawrence and

Lanzrein, 1993). Typically some developmental effects are induced by virus infection, although other wasp-derived factors such as teratocytes have important roles (Dahlman and Vinson, 1993). The virally induced developmental alterations primarily are associated with changes in host hormone titers. For example, *M. croceipes*-parasitized *H. virescens* larvae prematurely initiate metamorphosis, indicating reduced JH titers and occurrence of the ecdysteroid commitment peak, which typically reprograms larval cells for commitment in anticipation of the pupal molt (Webb and Dahlman, 1986). However, the second ecdysteroid peak, which induces synthesis of the pupal cuticle, does not occur in parasitized or virus-infected larvae. *Manduca sexta* larvae injected with *Cc*BV and *C. congregata* venom have elevated JH titers (Balgopal *et al.,* 1996), although not as high as occur in naturally parasitized larvae. This suggests involvement of other factors, including the parasite itself, in manipulation of host development (Lawrence and Lanzrein, 1993).

Rather than altering JH titers, *Cn*BV reduces ecdysteroid titer to arrest host development through suppression of prothoracic gland (PTG) synthesis of ecdysteroids (Pennachio *et al.,* 1997). Reduced phosporylation of proteins of the prothoracicotropic hormone–cAMP–prothoracic gland pathway is associated with reductions in ecdysone synthesis (Pennachio *et al.,* 1998a, b). *Cn*BV potentially encodes several genes with protein kinase inhibitor (PKI) activity, as a mouse PKI probe hybridized with viral genomic DNA (Varricchio *et al.,* 1999). At least one *Cn*BV transcript, *CnPDV1*, localizes to PTG cells with irregular morphology (Varricchio *et al.,* 1999). The transcript has a single putative N-linked glycosylation site and four potential phosphorylation sites and lacks a signal peptide. Although *CnPDV1* does not exhibit any similarities to PKI genes in the NCBI nr(nonredundant) database, indicating that it may not be the source of the PKI activity, it may be the source of observed cytological abnormalities.

The *Ci*BV–*Spodoptera littoralis* system is unusual in that *Chelonus inanitus* parasitizes the lepidopteran egg, a host stage lacking a functional immune system, but completes larval development in and emerges from the larval stage. Premature metamorphosis of *Ch. inanitus*-parasitized *S. littoralis* during the fifth larval instar requires precocious depression of host JH (Pfister-Wilhelm and Lanzrein, 1996; Steiner *et al.,* 1999). Introduction of calyx fluid (=*Ci*BV) and venom induces host developmental arrest, whereas injection of anti-*Ci*BV antisera prior to parasitization rescues development, suggesting that the virus is the factor that alters host development (Soller and Lanzrein, 1996). Viral transcripts are evident in parasitized eggs and increase in titer until the penultimate larval instar (Johner *et al.,* 1999), when parasitized *S. littoralis* prematurely metamorphose. *Ci*BV also inhibits encapsulation during the larval stage via unknown mechanisms (Stettler *et al.,* 1998).

V. Origin of the Polydnaviridae

Stoltz (1993), in considering PDV biology, defined viruses as parasitic nucleic acid genomes encapsidated in a protein coat which aids in entry into host cells. Two differences between viruses and other mobile nucleic acid elements are the presence of the viral protein capsid (Holland and Domingo, 1998) and possession by the virus of "an evolutionary history independent of that of its host" (Strauss *et al.,* 1996). Although PDVs possess the former characteristic, PDVs do not possess an "evolutionary history independent of its host" at present. If one includes as part of its evolutionary history their putative viral precursors, then PDVs arguably meet the latter definition of "virus." In closing this review, we consider the steps through which a viral or transposable element PDV antecedent would have to proceed. The two possibilities considered for PDV origin are (1) they are evolutionary derivatives of viruses which have become replication-defective, or (2) they are wasp genetic secretions, possibly related to transposable elements, that have acquired a virus-like morphology. We then will conclude by describing some research directions that may contribute to elucidating the evolutionary origin of PDVs.

A. Evolutionary Model I: A Viral Ancestor

If PDVs are viral derivatives, they should share characteristics of other DNA viruses. However, the dichotomous selection pressures imposed by the PDV life cycle could have major impact on genome structure and composition. The relationships between BV and baculovirus have been discussed, and some similarities exist between IV and ascovirus morphology as well (Miller, 1998). If these overall structural similarities are due to a common origin, then one would expect that at least some of the genes encoding structural proteins would be conserved. However, a thorough structural comparison of PDV structural proteins to other insect viruses is not possible because few genes encoding PDV structural proteins have been isolated and the genes encoding these proteins are not (always) encapsidated.

PDVs are the only segmented DNA viruses, a characteristic otherwise associated with RNA viruses. The similarity between segments and the existence of gene family members on several segments indicate that PDV genome segmentation may have resulted from repeated duplication of shorter sequences, rather than from simple fragmentation of a larger genome. We believe the selection pressures leading to this duplication and organization are imposed by the unusual PDV life cycle and result in the characteristics shared by IV and BV lineages. Diversity within lineage then would increase with wasp speciation, divergence, and exploitation of novel lepidopteran hosts.

Reduction in genome size is expected in a virus obligately associated with a single host through permanent genome integration. Functional redundancy between two symbiotic genomes should be reduced as the nuclear genome acquires or displaces essential symbiont genes (Berg and Kurland, 2000). Lateral gene transfer from endosymbiont to host also is predicted if the nuclear environment has a slower rate of change than the genome of the associate (Berg and Kurland, 2000). PDV structural genes therefore might be expected to be transfered to the nuclear genome to reduce the probability of inactivating mutations occurring in a genome, which otherwise appears to facilitate high rates of mutation and recombination. The removal of the viral sequences essential for virus replication and transmission from the PDV genome would relax selection on the encapsidated genome, thereby allowing relatively increased mutation rates and production of novel gene variants in the "viral" genome. Therefore, an integrated DNA virus, genetically passed by nonpathological vertical transmission, could exhibit higher rates of change and more effective coevolution with host immune responses in the pathological host.

PDV interactions with the lepidopteran host impose different evolutionary constraints on PDV genomes. As PDVs alter host immune and developmental physiology, they will be selected for compensatory mutations as the lepidopteran genome develops resistance to viral virulence genes. The ability of PDV genomes to generate novel mutations is enhanced by their large size, potential for recombination between and among segments, and highly repetitive sequence. These factors likely contribute to the relatively large PDV genome sizes and diversity within and among PDV gene families. Multiple copies of related genes may also contribute to gene dose effects, as is described for the *Cs*IV cys-motif proteins (Cui *et al.,* 2000). Multigene families also allow expansion of physiological function (Cooke *et al.,* 1997), and thus may potentiate host range expansion.

Genome structure and mode of replication may enhance the ability of PDVs to acquire genes from the host genome. The *Hd*IV and *Cs*IV vinnexin genes are clearly homologous to cellular genes over their entire ORF (M. W. Turnbull, A.-N. Volkoff, and B. A. Webb, unpublished). The high level of homology supports the idea that wasp-to-virus xenologous recombination may occur. Xenologous recombination is a common evolutionary mechanism exploited by large DNA viruses to acquire genes allowing manipulation of host immune responses and development (e.g., herpesvirus host genes; Raftery *et al.,* 2000). Specifically, insect viruses have been shown to acquire host genetic material for manipulation of host immune response (e.g., baculovirus alteration of the apoptotic response; Huang *et al.,* 2000) and development (e.g., baculovirus inactivation of host ecdysteroids by glycosylation; O'Reilly and Miller, 1989). Such gene transfer from wasp to PDV could rapidly expand pathological mechanisms. Similarities between Hymenoptera and Lepidoptera immune and developmental pathways are indicated by conservation of key immune response molecules and pathways (Gillespie *et al.,* 1997; Hughes, 1998). Therefore acquisition by PDVs of wasp regulatory genes may provide an effective means for disrupting lepidopteran immune responses. The selection pressures and mechanisms of diversification may rapidly distance PDVs from an ancestral viruses, complicating comparisons between PDVs and known viruses.

A final consideration relevant to the potential for free-living viruses to evolve toward mutualistic relationships is the documented beneficial relationships between the ascovirus *Dp*AV and a facultative symbiont, a parasitoid wasp, *Diadromus pulchrellus* (Bigot *et al.,* 1995). In this system, *Dp*AV is normally vectored by the *D. pulchrellus* with ascovirus infection having little adverse effect on the wasp. By contrast, a second parasitoid, *Itoplectis tunetana,* is adversely affected by the rapid replication of *Dp*AV in the parasitized lepidopteran host. Evidence suggest that *Dp*AV slows virus replication when in association with *D. pulchrellus* to the mutual benefit of the wasp and ascovirus even though this wasp–virus association is clearly not obligatory for either. The *Dp*AV–*D. pulchrellus* system is an intriguing ecological intermediate between a free-living virus that is vectored by parasitic wasps (e.g., all ascoviruses) and polydnaviruses, which are obligatory mutualists of parasitoids.

B. Evolutionary Model II: A Wasp-Encoded Episome

A second potential origin of viruses is the encapsidation of cellular material and eventual evolution of a replicating parasitic element (e.g., Sinkovics *et al.,* 1998). In this light PDVs may be early intermediates in the formation of a classical virus. PDVs exhibit some genetic characteristics more commonly associated with eukaryotes than viruses, such as large amounts of noncoding sequence, introns, and multigene families. We have suggested possible explanations for the origin and function of the repetition in the PDV genomes. However, few viruses possess such high ratios of intergenic nucleic acid. The *Cs*IV genome is approximately two-thirds noncoding sequence (Webb *et al.*, unpublished), and initial sequencing of the *Md*BV genome suggests that all genes may be present on one or two segments (K. Kadesh and M. R. Strand, personal communication). Introns also are fairly uncommon in viruses, but PDVs in both lineages have genes encoding intronic regions. The introns of *Cs*IV cys-motif genes are highly conserved relative to the flanking exons, though selection appears neutral (Dib-Hajj *et al.*, 1993; M. W. Turnbull and B. A. Webb, unpublished). A likely origin for introns is through incorporation of host genes. We have suggested that this occurs in PDVs, but the genetic structure of the vinnexins, the only homologous genes, differs from the genetic structure of the known xenologues in lacking introns, raising the possibility of recombination involving an mRNA intermediary. Because viruses typically do not encode introns, a parsimonious explanation for the presence of large nongenic regions is that they are introduced by encapsidation of wasp genetic material, but the repetitive nature of PDV segments may not be consistent with direct transfer of wasp noncoding sequence to PDVs.

Several immunological studies have detected extensive cross-reaction between anti-PDV antisera and host-encoded proteins. It is apparent that the *C. sonorensis* ovarian proteins and venom, for example, react with anti-*Cs*IV antibodies due to shared epitopes. Cross-reactivity of wasp and PDV components supports the use of wasp structural genes and proteins in capsid construction. Certainly, a striking difference between PDVs and other viruses is the dearth of encapsidated structural and replicative genes. If PDVs are of viral descent, then lateral transfer may explain this phenomenon. However, the intermediate states that would be predicted in cases of lateral transfer are not described, although this issue could be approached by comparing phylogenetically related viruses. Traits leading to excision and packaging of wasp genetic elements that enhance parasitoid larva survival would quickly fix in populations.

C. Possible Future Avenues of Research

PDV genome structure and genes are clearly unusual from a virological perspective, but studies that would directly address PDV evolution are limited because most work on PDVs has focused on transcript analysis or characterization of individual viral genes. Studies that will more directly address PDV origins and evolution are underway, but require comparative genome-level analyses to be truly informative. Perhaps as a result of the unusual selection pressures under which PDVs have evolved, traditional methods of comparative virology have not clearly identified antecedent viral groups. This appears to be because polydnavirus genes are unrelated to known viral genes and not closely related to any database proteins. Conversely, ubiquitous viral genes (e.g., polymerases) are not encoded in PDV genomes.

Comparative genomics is potentially a rich source for understanding PDV evolution, and has already led to the identification of two shared gene families from two unrelated IVs, *Cs*IV and *Hd*IV (B. A. Webb *et al.,* unpublished). Similarly, studies of EP1 genes indicate that this gene is present in most, but probably not all BVs (Whitfield, 2000). As PDV databases develop to a point that comparison of expressed sequence tag (EST) sequences and genomes is possible, viral genomes and gene families may be considered from a more informed evolutionary perspective. Such an approach will need to include both closely and more distantly related PDVs to track recombination among and within viral segments, as well as the rate of change of individual genes and gene families. As PDV genes with structural similarities to known genes are uncovered, gene origin may be determined through comparison with other insect and possibly virus genes. Potentially, directed analyses of more basal PDV-containing lineages may lead to analyses of informative intermediate PDVs and provide insight into reduction of viral genomes to the point of extinction (e.g., *V. canescens* VLPs) and potentially discover direct associations with conventional viruses.

Molecular analyses of associated wasp genes also may give insight into the PDV life cycle and viral gene evolution. Phylogenetic studies of the wasps have begun to outline viral phylogenies (Belshaw *et al.,* 1998; Whitfield, 1997), but a systematic sampling of wasp–virus association has not yet been undertaken. Genomic analyses of PDV-associated wasps may clearly identify wasp-derived PDV genetic material and identify regulatory genes controlling proviral segment excision and recombination. Comparative analyses of associated wasps will also allow study of wasp–virus coevolution and the impacts of viral gene diversificaton on wasp host range.

For the immediate future sequencing of viral genomes will continue to be an effective way to accumulate data relevant to the origins of PDVs. If PDVs are derivatives of conventional viruses, massive changes have occurred since the initial associations. However, some pathological and structural genes likely have been conserved, albeit possibly relocated to the wasp genome. In this context, the fact that baculoviruses and BVs encode genes that distort actin polymerization in the infected hosts may be relevant. Strategies designed to sequence viral genes and "wasp" genes expressed during virus replication seem likely to identify genetic homologies between PDVs, their ancestral viruses, and their insect hosts in the near future, which has potential to significantly improve our understanding of the origins of PDVs.

Cytogenetic studies may clarify the evolution of PDVs by identification of integration sites and modes of regulation. Analyses of integration location on wasp chromosomes have elucidated some of the similarities and dissimilarities of the two PDV lineages, and further studies should better inform us of PDV evolution at the gene, gene family, and genomic levels. Analyses of mechanisms of excision may elucidate the origin of PDV mobility as well. The regulatory mechanisms governing the viral life cycle, in both hosts, are relatively unclear, and may further identify similarities to possible viral ancestors and mobile elements.

VI. Conclusion

PDVs provide an interesting system for viral evolutionary studies. The unique life cycle involves coevolution with two hosts, one mutualistic and one pathogenic, thus imposing both reductive and diversifying selection pressures on viral genes. This probably has driven the convergent evolution of their highly modified genetic structure.

PDVs may represent genetic secretions evolving toward a classical virus, or an unusually extreme instance of convergent evolution of two virus–wasp mutualisms. PDVs are among the most fascinating biological associations known, but many aspects of their biology, required to better understand their evolutionary history, have yet to be examined. The presence of large multigene families in a virus is not unique, but the extent of diversity within *Cs*IV and other PDVs is exceptional. The evolutionary function of virus diversification is likely related to wasp diversification and host range expansion, but detailed studies examining this correlation have not been performed. Additional comparisons of PDV genome structure are required to adequately answer questions concerning the relationships between and among PDV segments and genes (e.g., the occurrence of only one gene family per segment in *Cs*IV and all BV genes occurring on a single segment). Finally, not only is the evolutionary origin of PDVs interesting from a basic perspective, but it also can provide novel insight into the evolution of virulence in classical viruses and other pathogens. There are surely few viral systems in which it is possible to approach virus evolution over a period of

60 million years. PDV biology will continue to offer interesting future research opportunities and unexpected challenges as their evolution and ancestry is elucidated.

References

Albrecht, U., Wyler, T., Pfister-Wilhelm, R., Gruber, A., Stettler, P., Heiniger, P., Kurt, E., Schumperli, D., and Lanzrein, B. (1994). PDV of the parasitic wasp *Chelonus inanitus* (Braconidae): Characterization, genome organization and time point of replication. *J. Gen. Virol.* **75,** 3353–3363.

Altschul, S. F., Madden, T. L., Schaffer, A. A., Zhang, J., Zhang, Z., Miller, W., and Lipman, D. J. (1997). Gapped BLAST and PSI-BLAST: A new generation of protein database search programs. *Nucleic Acids Res.* **25,** 3389–3402.

Andersson, S. G., and Kurland, C. G. (1998). Reductive evolution of resident genomes. *Trends Microbiol.* **6,** 263–268.

Asgari, S., and Schmidt, O. (1994). Passive protection of eggs from the parasitoid, *Cotesia rubecula,* in the host, *Pieris rapae*. *J. Insect Physiol.* **40,** 789–795.

Asgari, S., Hellers, M., and Schmidt, O. (1996). Host haemocyte inactivation by an insect parasitoid: Transient expression of a PDV gene. *J. Gen. Virol.* **77,** 2653–2662.

Asgari, S., Schmidt, O., and Theopold, U. (1997). A PDV-encoded protein of an endoparasitoid wasp is an immune suppressor. *J. Gen. Virol.* **78,** 3061–3070.

Asgari, S., Theopold, U., Wellby, C., and Schmidt, O. (1998). A protein with protective properties against the cellular defense reactions in insects. *Proc. Natl. Acad. Sci. USA* **95,** 3690–3695.

Avery, S. W., Jouvenaz, D. P., Banks, W. A., and Anthony, D. W. (1977). Virus-like particles in a fire ant, *Solenopsis* sp. (Hymenoptera: Formicidae) from Brazil. *Fla. Entomol.* **60,** 17–21.

Balgopal, M. M., Dover, B. A., Goodman, W. G., and Strand, M. R. (1996). Parasitism by *Microplitis demolitor* induces alterations in the juvenile hormone titers and juvenile hormone esterase activity of its host, *Pseudoplusia includens*. *J. Insect Physiol.* **42,** 337–345.

Barratt, B. I. P., Evans, A. A., Stoltz, D. B., Vinson, S. B., and Easingwood, R. (1999). Virus-like paricles in the ovaries of *Microtonus aeithiopoides* Loan (Hymenoptera: Braconidae), a parasitoid of adult weevils (Coleoptera: Curculionidae). *J. Invertbr. Pathol.* **73,** 182–188.

Beck, M., Theopold, U., and Schmidt, O. (2000). Evidence for serine protease inhibitor activity in the ovarian calyx fluid of the endoparasitoid *Venturia canescens*. *J. Insect Physiol.* **46,** 1275–1283.

Beckage, N. E. (1993). Games parasites play: The dynamic roles of peptides and proteins in the host–parasite interaction. *In* "Parasites and Paathogens of Insects" (N. E. Beckage, S. N., Thompson and B. A. Federici, Eds.), Vol. 1, pp. 25–58. Academic Press, New York.

Beckage, N. E., and Kanost, M. R. (1993). Effects of parasitism by the braconid wasp *Cotesia congregata* on host hemolymph proteins of the tobacco hornworm, *Manduca sexta*. *Insect Biochem. Mol. Biol.* **23,** 643–653.

Beckage, N. E., Templeton, T. J., Nielsen, B. D., Cook, D. I., and Stoltz, D. B. (1987). Parasitism-induced hemolymph polypeptides in *Manduca sexta* (L.) larvae parasitized by the braconid wasp *Cotesia congregata* (Say). *Insect Biochem.* **17,** 439–455.

Beliveau, C., Laforge, M., Cusson, M., and Bellemare, G. (2000). Expression of a *Tranosema rostrale* PDV gene in the spruce budworm, *Choristoneura fumiferana*. *J. Gen. Virol.* **81,** 1871–1880.

Belshaw, R., Fitton, M., Herniou, E., Gimeno, C., and Quicke, D. L. J. (1998). A phylogenetic reconstruction of the Ichneumonoidea (Hymenoptera) based on the D2 variable region of 28S ribosomal RNA. *Syst. Entomol.* **23,** 109–123.

Berg, O. G., and Kurland, C. G. (2000). Why mitochondrial genes are most often found in nuclei. *Mol. Biol. Evol.* **17,** 951–961.

Berg, R., Schuchmann-Feddersen, I., and Schmidt, O. (1988). Bacterial infection induces a moth (*Ephestia kuhniella*) protein which has antigenic similarity to virus-like particle proteins of a parasitoid wasp (*Venturia canescens*). *J. Insect Physiol.* **34,** 473–480.

Bevans, C. G., Kordel, M., Rhee, S. K., and Harris, A. L. (1998). Isoform composition of connexin channels determines selectivity among second messengers and uncharged molecules. *J. Biol. Chem.* **273,** 2808–2816.

Bigot, Y., Drezen, J. M., Sizaret, P. Y., Rabouille, A., Hamelin, M. H., and Periquet, G. (1995). The genome segments of *Dp*RV, a commensal reovirus of the wasp *Diadromus pulchellus* (Hymenoptera). *Virology* **210,** 109–119.

Blissard, G. W., Vinson, S. B., and Summers, M. D. (1986). Identification, mapping and *in vitro* translation of *Campoletis sonorensis* virus mRNAs from parasitized *Heliothis virescens* larvae. *J. Virol.* **57,** 318–327.

Blissard, G. W., Smith, O. P., and Summers, M. D. (1987). Two related viral genes are located on a single superhelical DNA segment of the multipartite *Campoletis sonorensis* virus genome. *Virology* **160,** 120–134.

Blissard, G. W., Theilmann, D. A., and Summers, M. D. (1989). Segment W of *Campoletis sonorensis* virus: Expression, gene products, and organization. *Virology* **169,** 78–89.

Brown, F. (1986). The classification and nomenclature of viruses: summary of results of meetings of the International Committee on Taxonomy of Viruses in Sendai. *Intervirology* **25,** 141–143.

Churchill, D., Coodin, S., Shivers, R. R., and Caveney, S. (1993). Rapid *de novo* formation of gap junctions between insect hemocytes *in vitro:* A freeze–fracture, dye-transfer and patch-clamp study. *J. Cell Sci.* **104,** 763–772.

Clark, K. D., Witherell, A., and Strand, M. R. (1998). Plasmatocyte spreading peptide is encoded by an mRNA differentially expressed in tissues of the moth *Pseudoplusia includens*. *Biochem. Biophys. Res. Commun.* **250,** 479–485.

Conticello, S., Gilad, Y., Avidan, N., Ben-Asher, E., Levy, Z., and Fainzilber, M. (2001). Mechanisms for evolving hypervariability: The case of conopeptides. *Mol. Biol. Evol.* **18,** 120–131.

Cook, D., and Stoltz, D. B. (1983). Comparative serology of viruses isolated from ichneumonid parasitoids. *Virology* **130,** 215–220.

Cooke, J., Nowak, M. A., Boerlijst, M., and Maynard-Smith, J. (1997). Evolutionary origins and maintenance of redundant gene expression during metazoan development. *Trends Genet.* **13,** 360–364.

Cox-Foster, D. L., and Stehr, J. E. (1994). Induction and localization of FAD–glucose dehydrogenase (GLD) during encapsulation of abiotic implants in *Manduca sexta* larvae. *J. Insect Physiol.* **40,** 235–249.

Cui, L., and Webb, B. A. (1996). Isolation and characterization of a member of the cysteine-rich gene family from *Campoletis sonorensis* PDV. *J. Gen. Virol.* **77,** 797–809.

Cui, L., and Webb, B. A. (1997). Homologous sequences in the *Campoletis sonorensis* PDV genome are implicated in replication and nesting of the W segment family. *J Virol.* **71,** 8504–8513.

Cui, L., Soldevila, A., and Webb, B. A. (1997). Expression and hemocyte-targeting of a *Campoletis sonorensis* PDV cysteine-rich gene in *Heliothis virescens* larvae. *Arch. Insect. Biochem. Physiol.* **36,** 251–271.

Cui, L., Soldevila, A. I., and Webb, B. A. (2000). Relationships between PDV gene expression and host range of the parasitoid wasp *Campoletis sonorensis*. *J. Insect Physiol.* **46,** 1397–1407.

Curtin, K. D., Zhang, Z., and Wyman, R. J. (1999). *Drosophila* has several genes for gap junction proteins. *Gene* **232,** 191–201.

Cusson, M., Laforge, M., Miller, D., Cloutier, C., and Stoltz, D. (2000). Functional significance of parasitism-induced suppression of juvenile hormone esterase activity in developmentally delayed *Choristoneura fumiferana* larvae. *Gen. Comp. Endocrinol.* **117,** 343–354.

Dahlman, D. L., and Vinson, S. B. (1993). Teratocytes: Developmental and biochemical characteristics. *In* "Parasites and Pathogens of Insects: Parasites" (S. N. Thompson, B. A. Federici, and N. E. Beckage, Eds.), Vol. I, pp. 145–165. Academic Press, New York.

Davies, D. H., and Vinson, S. B. (1988). Interference with function of plasmatocytes of *Heliothis virescens in vivo* by calyx fluid of the parasitoid *Campoletis sonorensis*. *Cell Tiss. Res.* **251,** 467–475.

Davies, D. H., Strand, M. R., and Vinson, S. B. (1987). Changes in differential haemocyte count and *in vitro* behaviour of plasmatocytes from host *Heliothis virescens* caused by *Campoletis sonorensis* PDV. *J. Insect Physiol.* **33,** 143–153.

de Buron, I., and Beckage, N. E. (1992). Characterization of a polydnavirus (PDV) and virus-like filamentous particle (VLFP) in the braconid wasp *Cotesia congregata* (Hymenoptera: Braconidae). *J. Invertebr. Pathol.* **59,** 315–327.

Deng, L., and Webb, B. A. (1999). Cloning and expression of a gene encoding a *Campoletis sonorensis* PDV structural protein. *Arch. Insect Biochem. Physiol.* **40,** 30–40.

Deng, L., Stoltz, D. B., and Webb, B. A. (2000). A gene encoding a PDV structural polypeptide is not encapsidated. *Virology* **269,** 440–450.

Dib-Hajj, S. D., Webb, B. A., and Summers, M. D. (1993). Structure and evolutionary implications of a "cysteine-rich" *Campoletis sonorensis* PDV gene family. *Proc. Natl. Acad. Sci. USA* **90,** 3765–3769.

Doucet, D., and Cusson, M. (1996). Role of calyx fluid in alterations of immunity in *Choristoneura fumiferana* larvae parasitized by *Tranosema rostrale*. *Comp. Biochem. Physiol.* **114A,** 311–317.

Dover, B. A., Davies, D. H., Strand, M. R., Gray, R. S., Keeley, L. L., and Vinson, S. B. (1987). Ecdysteroid-titre reduction and developmental arrest of last-instar *Heliothis virescens* larvae by calyx fluid from the parasitoid *Campoletis sonorensis* PDV. *J. Insect Physiol.* **33,** 333–338.

Dover, B. A., Davies, D. H., and Vinson, S. B. (1988a). Degeneration of last instar *Heliothis virescens* prothoracic glands by *Campoletis sonorensis* PDV. *J. Invertebr. Pathol.* **51,** 80–91.

Dover, B. A., Davies, D. H., and Vinson, S. B. (1988b). Dose-dependent influence of *Campoletis sonorensis* PDV on the development and ecdysteroid titers of last-instar *Heliothis virescens* larvae. *Arch. Insect Biochem. Physiol.* **8,** 113–126.

"Dupas, S., Turnbull, M. W., and Webb, B. A. (submitted). Diversifying selection in a parasitoid's symbiotic virus among genes involved in inhibiting host immunity."

Fleming, J. G., and Summers, M. D. (1986). *Campoletis sonorensis* endoparasitic wasps contain forms of *C. sonorensis* virus DNA suggestive of integrated and extrachromosomal PDV DNAs. *J. Virol.* **57,** 552–562.

Fleming, J. G., and Summers, M. D. (1991). PDV DNA is integrated in the DNA of its parasitoid wasp host. *Proc. Natl. Acad. Sci. USA* **88,** 9770–9774.

Foote, C. I., Zhou, L., Zhu, X., and Nicholson, B. J. (1998). The pattern of disulfide linkages in the extracellular loop regions of connexin 32 suggests a model for the docking interface of gap junctions. *J. Cell. Biol.* **140,** 1187–1197.

Ganfornina, M. D., Sanchez, D., Herrera, M., and Bastiani, M. J. (1999). Developmental expression and molecular characterization of two gap junction channel proteins expressed during embryogenesis in the grasshopper *Schistocerca americana*. *Dev. Genet.* **24,** 137–150.

Gardiner, E. M. M., and Strand, M. R. (1999). Monoclonal antibodies bind distinct classes of hemocytes in the moth *Pseudoplusia includens*. *J. Insect Physiol.* **45,** 113–126.

Gillespie, J. P., Kanost, M. R., and Trenczek, T. (1997). Biological mediators of insect immunity. *Annu. Rev. Entomol.* **42,** 611–643.

Gouranton, J. (1972). Development of an intranuclear non-occluded rod-shaped virus in some midgut cells of an adult insect, *Gyrinus notator* L. (Coleoptera). *J. Ultrastruct. Mol. Struct. Res.* **39,** 281–294.

Gruber, A., Stettler, P., Heiniger, P., Schumperli, D., and Lanzrein, B. (1996). PDV DNA of the braconid wasp *Chelonus inanitus* is integrated in the wasp's genome and excised only in later pupal and adult stages of the female. *J. Gen. Virol.* **77,** 2873–2879.

Gundersen-Rindal, D., and Dougherty, E. M. (2000). Evidence for integration of *Glyptapanteles indiensis* PDV DNA into the chromosome of *Lymantria dispar in vitro*. *Virus Res.* **66,** 27–37.

Gundersen-Rindal, D., Lynn, D. E., and Dougherty, E. M. (1999). Transformation of lepidopteran and coleopteran insect cell lines by *Glyptapanteles indiensis* PDV DNA. *In Vitro Cell. Dev. Biol. Anim.* **35,** 111–114.

Haren, L., Ton-Hoang, B., and Chandler, M. (1999). Integrating DNA: Transposases and retroviral integrases. *Annu. Rev. Microbiol.* **53,** 245–281.

Harwood, S. H., Grosovsky, A. J., Cowles, E. A., Davis, J. W., and Beckage, N. E. (1994). An abundantly expressed hemolymph glycoprotein isolated from newly parasitized *Manduca sexta* larvae is a PDV gene product. *Virology* **205,** 381–392.

Hayakawa, Y., Yazaki, K., Yamanaka, A., and Tanaka, T. (1994). Expression of PDV genes from the parasitoid wasp *Cotesia kariyai* in two noctuid hosts. *Insect Mol. Biol.* **3,** 97–103.

Hilgarth, R. (1995). *Physiological effects of Campoletis sonorensis calyx fluid on Heliothis virescens*. M. S. Thesis, Kansas State University, Manhattan, Kansas.

Hilgarth, R., and Webb, B. A. Characterization of *Campoletis sonorensis* Segment I Genes as Members of the *repeat element* Gene Family. *J. Gen. Virol.,* in press.

Holland, J., and Domingo, E. (1998). Origin and evolution of viruses. *Virus Genes* **16,** 13–21.

Huang, Q., Deveraux, Q. L., Maeda, S., Salvesen, G. S., Stennicke, H. R., Hammock, B. D., and Reed, J. C. (2000). Evolutionary conservation of apoptosis mechanisms: Lepidopteran and baculoviral inhibitor of apoptosis proteins are inhibitors of mammalian caspase-9. *Proc. Natl. Acad. Sci. USA* **97,** 1427–1432.

Hughes, A. L. (1998). Protein phylogenies provide evidence of a radical discontinuity between arthropod and vertebrate immune systems. *Immunogenetics* **47,** 283–296.

Jiang, J. X., and Goodenough, D. A. (1996). Heteromeric connexons in lens gap junction channels. *Proc. Natl. Acad. Sci. USA* **93,** 1287–1291.

Johner, A., Stettler, P., Gruber, A., and Lanzrein, B. (1999). Presence of PDV transcripts in an egg-larval parasitoid and its lepidopterous host. *J. Gen. Virol.* **80,** 1847–1854.

Kim, M. K., Sisson, G., and Stoltz, D. B. (1996). Ichnovirus infection of an established gypsy moth cell line. *J. Gen. Virol.* **77,** 2321–2328.

Kinuthia, W., Li, D., Schmidt, O., and Theopold, U. (1999). Is the surface of endoparasitic wasp eggs and larvae covered by a limited coagulation reaction? *J. Insect Physiol.* **45,** 501–506.

Kotani, E., Yamakawa, M., Iwamoto, S., Tashiro, M., Mori, H., Sumida, M., Matsubara, F., Taniai, K., Kadono-Okuda, K., Kato, Y., *et al.* (1995). Cloning and expression of the gene of hemocytin, an insect humoral lectin which is homologous with the mammalian von Willebrand factor. *Biochim. Biophys. Acta* **1260,** 245–258.

Kramerov, A. A., Arbatsky, N. P., Rozovsky, Y. M., Mikhaleva, E. A., Polesskaya, O. O., Gvozdev, V. A., and Shibaev, V. N. (1996). Mucin-type glycoprotein from *Drosophila melanogaster* embryonic cells: Characterization of carbohydrate component. *FEBS Lett.* **378,** 213–218.

Krell, P. J., Summers, M. D., and Vinson, S. D. (1982). A virus with a multipartite superhelical DNA genome from the ichneumonid parasitoid *Campoletis sonorensis*. *J. Virol.* **43,** 859–870.

Kunzelmann-Marche, C., Freyssinet, J. M., and Martinez, M. C. (2000). Regulation of phosphatidylserine transbilayer redistribution by store-operated Ca^{2+} entry: Role of actin cytoskeleton. *J. Biol. Chem.* **276,** 5134–5139.

Labrador, M., and Corces, V. G. (1997). Transposable element–host interactions: Regulation of insertion and excision. *Annu. Rev. Genet.* **31,** 381–404.

Landesman, Y., White, T. W., Starich, T. A., Shaw, J. E., Goodenough, D. A., and Paul, D. L. (1999). Innexin-3 forms connexin-like intercellular channels. *J. Cell Sci.* **112,** 2391–2396.

Lavine, M. D., and Beckage, N. E. (1996). Temporal pattern of parasitism-induced immunosuppression in *Manduca sexta* larvae parasitized by *Cotesia congregata*. *J. Insect Physiol.* **42,** 41–51.

Lawrence, P. O., and Akin, D. (1990). Virus-like particles from the poison glands of the parasitic wasp *Biosteres longicaudatus* (Hymenoptera, Braconidae). *Can. J. Zool.* **68,** 539–546.

Lawrence, P. O., and Lanzrein, B. (1993). Hormonal interactions between insect endoparasites and their host insects. *In* "Parasites and Pathogens of Insects: Parasites" (S. N. Thompson, B. A. Federici, and N. E. Beckage, Eds.), Vol. I, pp. 59–86. Academic Press, New York.

Li, J., Zhao, X., and Christensen, B. M. (1994). Dopachrome conversion activity in *Aedes aegypti:* Significance during melanotic encapsulation of parasites and cuticular tanning. *Insect Biochem. Mol. Biol.* **24,** 1043–1049.

Li, X., and Webb, B. A. (1994). Apparent functional role for a cysteine-rich PDV protein in suppression of the insect cellular immune response. *J. Virol.* **68,** 7482–7489.

Lovallo, N., and Cox-Foster, D. L. (1999). Alteration in FAD–glucose dehydrogenase activity and hemocyte behavior contribute to initial disruption of *Manduca sexta* immune response to *Cotesia congregata* parasitoids. *J. Insect Physiol.* **45,** 1037–1048.

Luckhart, S., and Webb, B. A. (1996). Interaction of a wasp ovarian protein and PDV in host immune suppression. *Dev. Comp. Immunol.* **20,** 1–21.

Mckelvey, T. A., Lynn, D. E., Gundersen-Rindal, D., Guzo, D., Stoltz, D., Guthrie, K. P., Taylor, P., and Dougherty, E. M. (1996). Transformation of gypsy moth (*Lymantria dispar*) cell lines by infection with *Glyptapanteles indiensis* polydnavirus. *Biochem. Biophys. Res. Commun.* **225,** 764–770.

Miller, L. K. (1996). Insect viruses. *In* "Fundamental Virology" (B. N. Fields, D. M. Knipe, and P. M. Howley, Eds.), pp. 401–424. Lippincott-Raven, Philadelphia.

Miller, L. K. (1998). Ascoviruses. *In* "The Insect Viruses" (L. K. Miller and L. A. Ball, eds.), pp. 91–103. Plenum Publishing Corporation, New York.

Nicolas, E., Nappi, A. J., and Lemaitre, B. (1996). Expression of antimicrobial peptide genes after infection by parasitoid wasps in *Drosophila*. *Dev. Comp. Immunol.* **20,** 175–181.

Norton, W. N., and Vinson, S. B. (1983). Correlating the initiation of virus replication with a specific pupal developmental phase of an ichneumonid parasitoid. *Cell Tiss. Res.* **231,** 387–398.

Oliver, M. G., and Specian, R. D. (1990). Cytoskeleton of intestinal goblet cells: Role of actin filaments in baseline secretion. *Am. J. Physiol.* **259**(6, Part 1), G991–G997.

O'Reilly, D. R., and Miller, L. K. (1989). A baculovirus blocks insect molting by producing ecdysteroid UDP-glucosyl transferase. *Science* **245,** 1110–1112.

Pallaghy, P. K., Nielsen, K. J., Craik, D. J., and Norton, R. S. (1994). A common structural motif incorporating a cystine knot and a triple-stranded beta-sheet in toxic and inhibitory polypeptides. *Protein Sci.* **3,** 1833–1839.

Pech, L. L., and Strand, M. R. (1996). Granular cells are required for encapsulation of foreign targets by insect haemocytes. *J. Cell Sci.* **109,** 2053–2060.

Pennacchio, F., Sordetti, R., Falabella, P., and Vinson, S. B. (1997). Biochemical and ultrastructural alterations in prothoracic glands of *Heliothis virescens* (F.) (Lepidoptera: Noctuidae) last instar larvae parasitized by *Cardiochiles nigriceps* Viereck (Hymenoptera: Braconidae). *Insect Biochem. Mol. Biol.* **27,** 439–450.

Pennacchio, F., Falabella, P., Sordetti, R., Varricchio, P., Malva, C., and Vinson, S. B. (1998a). Prothoracic gland inactivation in *Heliothis virescens* (F.) (Lepidoptera: Noctuidae) larvae parasitized by *Cardiochiles nigriceps* Viereck (Hymenoptera: Braconidae). *J. Insect Physiol.* **44,** 845–857.

Pennacchio, F., Falabella, P., and Vinson, S. B. (1998b). Regulation of *Heliothis virescens* prothoracic glands by *Cardiochiles nigriceps* polydnavirus. *Arch. Insect Biochem. Physiol.* **38,** 1–10.

Pfister-Wilhelm, R., and Lanzrein, B. (1996). Precocious induction of metamorphosis in *Spodoptera littoralis* (Noctuidae) by the parasitic wasp *Chelonus inanitus* (Braconidae): Identification of the parasitoid larva as the key regulatory element and the host corpora allata as the main targets. *Arch. Insect Biochem. Physiol.* **32,** 511–525.

Phelan, P., Bacon, J. P., Davies, J. A., Stebbings, L. A., Todman, M. G., Avery, L., Baines, R. A., Barnes, T. M., Ford, C., Hekimi, S., Lee, R., Shaw, J. E., Starich, T. A., Curtin, K. D., Sun, Y. A., and Wyman, R. J. (1998a). Innexins: A family of invertebrate gap-junction proteins. *Trends Genet.* **14,** 348–349.

Phelan, P., Stebbings, L. A., Baines, R. A., Bacon, J. P., Davies, J. A., and Ford, C. (1998b). *Drosophila* Shaking-B protein forms gap junctions in paired *Xenopus oocytes*. *Nature* **391,** 181–184.

Prevost, G., Davies, D. H., and Vinson, S. B. (1990). Evasion of encapsulation by parasitoid correlated with the extent of host hemocyte pathology. *Entomol. Exp. Appl.* **55,** 1–10.

Raftery, M., Muller, A., and Schonrich, G. (2000). Herpesvirus homologues of cellular genes. *Virus Genes* **21,** 65–75.

Rotheram, S. M. (1967). Immune surface of eggs of a parasitic insect. *Nature* **214,** 700.

Saez, J. C., Branes, M. C., Corvalan, L. A., Eugenin, E. A., Gonzalez, H., Martinez, A. D., and Palisson, F. (2000). Gap junctions in cells of the immune system: Structure, regulation and possible functional roles. *Braz. J. Med. Biol. Res.* **33,** 447–455.

Savary, S., Beckage, N., Tan, F., Periquet, G., and Drezen, J. M. (1997). Excision of the PDV chromosomal integrated EP1 sequence of the parasitoid wasp *Cotesia congregata*

(Braconidae, Microgastinae) at potential recombinase binding sites. *J. Gen. Virol.* **78,** 3125–3134.

Savary, S., Drezen, J. M., Tan, F., Beckage, N. E., and Periquet, G. (1999). The excision of PDV sequences from the genome of the wasp *Cotesia congregata* (Braconidae, microgastrinae) is developmentally regulated but not strictly restricted to the ovaries in the adult. *Insect Mol. Biol.* **8,** 319–327.

Schepers, E. J., Dahlman, D. L., and Zhang, D., (1998 *Microplitis croceipes* teratocytes: *In vitro* culture and biological activity of teratocyte secretory protein. *J. Insect Physiol.* **44,** 767–777.

Shelby, K. S., and Webb, B. A. (1994). PDV infection inhibits synthesis of an insect plasma protein, arylphorin. *J. Gen. Virol.* **75,** 2285–2292.

Shelby, K. S., and Webb, B. A. (1997). PDV infection inhibits translation of specific growth-associated host proteins. *Insect Biochem. Mol. Biol.* **27,** 263–270.

Shelby, K. S., and Webb, B. A. (1999). PDV-mediated suppression of insect immunity. *J. Insect Physiol.* **45,** 507–514.

Shelby, K. S., Cui, L., and Webb, B. A. (1998). PDV-mediated inhibition of lysozyme gene expression and the antibacterial response. *Insect Mol. Biol.* **7,** 265–272.

Shelby, K. S., Adeyeye, O. A., Okot-Kotber, B. M., and Webb, B. A. (2000). Parasitism-linked block of host plasma melanization. *J. Invertebr. Pathol.* **75,** 218–225.

Sinkovics, J., Horvath, J., and Horak, A. (1998). The origin and evolution of viruses (a review). *Acta Microbiol. Immunol. Hung.* **45,** 349–390.

Soldevila, A. I., Heuston, S., and Webb, B. A. (1997). Purification and analysis of a PDV gene product expressed using a poly-histidine baculovirus vector. *Insect Biochem. Mol. Biol.* **27,** 201–211.

Soller, M., and Lanzrein, B. (1996). PDV and venom of the egg-larval parasitoid *Chelonus inanitus* (Braconidae) induce developmental arrest in the prepupa of its host *Spodoptera littoralis* (Noctuidae). *J. Insect Physiol.* **42,** 471–481.

Starich, T. A., Lee, R. Y., Panzarella, C., Avery, L., and Shaw, J. E. (1996). eat-5 and unc-7 represent a multigene family in *Caenorhabditis elegans* involved in cell–cell coupling. *J. Cell Biol.* **134,** 537–548.

Stauffer, K. A. (1995). The gap junction proteins beta 1-connexin (connexin-32) and beta 2-connexin (connexin-26) can form heteromeric hemichannels. *J. Biol. Chem.* **270,** 6768–6772.

Steiner, B., Pfister-Wilhelm, R., Grossniklaus-Buergin, C., Rembold, H., Treiblmayr, K., and Lanzrein, B. (1999). Titres of juvenile hormone I, II and III in *Spodoptera littoralis* (Noctuidae) from the egg to the pupal moult and their modification by the egg-larval parasitoid *Chelonus inanitus* (Braconidae). *J. Insect Physiol.* **45,** 401–413.

Stettler, P., Trenczek, T., Wyler, T., Pfister-Wilhelm, R., and Lanzrein, B. (1998). Overview of parasitism associated effects on host haemocytes in larval parasitoids and comparison with effects of the egg-larval parasitoid *Chelonus inanitus* on its host *Spodoptera littoralis*. *J. Insect Physiol.* **44,** 817–831.

Stoltz, D. B. (1993). The PDV life cycle. *In* "Parasites and Pathogens of Insects: Parasites" (S. N. Thompson, B. A. Federici, and N. E. Beckage, Eds.), Vol. I, pp. 167–187. Academic Press, New York.

Stoltz, D. B., and Cook, D. I. (1983). Inhibition of host phenoloxidase activity by parasitoid Hymenoptera. *Experientia* **39,** 1022–1024.

Stoltz, D. B., and Faulkner, G. (1978). Apparent replication of an unusual virus-like particle in both a parasitoid wasp and its host. *Can. J. Microbiol.* **24,** 1509–1514.

Stoltz, D. B., and Makkay, A. (2000). Co-replication of a reovirus and a polydnavirus in the ichneumonid parasitoid *Hyposoter exiguae*. *Virology* **278,** 266–275.

Stoltz, D. B., and Vinson, S. B. (1977). Baculovirus-like particles in the reproductive tracts of female parasitoid wasps. II. The genus *Apanteles*. *Can. J. Microbiol.* **23,** 28–37.

Stoltz, D. B., and Vinson, S. B. (1979a). Penetration into caterpillar cells of virus-like particles injected during oviposition by parasitoid ichneumonid wasps. *Can. J. Microbiol.* **25,** 207–216.

Stoltz, D. B., and Vinson, S. B. (1979b). Viruses and parasitism in insects. *Adv. Virus Res.* **24,** 125–171.

Stoltz, D. B., and Whitfield, J. B. (1992). Viruses and virus-like entities in the parasite Hymenoptera. *J. Hymenop. Res.* **1,** 125–139.

Stoltz, D. B., and Xu, D. (1990). Polymorphism in PDV genomes. *Can. J. Microbiol.* **36,** 538–543.

Stoltz, D. B., Vinson, S. B., and MacKinnon, E. A. (1976). Baculovirus-like particles in the reproductive tracts of female parasitoid wasps. *Can. J. Microbiol.* **22,** 1013–1023.

Stoltz, D. B., Krell, P. J., and Vinson, S. B. (1981). Polydisperse viral DNA's in ichneumonid ovaries: A survey. *Can. J. Microbiol.* **27,** 123–130.

Stoltz, D. B., Guzo, D., and Cook, D. (1986). Studies on PDV transmission. *Virology* **155,** 120–131.

Stoltz, D. B., Beckage, N. E., Blissard, G. W., Fleming, J. G. W., Krell, P. J., Theilmann, D. A., Summers, M. D., and Webb, B. A. (1995). Polydnaviridae. *In* "Virus Taxonomy" (F. A. Murphy, C. M. Fauquet, D. H. L. Bishop, S. A. Ghabrial, A. W. Jarvis, G. P. Martelli, M. A. Mayo, and M. D. Summers, Eds.), pp. 143–147. Springer-Verlag, New York.

Strand, M. R. (1994). *Microplitis demolitor* PDV infects and expresses in specific morphotypes of *Pseudoplusia includens* haemocytes. *J. Gen. Virol.* **75,** 3007–3020.

Strand, M. R., and Noda, T. (1991). Alterations in the haemocytes of *Pseudoplusia includens* after parasitism by *Microplitis demolitor*. *J. Insect Physiol.* **37,** 839–850.

Strand, M. R., and Pech, L. L. (1995). Immunological basis for compatibility in parasitoid–host relationships. *Annu. Rev. Entomol.* **40,** 31–56.

Strand, M. R., McKenzie, D. I., Grassl, V., Dover, B. A., and Aiken, J. M. (1992). Persistence and expression of *Microplitis demolitor* PDV in *Pseudoplusia includens*. *J. Gen. Virol.* **73,** 1627–1635.

Strand, M. R., Witherell, R. A., and Trudeau, D. (1997). Two *Microplitis demolitor* PDV mRNAs expressed in hemocytes of *Pseudoplusia includens* contain a common cysteine-rich domain. *J. Virol.* **71,** 2146–2156.

Strauss, E. G., Strauss, J. H., and Levine, A. J. (1996). Virus evolution. *In* "Fundamental Virology" (B. N. Fields, D. M. Knipe, and P. M. Howley, Eds.), pp. 141–159. Lippincott-Raven, Philadelphia.

Sugumaran, M. (1996). Role of insect cuticle in immunity. *In* "New Directions in Invertebrate Immunology" (K. Soderhall and S. Iwanaga, Eds.), pp. 355–374. SOS, Fair Haven, New Jersey.

Theilmann, D. A., and Summers, M. D. (1986). Molecular analysis of *Campoletis sonorensis* virus DNA in the lepidopteran host *Heliothis virescens*. *J. Gen. Virol.* **67,** 1961–1969.

Theilmann, D. A., and Summers, M. D. (1987). Physical analysis of the *Campoletis sonorensis* virus multipartite genome and identification of a family of tandemly repeated elements. *J. Virol.* **61,** 2589–2598.

Theilmann, D. A., and Summers, M. D. (1988). Identification and comparison of *Campoletis sonorensis* virus transcripts expressed from four genomic segments in the insect hosts *Campoletis sonorensis* and *Heliothis virescens*. *Virology* **167,** 329–341.

Theopold, U., and Schmidt, O. (1997). *Helix pomatia* lectin and annexin V, two molecular probes for insect microparticles: Possible involvement in hemolymph coagulation. *J. Insect Physiol.* **43,** 667–674.

Trudeau, D., and Strand, M. R. (1998). A limited role in parasitism for *Microplitis demolitor* PDV. *J. Insect Physiol.* **44,** 795–805.

Trudeau, D., Witherell, R. A., and Strand, M. R. (2000). Characterization of two novel *Microplitis demolitor* PDV mRNAs expressed in *Pseudoplusia includens* haemocytes. *J. Gen. Virol.* **81,** 3049–3058.

Trudeau, D., Washburn, J. O., and Volkman, L. E. (2001). Central role of hemocytes in *Autographa californica* M nucleopolyhedrovirus pathogenesis in *Heliothis virescens* and *Helicoverpa zea*. *J. Virol.* **75,** 996–1003.

Varricchio, P., Falabella, P., Sordetti, R., Graziani, F., Malva, C., and Pennacchio, F. (1999). *Cardiochiles nigriceps* PDV: Molecular characterization and gene expression in parasitized *Heliothis virescens* larvae. *Insect Biochem. Mol. Biol.* **29,** 1087–1096.

Vinson, S. B. (1990). Physiological interactions between the host genus *Heliothis* and its guild of parasitoids. *Arch. Insect Biochem. Physiol.* **13,** 63–81.

Vinson, S. B., and Iwantsch, G. F. (1980). Host regulation by insect parasitoids. *Q. Rev. Biol.* **55,** 143–165.

Vinson, S. B., and Scott, J. R. (1975). Particles containing DNA associated with the oocyte of an insect parasitoid. *J. Invertebr. Pathol.* **25,** 375–378.

Vinson, S. B., Edson, K. M., and Stoltz, D. B. (1979). Effect of a virus associated with the reproductive system of the parasitoid wasp *Campoletis sonorensis,* on host weight gain. *J. Invertebr. Pathol.* **34,** 133–137.

Volkoff, A.-N., Ravallec, M., Bossy, J., Cerutti, P., Rocher, J., Cerutti, M., and Devauchelle, G. (1995). The replication of *Hyposoter didymator* PDV: Cytopathology of the calyx cells in the parasitoid. *Biol. Cell* **83,** 1–13.

Volkoff, A.-N., Cerutti, P., Rocher, J., Ohresser, M. C., Devauchelle, G., and Duonor-Cerutti, M. (1999). Related RNAs in lepidopteran cells after *in vitro* infection with *Hyposoter didymator* virus define a new polydnavirus gene family. *Virology* **263,** 349–363.

Volkoff, A.-N., Rocher, J., Cerutti, P., Ohressor, M. C. P., d'Aubenton-Carafa, Y., Devauchelle, G., and Duoner-Cerutti, M. (2001). Persistent expression of a newly characterized *Hyposoter didymator* PDV gene in long-term infected lepidopteran cell lines. *J. Gen. Virol.* **82,** 963–969.

Volkoff, A.-N., Béliveau, C., Rocher, J., Hilgarth, R., Levasseur, A., Duonor-Cérutti, M., Cusson, M., and Webb, B. A. Evidence for a conserved polydnavirus gene family: Ichnovirus homologs of the *Cs*IV *repeat element* gene family. *Virology,* in press.

Washburn, J. O., Kirkpatrick, B. A., and Volkman, L. E. (1995). Comparative pathogenesis of *Autographa californica* M nuclear polyhedrosis virus in larvae of *Trichoplusia ni* and *Heliothis virescens*. *Virology* **209,** 561–568.

Washburn, J. O., Haas-Stapleton, E. J., Tan, F. F., Beckage, N. E., and Volkman, L. E. (2000). Co-infection of *Manduca sexta* larvae with PDV from *Cotesia congregata* increases susceptibility to fatal infection by *Autographa californica* M nucleopolyhedrovirus. *J. Insect Physiol.* **46,** 179–190.

Webb, B. A. (1998). PDV biology, genome structure, and evolution. *In* "The Insect Viruses" (L. K. Miller and L. A., Ball, Eds.), pp. 105–139. Plenum Press, New York.

Webb, B. A., and Cui, L. (1998). Relationships between PDV genomes and viral gene expression. *J. Insect Physiol.* **44,** 785–793.

Webb, B. A., and Dahlman, D. (1986). Ecdysteroid influence on the development of the host *Heliothis virescens* and its endoparasite *Microplitis croceipes*. *J. Insect Physiol.* **32,** 339–345.

Webb, B. A., and Luckhart, S. (1994). Evidence for an early immunosuppressive role for related *Campoletis sonorensis* venom and ovarian proteins in *Heliothis virescens*. *Arch. Insect Biochem. Physiol.* **26,** 147–163.

Webb, B. A., and Summers, M. D. (1990). Venom and viral expression products of the endoparasitic wasp *Campoletis sonorensis* share epitopes and related sequences. *Proc. Natl. Acad. Sci. USA* **87,** 4961–4965.

Webb, B. A., and Summers, M. D. (1992). Stimulation of PDV replication by 20-hydroxyecdysone. *Experientia* **48,** 1018–1022.

Webb, B. A., Beckage, N. E., Hayakawa, Y., Krell, P. J., Lanzrein, B., Stoltz, D. B., Strand, M. R., and Summers, M. D. (2000). Polydnaviridae. *In* "Virus Taxonomy: The Classification and Nomenclature of Viruses. The Seventh Report of the International Committee on Taxonomy of Viruses" (M. H. V. van Regenmortel, C. M. Fauquet, D. H. L. Bishop, E. B. Carstens, M. K. Estes, S. M. Lemon, J. Maniloff, M. A. Mayo, D. J. McGeoch, C. R. Pringle, and R. B. Wickner, Eds.), p. 1167. Academic Press, San Diego, California.

Webb, B. A., Dahlman, D. L., and Rana, R. L. (2001). Endoparasite mediated disruption of host endocrine systems: Common themes through uncommon means. *In* "Manipulating Hormonal Control" (J. E. Edwards, Ed.), pp. 88–93. Bios Scientific Publishers, Oxford.

Whitfield, J. B. (1997). Molecular and morphological data suggest a single origin of the PDVs among braconid wasps. *Naturwissenschaften* **84,** 502–507.

Whitfield, J. B. (2000). Phylogeny of microgastroid braconid wasps, and what it tells us about polydnavirus evolution. *In* "The Hymenoptera: Evolution, Biodiversity and Biological Control" (A. D. Austin and M. Dowton, Eds.), pp. 97–105. CSIRO, Melbourne, Australia.

Willott, E., Trenczek, T., Thrower, L. W., and Kanost, M. R. (1994). Immunochemical identification of insect hemocyte populations: Monoclonal antibodies distinguish four major hemocyte types in *Manduca sexta*. *Eur. J. Cell Biol.* **65,** 417–423.

Xu, D., and Stoltz, D. (1993). PDV genome segment families in the ichneumonid parasitoid *Hyposoter fugitivus*. *J. Virol.* **67,** 1340–1349.

ADVANCES IN VIRUS RESEARCH, VOL. 58

BACTERIOPHAGE ϕ29 DNA PACKAGING

Shelley Grimes,* Paul J. Jardine,* and Dwight Anderson*,†

*Department of Oral Science and †Department of Microbiology
University of Minnesota, Minneapolis, Minnesota 55455

I. Introduction
II. Overview of ϕ29 DNA Packaging
III. Components of the Defined ϕ29 DNA Packaging System
A. DNA–gp3
B. The Prohead
C. The Head–Tail Connector
D. Prohead RNA
E. The DNA-Translocating ATPase gp16
IV. The DNA Packaging Process
A. DNA Maturation as a Prerequisite for Packaging
B. Initiation of DNA Packaging
C. DNA–gp3 Translocation
V. Aims and Prospects
References

I. Introduction

How the double-stranded (ds) DNA bacteriophages package their genomes into protein capsids is a mystery and a marvel. The length of DNA to be packaged is on the order of 150 times the width of the capsid, and the DNA is packed to near crystalline density (Earnshaw and Casjens, 1980). If the protein shell of bacteriophage ϕ29 of *Bacillus subtilis* is scaled to a 1-in. structure, the DNA at this scale is more than 13 feet long! Also, inherent in the packaging process are the electrostatic, entropic, and bending energies of the DNA to overcome. Moreover, there are topological considerations, for the DNA that goes in must exit freely during ejection. No knots or tangles are allowed, and the exit channel is not wide enough for two strands of DNA. It is a wonder so unique that investigators first assumed that the viral protein shell must be assembled around the condensed DNA (e.g., see Kellenberger *et al.,* 1959).

The tailed dsDNA phages share many common features of the packaging process and probably use a common mechanism (for reviews see Black, 1989; Fujisawa and Morita, 1997). They all package their viral DNA unidirectionally into a preformed capsid precursor called the prohead. Situated at a unique vertex in the prohead is the dodecameric

0065-3527/02 $35.00

head–tail connector with an axial channel that is assumed to accommodate DNA passage during packaging and ejection. The DNA substrate for packaging is either a concatemer or a unit-length molecule, depending on the replication strategy of the particular phage. In addition, the phages have two packaging proteins, which are responsible for the maturing and the translocation of the DNA. The smaller protein of the pair binds to the DNA, while the larger protein interacts with the prohead and binds the smaller subunit to link the prohead and DNA. Many of the large subunits have ATP-binding sites and are probably the DNA-translocating ATPases. In ϕ29, RNA also plays an essential role in packaging (see below). A prevailing theory of the mechanism is that DNA packaging utilizes the symmetry mismatch between the 5-fold-symmetric icosahedral shell and the 12-fold-symmetric head–tail connector (Hendrix, 1978). This mismatch allows the connector to fit loosely in the capsid shell, but the idea that the connector rotates with respect to the shell with the aid of ATP hydrolysis to drive packaging is yet to be tested. Although years of genetic and biochemical analyses on the dsDNA phages have led to identification of the macromolecular components in the packaging process, knowledge of the mechanism is still rudimentary. Recent progress in delineating the structure and function of the ϕ29 packaging motor offers renewed hope for understanding the packaging mechanism.

II. Overview of ϕ29 DNA Packaging

The *B. subtilis* phage ϕ29 manifests the most efficient *in vitro* DNA packaging system of the phages currently being studied. ϕ29 has been on the leading edge of research on packaging due to its relatively simple composition, high particle yield per cell, and high efficiency of *in vitro* assembly. Simplicity is inherent in its DNA of 19.3 kbp, which encodes about 20 proteins, yet the ϕ29 structure has complexity typical of larger dsDNA phages (Fig. 1) (Anderson *et al.,* 1966). In contrast, the 170-kbp genome of T4 encodes more than 200 proteins (Carlson and Miller, 1994). The ϕ29 burst size is more than 1000, and mutant phage infection yields 5000 proheads per cell. By comparison, the yield of phage lambda proheads is a few hundred per cell. The active ϕ29 prohead may contain only two proteins, the connector and the major capsid protein. Nearly every prohead is active in the completely defined *in vitro* packaging system, and every DNA can be packaged, making the *in vitro* system an analogue of *in vivo* assembly. These traits—simplicity, yield, and efficiency—make the ϕ29 packaging system uniquely amenable to biophysical analysis as a complement to genetic and biochemical analyses.

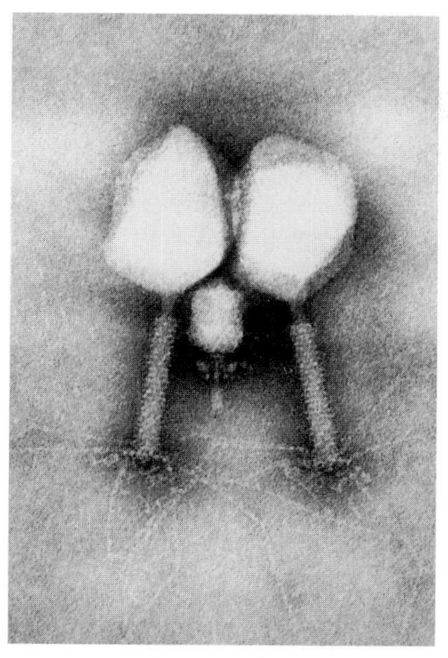

FIG. 1. Family portrait of phages ϕ29 and T2. Magnification, ×220,000.

A schematic of the ϕ29 assembly pathway is shown in Fig. 2. The prolate prohead, composed of the head–tail connector (gene product 10, gp10), the scaffolding protein (gp7), and the major capsid protein (gp8), assembles with the aid of a host chaperonin (Rajagopal *et al.,* 1993). A preassembled multimer of the ϕ29-encoded prohead RNA (pRNA) binds the connector to complete assembly of the prohead. The DNA–gene product 3 (DNA–gp3) complex is packaged into the prohead with the aid of the large packaging protein gp16 and ATP to yield a filled head. Early in packaging the scaffolding protein gp7 exits, and the shell undergoes a transformation from a rounded profile to an angular, less-charged structure (Bjornsti *et al.,* 1983). Unlike most of the dsDNA phages, the ϕ29 shell does not undergo an expansion during packaging (Tao *et al.,* 1998). The lower collar (gp11), tail (gp9), and appendages (gp12*) are then assembled onto the DNA-filled head to complete the mature phage particle. All steps in the pathway can be done *in vitro,* except for prohead assembly, either by use of extracts as a source of proteins or by use of purified proteins.

We define DNA packaging as the production of DNA-filled heads. Our *in vitro* DNA packaging assay consists of mixing purified proheads, DNA–gp3, the ATPase gp16, and ATP and assaying for head filling by DNase protection (Guo *et al.,* 1986). After 10 min of incubation, DNase I

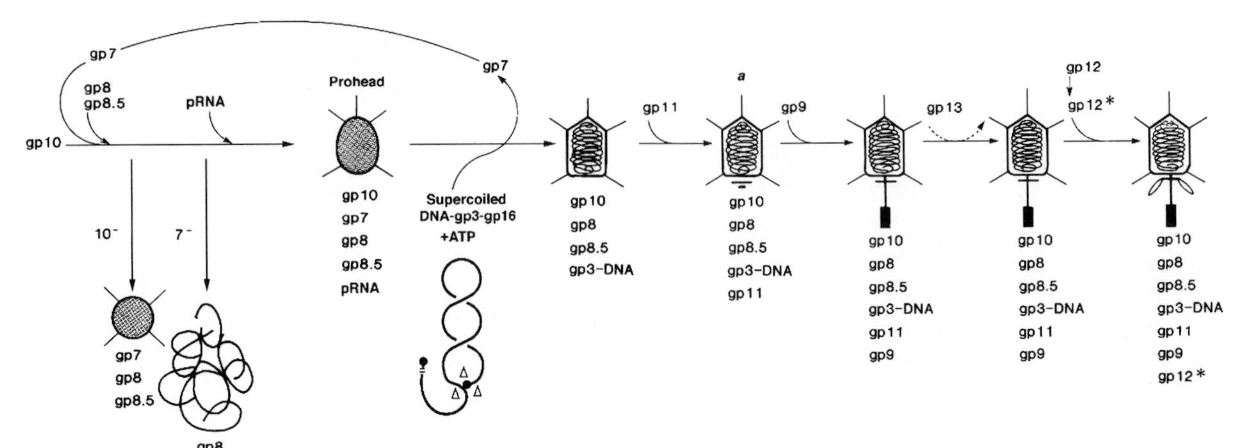

FIG. 2. Pathway of $\phi 29$ morphogenesis. The symbol "gp" before the gene number refers to the protein product of that gene. Proteins making up a structure are listed below it. The prohead and the 11^-, 12^-, and 13^- particles are true intermediates in the pathway. The 10^- and 7^- structures shown below the main pathway are abortive structures. (a): This particle has only been observed in the restrictive *sus* 12(716) infection together with the 12^- particle; its composition is inferred from morphology.

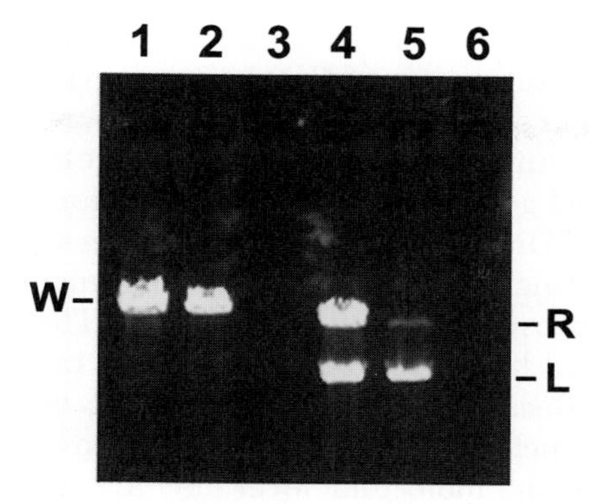

FIG. 3. DNA packaging in the defined system. DNA–gp3, gp16, proheads, and ATP are mixed and incubated for packaging. DNase I is then added to digest the unpackaged DNA, the DNase is inhibited, and the packaged DNA is extracted and quantified by agarose gel electrophoresis. Input DNA is the standard for quantification of packaging efficiency. ATP is omitted from the negative control lanes and shows that unpackaged DNA is sensitive to DNase I. The lanes contained (1) whole-length DNA input, (2) packaged whole-length DNA, (3) negative control, (4) *Cla I* DNA input, (5) packaged *Cla I* DNA, and (6) negative control. W, Whole-length; R, L, right and left ends, respectively.

is added to digest the unpackaged DNA, while the packaged DNA is protected in the head. The DNase is inhibited, the packaged DNA extracted, and the efficiency of packaging quantified by agarose gel electrophoresis (Grimes and Anderson, 1989a). Figure 3 illustrates efficient packaging of whole-length DNA–gp3 relative to the input DNA standard (lanes 1–2), and the negative control shows that all the DNA is sensitive to DNase in the absence of ATP (lane 3). Packaging of restriction digests demonstrates that the left end of DNA–gp3 is packaged preferentially, and thus the left end is the first end to enter the head (lanes 4–6) (Bjornsti *et al.,* 1983; Grimes and Anderson, 1989a). Alternatively, the filled heads can be converted to phage by the addition of the lower collar, appendage, tail, and assembly proteins produced from cloned genes (Lee and Guo, 1994, 1995). Packaging is completed in about 4 min, and about one ATP is hydrolyzed for every 2 bp packaged (Guo *et al.,* 1987c).

To initiate packaging the prohead binds to supercoiled DNA–gp3–gp16, and gp16 of this complex joins pRNA and the connector to complete assembly of the packaging motor. Thus the motor, which is set in the base of the prohead, comprises three multimers, the dodecameric connector, the multimer of pRNA, and the gp16. The intact motor has been visualized on partially packaged particles in cryo-electron microscopic (EM) three-dimensional (3D) reconstructions (Simpson *et al.,*

2000). Recently, it has been shown by single-molecule packaging studies that the $\phi 29$ motor is one of the most powerful molecular motors ever reported, with a stall force of about 57 pN (Smith *et al.,* 2001). The information base is sufficient to model and test the $\phi 29$ packaging mechanism by integrated genetic, biochemical, and biophysical approaches.

If the plethora of DNA, RNA, and protein interactions and the energetics inherent in this process are not enticing enough, there is on the practical side the prospect of drug development. The packaging process is a step in the viral life cycle with no analogue in the host cell (Wood and King, 1979). Because phages are excellent models for assembly of dsDNA viruses such as herpesvirus and adenovirus, understanding DNA packaging on the molecular level may aid in the design and development of antiviral therapies.

III. Components of the Defined $\phi 29$ DNA Packaging System

A. *DNA–gp3*

The genome of $\phi 29$ is a dsDNA of 19,285 bp, encoding about 20 genes. This is about 40% the size of the phage lambda genome and a bit more than 10% the size of phage T4 DNA. The small size of the $\phi 29$ DNA affords the opportunity to uncover the function of every gene. The early genes needed in DNA replication are found on the ends of the DNA, and the late genes encoding structural and morphogenetic factors are expressed from the opposite strand in the middle of the genome (Mosharrafa *et al.,* 1970; Schachtele *et al.,* 1972). $\phi 29$ is unusual among the well-studied dsDNA phages in that a protein molecule is covalently bound to each 5′ end of the DNA (Salas *et al.,* 1978; Harding *et al.,* 1978; Yehle, 1978). This protein, gp3, is bound to the first base A on each 5′ end through serine 232 (Hermoso *et al.,* 1985). gp3 primes DNA replication by a strand-displacement mechanism, where a free gp3–A interacts with the covalently bound parental gp3 on DNA to initiate bidirectional replication (for reviews see Salas, 1991, 1999; Meijer *et al.,* 2001).

The gp3 terminal protein is also essential for efficient DNA packaging, as partial digestion of the gp3 with trypsin or proteinase K reduces packaging by 50% and 75%, respectively (Bjornsti *et al.,* 1983; Grimes and Anderson, 1989a). Also, restriction fragments containing gp3 are packaged at least an order of magnitude more efficiently than fragments without gp3 (Grimes and Anderson, 1989a). Thus, the small packaging protein gp3 is covalently linked to the DNA, packaged, and injected upon infection to initiate a new replication cycle.

B. The Prohead

The ϕ29 prohead is a prolate icosahedron that is assembled from as few as three proteins. The prohead contains 12 copies of the connector protein gp10, at least 150 copies of the scaffolding protein gp7, and 235 copies of the major capsid protein gp8 (Tao *et al.,* 1998; Peterson *et al.,* 2001). All three proteins are essential for prohead assembly, as mutants in the scaffold or connector yield aberrant structures or isometric particles (for a review, see Anderson and Reilly, 1993). Further, an unidentified host protein is required for prohead assembly (Rajagopal *et al.,* 1993). Addition of the nonessential head fibers and a preformed multimer of pRNA (Guo *et al.,* 1991b, 1998; F. Zhang *et al.,* 1998; Chen *et al.,* 2000) completes assembly of the prohead. The scaffolding protein exits during DNA packaging and recycles to form new proheads (Nelson *et al.,* 1976; Bjornsti *et al.,* 1983). Only the connector and the major capsid protein may be needed for DNA packaging, as particles with only 15% of the normal complement of scaffolding protein are fully active (S. Grimes, unpublished).

Structures of the ϕ29 prohead and virion have been solved to 27-Å resolution by cryo-EM 3D reconstruction (Fig. 4) (Tao *et al.,* 1998). This is the first prolate virus structure solved by this method and also the first visualization of a connector *in situ* in the head. The prohead is

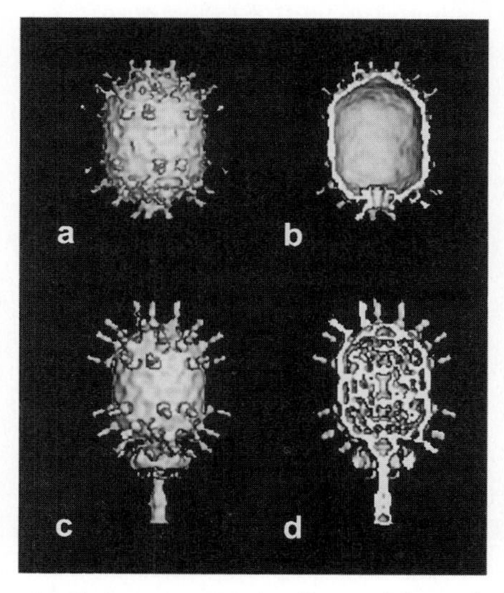

FIG. 4. Three-dimensional image reconstructions of bacteriophage ϕ29 particles. (a) Fibered prohead and (c) fibered mature phage. (b, d) Cross section of the respective particles, showing features such as the thin capsid wall, the dsDNA, and the structures of the connector and tail assemblies. (Adapted from Tao *et al.,* 1998, Fig. 2, p. 432.)

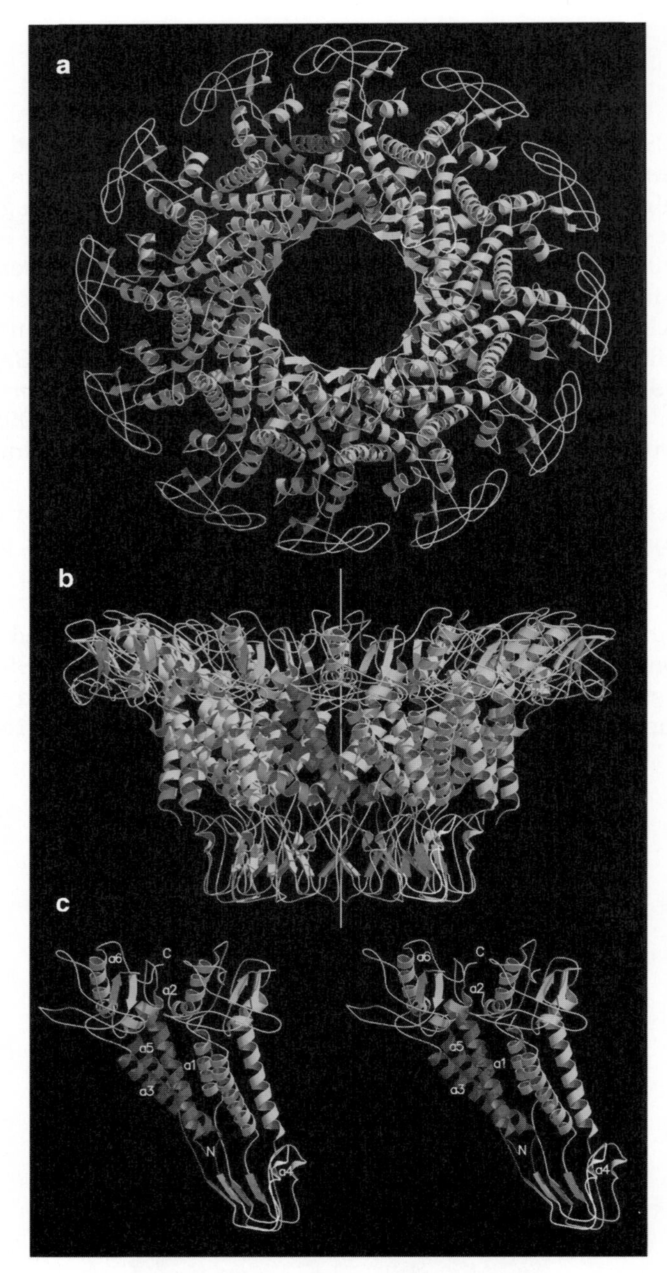

FIG. 5. Connector structure ribbon diagrams. (a) The dodecameric connector seen from the tail looking toward the head; (b) a side view with the pRNA-binding site at the bottom, showing the conical structure of the connector and the helical twist of each subunit around the 12-fold axis (*white*); (c) a stereo diagram of a pair of monomers. One monomer is colored

an extended $T = 3$ structure with the dimensions of 420×540 Å and a shell thickness of only 16 Å. The prolate prohead contains 30 hexamers of gp8, compared to 20 hexamers of gp8 in the isometric particle. The proximal and the distal caps of the head contain six and five pentamers of gp8, respectively, with the missing pentamer replaced by the connector. The dimeric head fibers (Carrascosa *et al.,* 1981) bind to gp8 pentamers at the quasi-threefold axes and exhibit partial occupation of the 55 binding sites on the prohead and virion (Tao *et al.,* 1998; Peterson *et al.,* 2001).

C. The Head–Tail Connector

The ϕ29 head–tail connector is a cone-shaped dodecamer of gp10 with a central channel through which the DNA may pass during packaging and ejection. Connectors of phages including T4, T3, and lambda are morphologically similar to the ϕ29 connector, although no amino acid sequence homology exists among them (for a review, see Valpuesta and Carrascosa, 1994). The ϕ29 connector is required in prolate head assembly (Hagen *et al.,* 1976; Camacho *et al.,* 1977; Guo *et al.,* 1991a) and is the crux of the DNA packaging machine. It has both RNA-binding (Valle *et al.,* 1999; Simpson *et al.,* 2000) and DNA-binding (Donate *et al.,* 1992) domains, which reflect its central position in the ϕ29 packaging motor. The ϕ29 connector has recently been solved to 3.2-Å resolution by X-ray crystallography, the first connector structure solved at near-atomic resolution (Fig. 5; also see color insert) (Simpson *et al.,* 2000, 2001). In side view, the conical profile has three distinct domains with heights of 22, 28, and 25 Å and external radii of 69, 47, and 33 Å, respectively. The wide end of the connector resides within the viral head, and the narrow end protrudes from the head and contains the RNA-binding domain. Connecting the wide and narrow regions is the central region, whose striking feature is the presence of three long α helices in each

red in the central domain, green in the wide-end domain which resides inside the capsid, and yellow at the narrow-end domain. The other monomer is colored blue. The ordered part of the polypeptide starts with helix α 1 on the outside of the connector, going toward the wide end (residues 61–128 and 247–286). Helix α 3 (residues 129–157) returns the chain to the narrow end (residues 158–202). The tip of the connector at the narrow end is formed by residues 164–170 and 185–196. Helix α 5 (residues 208–226) returns the polypeptide to the wide end through the second disordered section. Drawn with the program MOLSCRIPT and RASTER3D. (Adapted from Simpson *et al.,* 2000, Fig. 2, p. 747.) Also see color insert.

subunit arranged at a 40° angle with respect to the 12-fold axis and inclined obliquely to the right-handed DNA helix that traverses its channel. The surface of each monomer is positively charged on one side and negatively charged on the other, presumably to stabilize the oligomer. The outside of the dodecamer is hydrophobic, a property that might facilitate connector rotation within the head. The central channel, with an internal diameter ranging from 36 Å at the narrow end to 60 Å in the head, carries a negative surface charge that may aid transit of the DNA during packaging. Comparisons of the structural differences between connectors in three crystal forms show that the structure is relatively flexible, particularly the domain that is thought to rotate against the pentagonal portal of the prohead (Simpson *et al.,* 2001).

D. Prohead RNA

1. General Properties

We were astonished that ϕ29 *in vitro* DNA packaging was sensitive to ribonuclease, and this finding led to the discovery of the 174-base, ϕ29-encoded RNA that is essential for DNA packaging (Guo *et al.,* 1987a). The RNA is bound to the connector in the prohead (Guo *et al.,* 1987d; Garver and Guo, 1997), hence the designation prohead RNA (pRNA). pRNA is a transcript from a promoter at the extreme left end of the genome (bases 320–147) (Wichitwechkarn *et al.,* 1989; Guo *et al.,* 1987a). Addition of pRNA to the prohead is the last step in prohead assembly, as pRNA is not needed to assemble the prohead from products of the cloned genes 7, 8, and 10 (Guo *et al.,* 1991b). Five to six copies of pRNA are bound to the prohead in a Mg^{2+}-dependent reaction as determined by biochemical analyses (Wichitwechkarn *et al.,* 1989; Reid *et al.,* 1994a; Trottier and Guo, 1997). The essential role of pRNA in packaging is transitory, as pRNA is absent in the mature phage (Guo *et al.,* 1987a).

The secondary structure of pRNA was determined by phylogenetic analysis of ϕ29 pRNA and three phage relatives with distinct prohead RNAs ranging in size from 161 to 174 bases (Bailey *et al.,* 1990). ϕ29 pRNA has two domains: domain I of 117 bases contains the 5′ end and is separated by a single-stranded region of 13 bases from the 44-base domain II (Fig. 6a). Although pRNAs of ϕ29 relatives share less than 50% sequence homology, the secondary structures are very similar, and 19 conserved bases are found in specific elements among the relatives. None of the pRNAs of relatives can substitute for the ϕ29 pRNA in DNA packaging. Initially pRNA was isolated as a 120-base entity containing

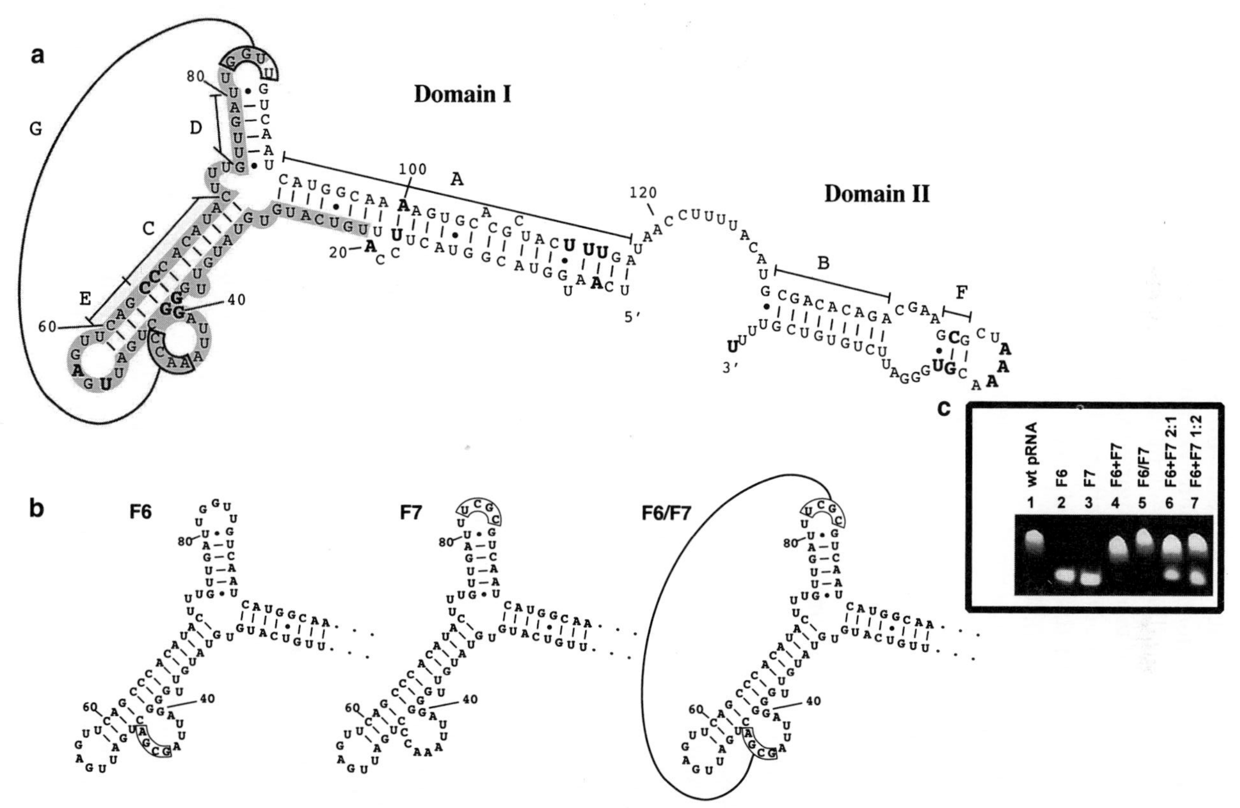

FIG. 6. Secondary structure of pRNA. (a) Secondary structure of the 174-base form of wild-type pRNA. The line shows the proposed tertiary interaction. Helices are designated A–G. The prohead-binding domain is marked by shading. (b) Sequence changes of mutants F6, F7, and F6/F7 are boxed, and the line shows the tertiary interaction. (c) Native polyacrylamide gel electrophoresis of wild-type (wt) or mutant pRNAs. (Adapted from F. Zhang *et al.*, 1998, Figs. 1 and 2, p. 142.)

domain I that was cut from the 174-base transcript by adventitious ribonucleases in prohead isolation (Wichitwechkarn *et al.,* 1989). This truncated form is totally sufficient in packaging and is the commonly used substrate for *in vitro* packaging experiments. The function of the 44-base domain II *in vivo* is unknown, but presence of its counterpart in all of the prohead RNAs of relatives suggests that it has biological relevance (Bailey *et al.,* 1990).

pRNA forms a multimeric ring that is sandwiched between the connector and a multimer of the ATPase gp16 to constitute the packaging motor. Why does the packaging motor use RNA as a component? Is its inherent flexibility needed to mediate protein conformational change? The following emphasizes the novel multiple roles of pRNA in ϕ29 assembly and DNA packaging, and reflects a large effort to relate structural elements and their sequences to biological activity.

2. pRNA Mutants

pRNA can be removed from the prohead with loss of DNA packaging activity, and the particle can be reconstituted with free RNA to full activity (Guo *et al.,* 1987d). Reconstitution of RNA-free proheads permits convenient evaluation of the activity of mutant RNAs. Early experiments with bisulfite mutants show that only those pRNA mutants predicted to fold by the M-fold algorithm (Zucker, 1989) survive in the cell (Wichitwechkarn *et al.,* 1992). Thus, the predicted secondary structure is credible, and subsequent pRNA mutant design was based on tests of folding of hypothetical mutants. Mutagenesis tested the role of conserved bases and confirmed and tested the role of structural elements such as helices, loops, and bulges. At least a few hundred RNA mutants have been constructed and tested in Minneapolis or in the laboratory of Peixuan Guo at Purdue University, so only highlights of these studies are presented here. Early experiments established two functional regions of the pRNA, a prohead-binding region (see shaded area in Fig. 6a) and a region involved in DNA translocation (Reid *et al.,* 1994b; C. Zhang *et al.,* 1994, 1995). Mutants in these regions are considered in turn.

The prohead-binding domain, bases 22–84, was identified by ribonuclease footprinting and confirmed by mutant studies and competitive binding experiments (Fig. 6a) (Reid *et al.,* 1994a–c). This area contains the C, E, and D helices and the C–E, E, and D loops. All of the helices and loops are essential, and sequence in these elements is also important to activity. Deletion of all or part of the C–E loop (bases 40–48) results in very low prohead-binding competitive activity, and the mutants are inactive in DNA packaging (Reid *et al.,* 1994b).

Mutation of the conserved base U54 or A56 in the E loop causes a modest reduction in packaging, but an RNA with all four bases of the E loop changed is inactive for packaging (Wichitwechkarn *et al.,* 1992; Reid *et al.,* 1994b). Changing the conserved bases 38GGG40/63CC64 to AAA/UU maintains the predicted structure, but the mutant has only one-half the activity of wild-type pRNA (Wichitwechkarn *et al.,* 1992). Mutants that disrupt the E helix are inactive, as is a mutant that inverts the wild-type sequence 49CUGA52/62GACU59 to AGUC/UCAG (Reid *et al.,* 1994b). Though the latter RNA is predicted to fold correctly, the binding to proheads is low, and the RNA is inactive in DNA packaging. Deletion of bulges U29 or U36 has little effect, but deletion of the 72UUU74 bulge, which forms a hinge at the junction of helices C and D, results in loss of packaging, perhaps by reducing the flexibility of the RNA molecule for tertiary interactions (Reid *et al.,* 1994c). Supporting this idea of flexibility is the activity of a circularly permuted RNA that contains the 72UUU74 deletion, but also new 5′ and 3′ ends that create an opening at this site (C. Zhang *et al.,* 1997). Deletion mutant studies show that the smallest pRNA with prohead-binding activity is a 62-base RNA (bases 30–91) (Reid *et al.,* 1994c) or a 64-base RNA (bases 28–91) (Garver and Guo, 1997). These truncated RNAs containing the prohead-binding region are ideal candidates for X-ray crystallography and nuclear magnetic resonance studies.

Residues 1–20 and 92–117 of pRNA constitute a large segment of the A helix, which functions in DNA translocation (Fig. 6a). Mutants in this region have prohead-binding capabilities near that of wild-type pRNA, but are defective in DNA packaging (Reid *et al.,* 1994b; C. Zhang *et al.,* 1994, 1995). The exact role of the A helix is unknown, but it probably interacts with the gp16 ATPase. Mutagenesis of the 5′/3′ end of the A helix has a marked effect on packaging activity, where both helix maintenance and sequence are important. The helix formed by residues 1–4 of the 5′ end and 117–114 of the 3′ end is critical for DNA packaging activity (C. Zhang *et al.,* 1994, 1995). A mispaired base at residue 1 is allowed, mismatches in positions 2, 3, or 4 are lethal, and the conserved base A3 is absolutely required for wild-type activity. Compensatory changes confirm the helical nature of the A region at positions 7–9/112–110 and 14–16/103–99, and helix maintenance is more important than sequence (C. Zhang *et al.,* 1995, 1997). The bulges U5, U35, A106, and C109 are dispensable for DNA packaging, whereas the CCA bulge (residues 18–20) is required (Reid *et al.,* 1994b; C. Zhang *et al.,* 1997). pRNA with the CCA bulge deleted cannot support DNA packaging, yet binds the prohead with wild-type efficiency; substitutions at all three positions, including the conserved A20, give wild-type activity (C. Zhang *et al.,* 1997). Most of the mutations of conserved single bases are tolerated in packaging, thus their conservation may be more important *in vivo* than *in vitro*. Construction of a ϕ29 strain with a deletion of the pRNA gene will permit assay of plasmid-encoded mutant RNAs *in vivo*. Also, activity of the 120-base versus 174-base RNA can be tested *in vivo* in hopes of uncovering the function of domain II.

3. Intermolecular Pseudoknot

Identical pRNA subunits form a novel hexamer by intermolecular base pairing, which is required for packaging (F. Zhang *et al.,* 1998; Guo *et al.,* 1998). Bases 45AACC48 of the C–E loop have the potential to base-pair with the 85UUGG82 of the D loop (Fig. 6a) (Reid *et al.,* 1994c). Similar potential base pairing exists in the predicted secondary structures of the prohead RNAs of the ϕ29 relatives examined (Bailey *et al.,* 1990; Reid *et al.,* 1994c). To test the biological significance of the pseudoknot, the AACC was changed to GCGA in mutant F6, and the UUGG was changed to CGCU in mutant F7 (Fig. 6b). Neither F6 nor F7 can form the pseudoknot, and both mutants are inactive in DNA packaging. The double mutant F6/F7, which restores the ability to form the pseudoknot, has full biological activity. However, a mixture of pRNAs F6 and F7, tested later, also has full biological activity (F. Zhang *et al.,* 1998), demonstrating that the pseudoknot is intermolecular and not intramolecular! Mobility shift in native gel electrophoresis demonstrates the Mg^{2+}-dependent physical interaction between the F6 and F7 mutants (Fig. 6c). The RNA form in the F6 + F7 mixture is slower than the individual F6 and F7 RNAs and comigrates with RNAs of the wild-type and the double mutant F6/F7, demonstrating multimers. If the ratio of either mutant is altered to 2 : 1, the same slower form is seen, along with excess RNA at the monomer position of the individual mutant, clearly showing a 1 : 1 interaction of F6 and F7 RNAs to form a multimer, which is likely a dimer.

Whereas complementations of two mutants in packaging assays suggest that the number of subunits of the packaging oligomer is a multiple of two, three-way mutant complementations exclude tetramers and octamers as active multimers and demonstrate that proheads use six copies of pRNA (F. Zhang *et al.,* 1998; Guo *et al.,* 1998). To this end, Guo *et al.* (1998) constructed six different RNAs, each with the ability to form an intermolecular link with only two other molecules of the set, and this combination has full biological activity.

Mutagenesis/native gel electrophoresis/packaging experiments show that dimers are intermediates in the formation of a hexameric RNA (F. Zhang *et al.,* 1998; Chen *et al.,* 2000). DNA packaging assays with mutant RNAs reveal a hierarchy of RNA–RNA interactions (F. Zhang *et al.*, 1998). Those mutant RNA molecules incapable of forming a dimer are inactive in packaging. RNAs requiring a noncanonical intermolecular base pairing for dimer formation, or dimers that cannot form higher multimers, have about 10% packaging activity. RNAs that can form dimers and then higher multimers using noncanonical base pairings have 40–60% packaging activity. Only when multimers can form by canonical base pairing is full biological activity seen. Chen *et al.* (2000) also report that dimers capable of forming a higher multimer are the most active RNA forms in prohead binding and in competition and inhibition studies of phage assembly. Analytical ultracentrifugation also suggests that dimers are intermediates in formation of a pRNA hexamer (F. Zhang *et al.*, 1998). Dimers and hexamers are observed in the wild-type pRNA and in the complementation of mutants F6 and F7, but no monomers, trimers, tetramers, or pentamers are found; only monomers are found in the individual mutant F6 and F7 controls.

Mutant studies of the intermolecular pseudoknot show that the 47CC48/83GG82 base pairs are sufficient for multimerization and biological activity (F. Zhang *et al.*, 1998; Chen *et al.*, 1999). At least one G/C pair is needed at these positions, but mutation of all four positions of the pseudoknot to G/C base pairs produces a dramatic decrease in packaging activity. Cross-linking studies confirm an interaction of G82 and bases flanking the C–E loop in tertiary structure, although the required C48 and G82 bases are not cross-linked (Garver and Guo, 2000). A 75-base RNA (residues 23–97) is sufficient for formation of dimers, and these are able to assemble into hexamers to inhibit phage assembly (Chen *et al.*, 1999). The pRNA hexamer is the first nucleic acid found to form a multimer by intermolecular base pairing of identical subunits, and this structure is essential for efficient DNA packaging. The pRNA hexamer thus represents a new structural motif in the RNA world.

4. SELEX Studies

In vitro selection of pRNA aptamers for prohead binding from a vast repertoire of RNA produced *in vitro* was done to test the structural elements and sequences needed for prohead binding, intermolecular linkage, and DNA packaging. The SELEX (Systematic Evolution of Ligands by Exponential Enrichment) method utilizes variation, selection, and replication to create variant RNA molecules capable of performing the same function as wild-type RNA (Tuerk and Gold, 1990). A partially randomized DNA template of the 62-base region (residues 30–91) (see Fig. 6a) shown to be sufficient for prohead binding was used to produce an initial RNA pool with a repertoire of 5×10^{14} molecules

(F. Zhang and Anderson, 1998). Seven consecutive rounds of binding to proheads, isolation of bound RNA, amplification as DNA, and *in vitro* transcription produces RNAs with ever-increasing affinity for proheads. After five rounds of selection the RNA pool is found to be equivalent to wild-type pRNA as a competitor of pRNA binding. Whereas RNAs from the third round show changes at nearly every base, by the fifth round, regions of the RNA are nearly invariant and wild-type. For example, RNAs from the fifth round show infrequent changes in residues 46–48 of the C–E loop, and most of the residues of the E stem and bases 53–56 of the E loop are wild-type. In addition, covariation was found in six of nine positions in the C helix and all six positions of the D helix, confirming the predicted helical structures.

A SELEX II series using the 120-base pRNA was designed to focus further on important prohead-binding elements and the intermolecular pseudoknot (F. Zhang and Anderson, 1998). Use of the 120-base RNA affords the opportunity to measure the DNA packaging capabilities of the various mutant pRNAs, which was not possible in the original SELEX design. A DNA template was constructed with total randomization of the bases 45–62 (the C–E loop, the E stem, and the E loop) and residues 81–84 of the D loop (see Fig. 6a). Again, by the fifth round the RNA pool is equivalent to wild-type pRNA in competitive binding experiments. Sequencing of the RNA pools shows a dramatic progression to wild-type between rounds 3 and 4 (Fig. 7). Sequence analysis of individual RNAs from the fourth and fifth rounds shows that residues

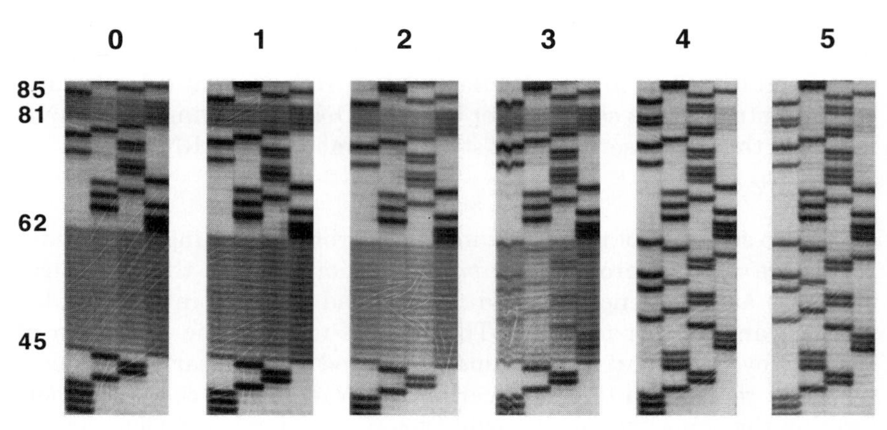

FIG. 7. Sequencing gel of pools of RNA from the initial selection through the fifth cycle of SELEX II. Numbers above the lanes refer to the selection cycles, and numbers at the left refer to positions in pRNA. Emergence of the wild-type is clear in selection cycle 4. (From F. Zhang and Anderson, 1998, Fig. 5, p. 2952.)

45–62 are wild-type in 22 of 31 sequences, whereas in contrast the D loop has only 3 of 31 RNAs that are wild-type. Chen *et al.* (2000) report that dimers are required for prohead binding, although mutant F6, in which dimers are not detected, has 20% prohead-binding competitor activity (Reid *et al.,* 1994c) and the high frequency of change in the D loop involved in intermolecular interaction still permits prohead binding. DNA packaging assays using the 31 RNAs reveal that only those RNAs having the wild-type GG at positions 83 and 84 are functional, mirroring the mutagenesis results (see pseudoknot above). Mutations in positions 81 and 82 of the pseudoknot are tolerated, confirming that the two 47, 48CC/GG84, 83 base pairs are sufficient for packaging and multimerization. The SELEX process recapitulates the natural evolution of pRNA, as no new RNA sequences emerged as a variant to the wild-type. Apparently the pRNA is already optimally designed for its biological functions.

5. *Multiple Functions of pRNA*

pRNA has multiple roles in the DNA packaging process. First, it is hypothesized to bind gp16 of supercoiled DNA–gp3–gp16 to link the prohead and DNA, thus completing the assembly of the packaging motor. Second, it has a role in recognition of the left end of the genome, the end that encodes the pRNA and the end that enters the capsid first (Grimes and Anderson, 1989a). A third function is that pRNA stimulates the ATPase activity of gp16, and it is needed continuously for this stimulation (Grimes and Anderson, 1990) (see below). Consistent with this, pRNA remains on the filled head until displaced by neck/tail assembly, and it has been proposed that sequential pRNA conformational change drives packaging (Chen and Guo, 1997; Guo *et al.,* 1998).

The formation of hexamers from dimers by intermolecular base pairing is essential to the role of pRNA in packaging (F. Zhang *et al.,* 1998; Guo *et al.,* 1998; Chen *et al.,* 2000). Mutant pRNAs that form hexamers from dimers have wild-type activity, whereas mutant pRNAs that form dimers that are incapable of forming hexamers have lower biological activity. Why is there a difference in activity? Gaps between dimers in the putative closed hexameric ring may perturb the binding of gp16 or may interfere with communication between pRNA subunits needed for coordinated and sequential steps of the motor. pRNA subunits in the motor have been hypothesized to function sequentially because incorporation of one inactive mutant into a hexamer is sufficient to block DNA packaging (Chen and Guo, 1997; Trottier and Guo, 1997). Also, aside from pseudoknot mutants, no two mutants of low activity have been able to complement one another to higher biological activity

(Chen and Guo, 1997). Taken together, the results suggest that although a fully base-paired pRNA hexamer is optimal in packaging, each RNA subunit performs a function individually, and the intermolecular linkages may bridge this dichotomy.

E. The DNA-Translocating ATPase gp16

The 39-kDa gp16 is the larger of the two ϕ29 packaging proteins. It contains Walker A- and B-type ATP-binding consensus sequences and the predicted secondary structure of ATPases (Guo *et al.,* 1987c), and it functions as the DNA-translocating ATPase in concert with pRNA (Grimes and Anderson, 1990). gp16 by itself has weak ATPase activity, which is stimulated by RNA or DNA. However, gp16 is most active when bound to the pRNA-containing prohead, and this stimulation is specific for pRNA. The ATPase activity of RNA-free proheads plus gp16 is nil, below the low activity found in gp16 alone. The pRNA is needed continuously, for RNase treatment of the gp16–pRNA complex during the assay halts ATP hydrolysis, showing that pRNA is an intimate part of the ATPase. Thus the consummate ATPase is the whole prohead–pRNA–gp16 machine, where each component is present in its correct conformation.

gp16 binds to gp3 on DNA–gp3 to supercoil the packaging substrate as a prerequisite for packaging (Grimes and Anderson, 1997; see below). Then an RNA recognition motif on gp16 enables DNA–gp3–gp16 to bind to pRNA on the prohead to form the ternary DNA–gp3/gp16/pRNA–prohead packaging complex (Guo *et al.,* 1987b; Grimes and Anderson, 1990). The number of copies of gp16 in the packaging motor and their location relative to the pRNA multimer are major open questions.

IV. The DNA Packaging Process

The ϕ29 DNA packaging process is presented as the phases of DNA maturation, packaging initiation, and DNA translocation (Fig. 8). The demonstrations of supercoiling of the packaging substrate (Fig. 8a) and wrapping of supercoiled DNA around the connector (Fig. 8b) as prerequisites for packaging are unique to the ϕ29 system at this time, although we believe that these processes likely occur in other dsDNA phages as well. ϕ29 DNA is packaged as unit-length molecules, so there is no termination event of cleaving a genome length from a DNA concatemer as, for example, in phage lambda. Although specific details are

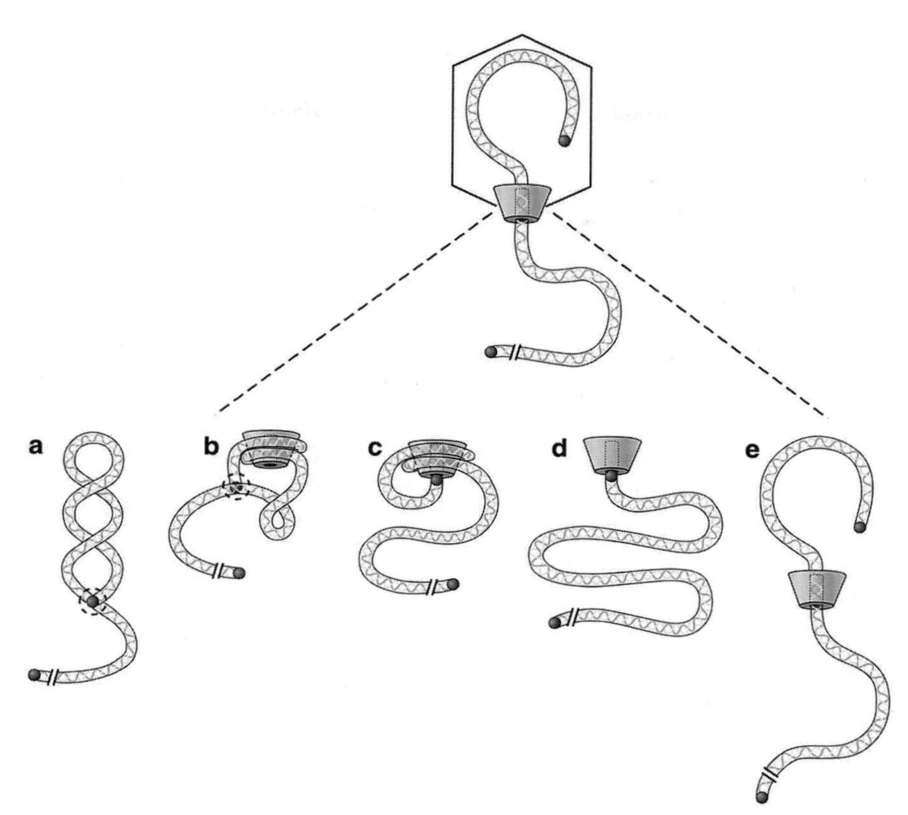

FIG. 8. ϕ29 DNA–gp3 packaging model. (a) gp3 covalently bound at the left end (*gray dot*) loops and forms a lariat by interaction with DNA, gp16 binds at the lariat loop junction (*dashed circle*), and gp3–gp16 produce a supercoiled packaging substrate (Grimes and Anderson, 1997); (b) supercoiled DNA wraps around the connector of the prohead; (c) the left DNA end is freed and binds the connector; (d) DNA is released from the outside of the connector; and (e) DNA is packaged as the connector expands, contracts, and rotates with the aid of ATP hydrolysis (see Figs. 17 and 18).

known about initiation, translocation, and termination in several of the dsDNA phages, the mechanism of DNA packaging is obscure.

A. DNA Maturation as a Prerequisite for Packaging

ϕ29 DNA–gp3 has its own maturation pathway prior to packaging (Grimes and Anderson, 1997). The ATPase gp16 acts in concert with the terminal protein gp3 to introduce supercoils into the DNA

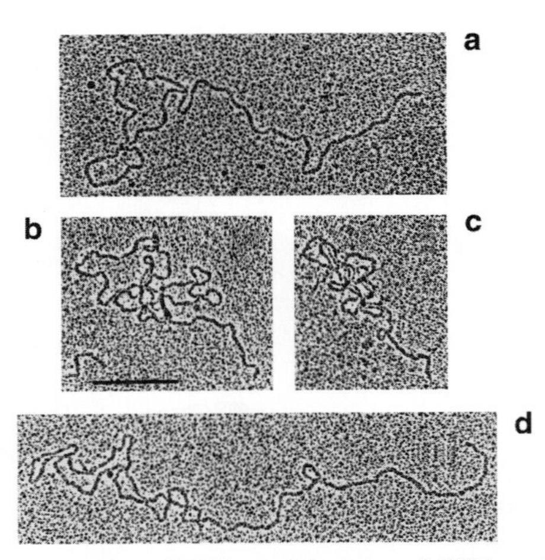

FIG. 9. Electron micrographs of DNA–gp3 lariat and DNA–gp3–gp16 coiled lariat complexes. *Hpa* I L-DNA–gp3 and *Hpa* I L-DNA–gp3–gp16 complexes were isolated on sucrose density gradients and prepared for microscopy using the polylysine method, followed by metal shadowing. (a) *Hpa* I L–DNA–gp3; (b–d) *Hpa* I L-DNA–gp3–gp16 complexes. The magnification bar represents 0.25 μm. (Adapted from Grimes and Anderson, 1997, Fig. 8, p. 909.)

(Fig. 8a). However, the terminal protein gp3 alone confers structure on the DNA. Whole-length DNA–gp3 isolated from the virion or from extracts of infected cells sediments as circles in sucrose density gradients, and proteinase K treatment to digest gp3 converts the circles to the slightly slower-sedimenting linear form. Sucrose density gradient and EM analyses show that both left-end and right-end DNA–gp3 restriction fragments form lariats, whereas internal fragments or proteinase K-treated fragments do not. The addition of gp16 causes these lariats to sediment faster and to become looped and coiled, as visualized by transmission electron microscopy (TEM) (Fig. 9) (Grimes and Anderson, 1997) and scanning transmission electron microscopy (STEM) (S. Grimes, M. Simon, and D. Anderson, unpublished). In some complexes gp16 is observed at the lariat loop junction. The coiled DNAs are packaged preferentially, showing biological relevance of the supercoiling. However, the mechanism of supercoiling is unknown. gp3–gp16 may track along the DNA like the *Eco*K restriction enzyme (Yuan *et al.*, 1980), or the protein complex may pass DNA through a double-stranded break like a gyrase. ATP dependence of supercoiling is ambiguous because both gp16 and gp3 bind and retain ATP, and the system has not been purged of ATP (Grimes and Anderson, 1997). Also, since both the left- and the right-end DNA–gp3 fragments are coiled, and this is a prerequisite for packaging, it is not clear how the specificity of packaging for the left end is determined. Does supercoiling of DNA–gp3 set ϕ29 apart from other dsDNA phages in terms of efficiency of packaging, or has the supercoiling of DNAs of other viruses gone undetected?

B. Initiation of DNA Packaging

It is likely that the supercoiling of DNA–gp3 is needed for efficient interaction of the prohead with DNA–gp3 to initiate packaging (Fig. 8b). Free connectors interact preferentially with supercoiled plasmid DNA, each connector wrapping DNA around it by slightly more than one turn to remove a negative supercoil (Turnquist *et al.,* 1992). Three lines of evidence support this conclusion: (1) The contour length of plasmid DNA is reduced by about 180 bp per connector bound (Fig. 10a), whereas open circular and linear DNA show no length reduction; (2) an increase in mass at the perimeter of the connector bound to the plasmid,

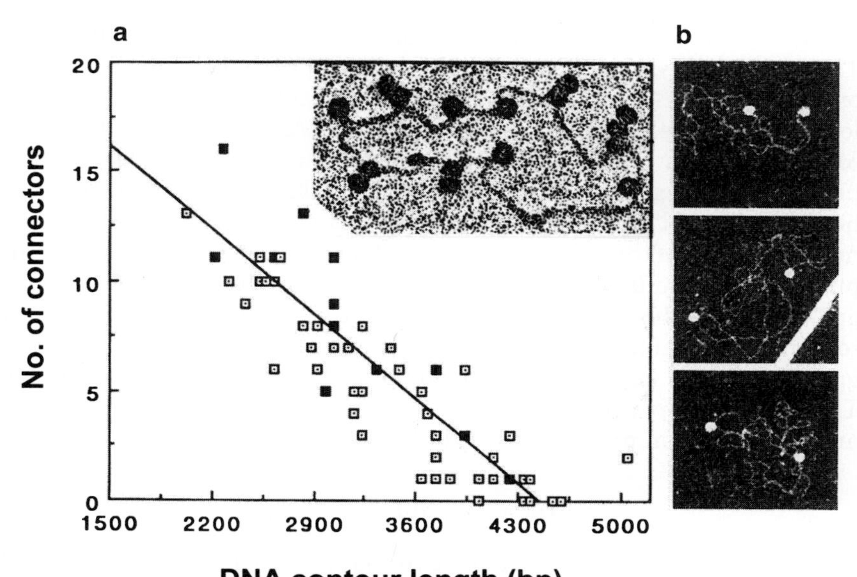

FIG. 10. (a) Supercoiled pBR322 wraps around connectors, reducing its contour length. Open symbols depict supercoiled complexes; closed symbols depict complexes nicked in preparation for electron microscopy. Inset: Relaxed complex with 16 bound connectors. Note the groups of two or three contiguous connectors. ×196,000. (b) STEM images of unstained, freeze-dried connector–pBR322 DNA complexes. ×84,000. (Adapted from Turnquist *et al.,* 1992, Figs. 2 and 3, pp. 10480–10481.)

consistent with about 170 bp of DNA, is detected by radial density measurements of STEM images (Fig. 10b); and (3) topoisomerase I treatment of plasmid–connector complexes to remove residual supercoils and seal any nicks, followed by deproteinization and gel electrophoresis, shows that supercoils equivalent to the number of connectors bound are restrained by the connectors, whereas the plasmid control is totally relaxed. Thus, connectors wrap supercoiled DNA *around the outside,* restraining negative supercoils. There are reports of connector binding preferentially to linear DNA over circular DNA, but it is difficult to compare the results with ours because there is no mention of the topological state of the circular DNA (Valpuesta *et al.,* 1992; Valle *et al.,* 1996). Our freshly purified connnectors, promptly frozen and stored at −70°C, bind supercoiled DNA 10 times more efficiently than linear or open circular DNA (Turnquist *et al.,* 1992).

Is the wrapping of supercoiled DNA around connectors biologically relevant? If so, the connector in the context of the prohead must also bind and wrap supercoiled DNA. Again, the connector in proheads is found to bind and wrap supercoiled plasmid DNA, reducing the contour length by about 100 base pairs for each connector bound as shown by TEM; conversely, the efficiency of binding of proheads to open circular or linear DNA is less than 1% that of supercoiled DNA, and contour length is not reduced (C. Peterson and D. Anderson, unpublished). Moreover, when proheads are incubated with supercoiled lariat complexes of DNA–gp3–gp16 and observed by TEM and STEM, the proheads are found on the coiled loops or at the lariat loop junction, but not on the tails of the lariats (S. Grimes, M. Simon, and D. Anderson, unpublished). The efficient binding of the prohead to the supercoiled DNA–gp3–gp16 complex probably positions multiple copies of gp16 on pRNA to complete assembly of the packaging motor, and in addition positions gp3 of the DNA left end at the opening of the motor channel (Fig. 8c). Then the DNA is released from the outside of the connector (Fig. 8d) and translocation proceeds with the aid of ATP hydrolysis (Fig. 8e).

C. DNA–gp3 Translocation

The consumption of one ATP per two base pairs of DNA–gp3 packaged (Guo *et al.,* 1987c) and the 4 min required for completion of packaging are known from bulk packaging experiments, but little additional information addresses the mechanism of translocation. How does the DNA get inside the head? Recent studies, described below, heighten prospects for approaching and understanding the mechanism. The connector has been visualized *in situ* in the prohead (Tao *et al.,* 1998). The connector structure has been solved to 3.2-Å resolution by X-ray crystallography and the structure fit into a cryo-EM 3D reconstruction of the motor (Simpson *et al.,* 2000). Laser-tweezer single-molecule packaging studies demonstrate the existence of a force-generating motor, and its force/velocity relationship has been determined (Smith *et al.,* 2001). These studies provide a new foundation and vista for progress.

1. *Motor Structure and Symmetry*

Cryo-EM 3D reconstructions of proheads and partially packaged particles, coupled with the first near-atomic-resolution structure of the connector, show that pRNA is bound at the narrow end of the connector and exhibits fivefold symmetry (Tao *et al.,* 1998; Simpson *et al.,* 2000). The electron density due to pRNA is demonstrated by a difference map of 3D reconstructions of proheads with pRNA and pRNA-free proheads (Figs. 11b–11d). pRNA is found as a ring containing five prominent spokes, and fivefold symmetry holds with or without

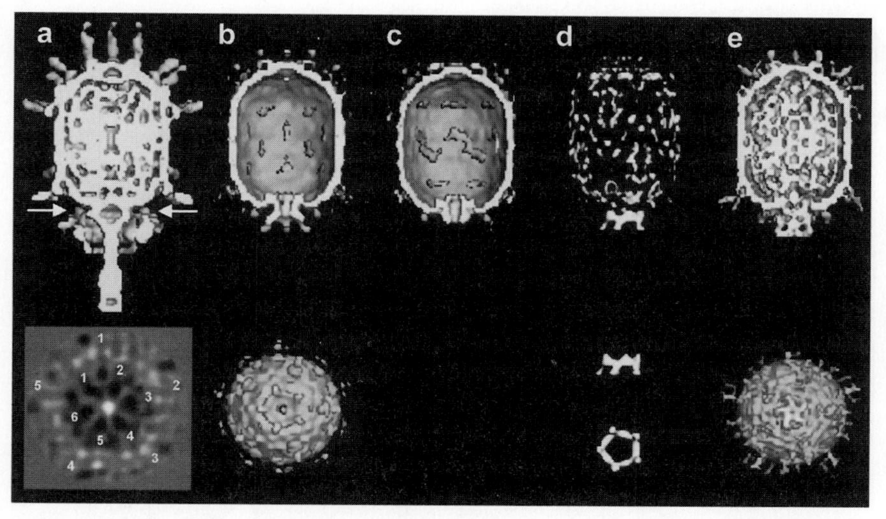

FIG. 11. Cryo-EM 3D reconstructions. Top and bottom rows: (a) mature ϕ29; (b) prohead +120-base pRNA; (c) prohead treated with RNase; (d) difference map between (b) and (c); and (e) partially packaged particle (with DNA in channel). End-on views, looking along the tail toward the head, are given for (b) and (e). Orthogonal difference pRNA densities are shown below the difference map in (d). The arrow in (a) shows the position of the section below, which was obtained by averaging particles without imposing fivefold symmetry. Note that both the fivefold symmetry of the downward-pointing head fibers (numbered 1–5) and the sixfold symmetry (numbered 1–6) of the lower collar are visible. (From Simpson *et al.,* 2000, Fig. 1, p. 746.)

imposition of fivefold symmetry in the reconstructions (Simpson *et al.,* 2000). Recently a method for 3D image reconstruction for symmetry-mismatched components in tailed phages was described, which illustrates the sixfold-symmetric neck and tail of ϕ29 that is attached to the fivefold-symmetric head and confirms the fivefold symmetry of pRNA (Morais *et al.,* 2001; see also Fig. 11a).

The initial 3D reconstructions revealed an area of contact between the icosahedral capsid shell and the pRNA ring, providing a rationale for the observed fivefold symmetry of the pRNA (Fig. 12; see also color

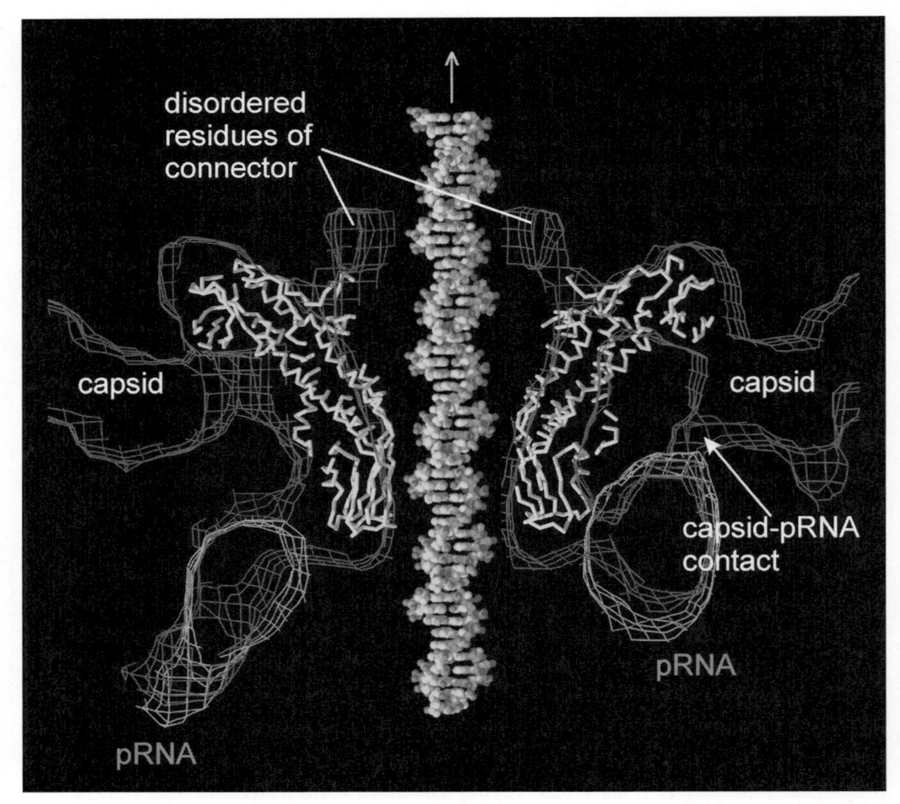

FIG. 12. Cryo-EM density fitted with atomic structures. Cross section of the cryo-EM prohead density (*red*) fitted with the Cα backbone of the connector (*yellow*) and the cryo-EM pRNA difference map (*green*). Shown also is a DNA molecule placed through the central channel of the connector. The prohead capsid, one of the five contacts between the pRNA with the capsid, and the partially disordered residues 229–246 and 287–309 in the connector are indicated. Drawn with the programs XTALVIEW and RASTER3D. (From Simpson *et al.,* 2000, Fig. 3, p. 748.) Also see color insert.

insert) (Simpson *et al.,* 2000). This is in direct conflict with the genetic studies and the analytical ultracentrifugation results (see pRNA section above), which show that pRNA is a hexamer (F. Zhang *et al.,* 1998; Guo *et al.,* 1998). Although a small region of pRNA is in contact with the shell in the reconstructions (Simpson *et al.,* 2000), the prohead-binding region comprises 62 bases of RNA protected from ribonuclease by the connector (Reid *et al.,* 1994a). In conflict with the report of Simpson *et al.* (2000), use of different methods of cryo-EM 3D reconstruction by Ibarra *et al.* (2000) shows the pRNA as a flat hexamer on the prohead, without spokes. Thus, the true symmetry of the pRNA on the prohead is controversial. The question is important because the pRNA symmetry bears directly on the structure of the packaging motor and its mechanism. If connector rotation is required in packaging (see mechanism below), a sixfold-symmetric pRNA may rotate with the connector, whereas fivefold symmetry for the pRNA lends credence to the proposal that it forms a stator with the shell, within which the connector rotates (Simpson *et al.,* 2000).

The crystal structure of the connector fits very nicely into the cryo-EM 3D reconstructions of proheads and partially packaged particles (Fig. 12) (Simpson *et al.,* 2000). The inner radius of the pRNA ring is 34 Å versus 33 Å for the narrow end of the connector, and thus the pRNA ring fits snugly over the connector. Partially packaged particles, flash-frozen and observed by cryo-EM 3D reconstruction, show additional mass, which joins the pRNA spokes to form a canister-like structure, the packaging motor, at the base of the head (Fig. 11e). The additional mass is interpreted as gp16, and its symmetry may be determined by pRNA. Finally, the channel of the connector in the partially packaged particle contains electron-dense material, likely DNA (Fig. 11e). There is a consensus that DNA passes through the connector channel in packaging, but there has been no formal demonstration of this.

2. Single-Molecule Studies of the ϕ29 DNA Packaging Motor

The use of laser optical tweezers to study single-molecule ϕ29 DNA packaging events shows that the packaging machine is a highly processive and force-generating motor (Smith *et al.,* 2001). Besides generating rate and force data, single-molecule studies offer the advantage of observing the nuances of the packaging process that are averaged out in population bulk studies. By tethering the packaging machine between two beads, the tweezers are able to measure the packaging rate in real time and the force of the motor. The packaging ternary complex of proheads, gp16, and DNA–gp3 is assembled in the test tube, and packaging is initiated by the addition of ATP. After 30 sec of packaging, the reaction is stalled by the addition of an excess of nonhydrolyzable γ-S-ATP to allow for attachment of the packaging complexes to beads and transfer to the laser trap. The stalled complexes are attached to streptavidin-coated beads via a biotin tag on the end of the unpackaged DNA, and this bead is caught in the laser trap. A second bead coated with antibody against the prohead is held by suction on a pipet tip and brought near the first bead to capture the prohead of the stalled packaging complexes and complete the hook-up (Fig. 13, left). ATP is restored, and packaging is analyzed using either the constant-force-feedback or the no-feedback mode of the tweezers.

Using the constant-force-feedback mode, the distance between the two beads is adjusted by a feedback mechanism to maintain a constant

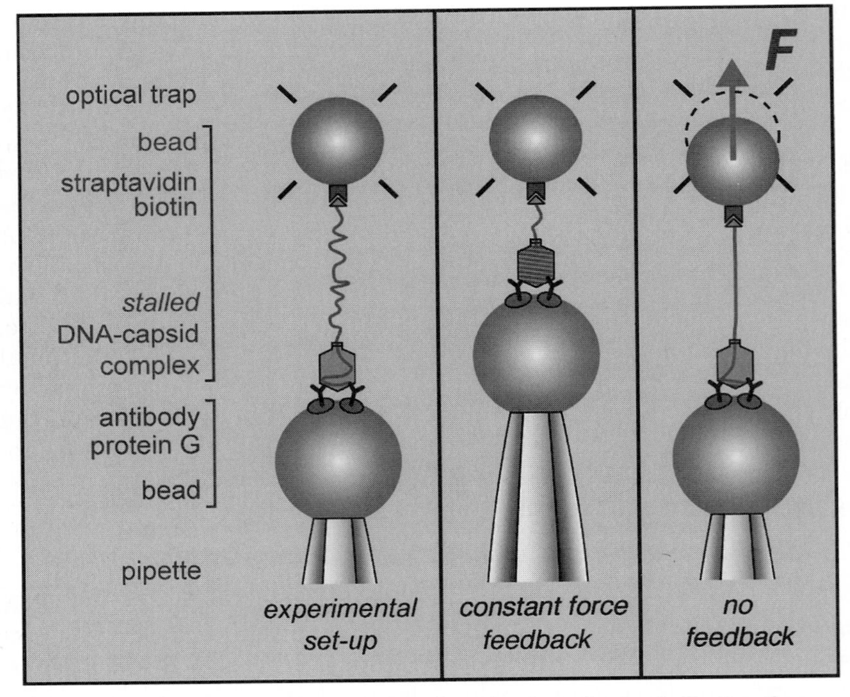

FIG. 13. Three schematic diagrams of single-molecule packaging indicating the experimental setup at the start of a measurement, constant-force-feedback-mode measurement, and no-feedback-mode measurement. A single ϕ29 packaging complex is tethered between two microspheres. Optical tweezers are used to trap one microsphere and measure the forces acting on it, while the other microsphere is held by a micropipet. To insure measurement on only one complex, the density of complexes on the microsphere is adjusted so that only about 1 of 5–10 microspheres yields hookups. Hookups break in one discrete step as the force is increased, indicating only one DNA molecule carries the load. (From Smith *et al.*, 2001, Fig. 1a.)

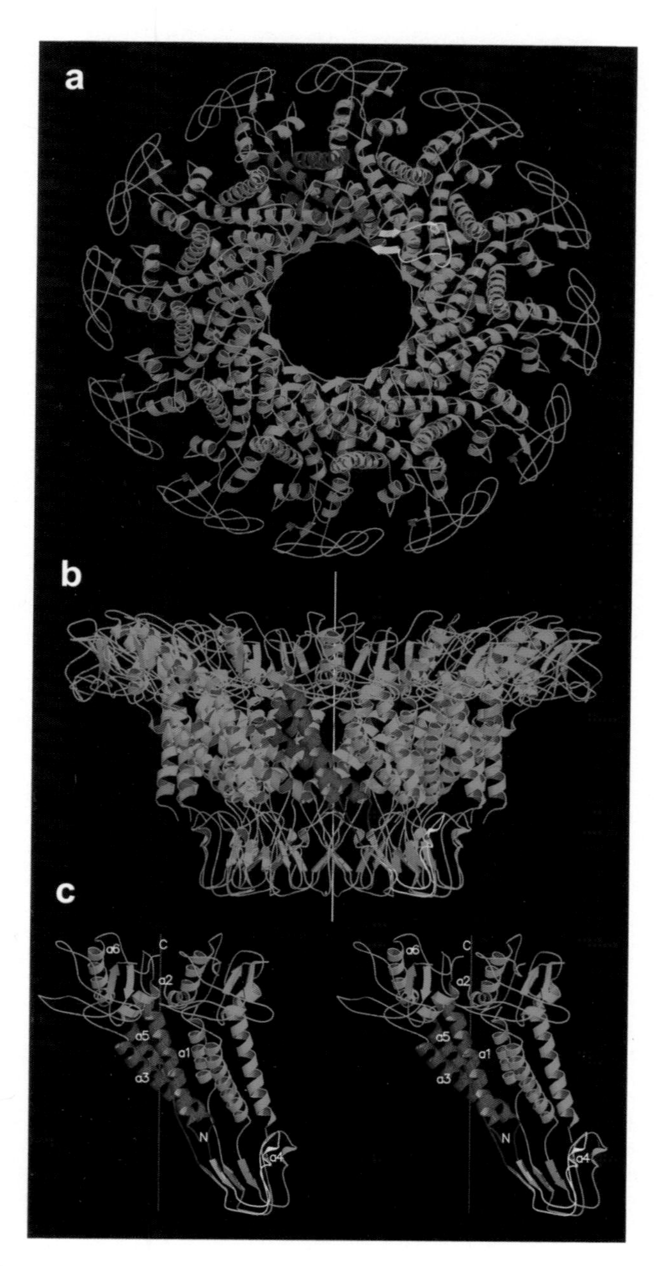

FIG 7.5 Connector structure ribbon diagrams. (a) The dodecameric connector seen from the tail looking toward the head; (b) a side view with the pRNA-binding site at the bottom, showing the conical structure of the connector and the helical twist of each subunit around the 12-fold axis (*white*); (c) a stereo diagram of a pair of monomers. One monomer is colored red in the central domain, green in the wide-end domain, which resides inside the capsid, and yellow at the narrow-end domain. The other monomer is colored blue. The ordered part of the polypeptide starts with helix α 1 on the outside of the connector, going toward the wide end (residues 61–128, and 247–286). Helix α 3 (residues 129–157) return the chain to the narrow end (residues 158–202). The tip of the connector at the narrow end is formed by residues 164–170, and 185–196. Helix α 5 (residues 208–226) returns the polypeptide to the wide end through the second disordered section. Drawn with the program MOLSCRIPT and RASTER3D. (Adapted from Simpson *et al.*, 2000, Fig. 2, p. 747.)

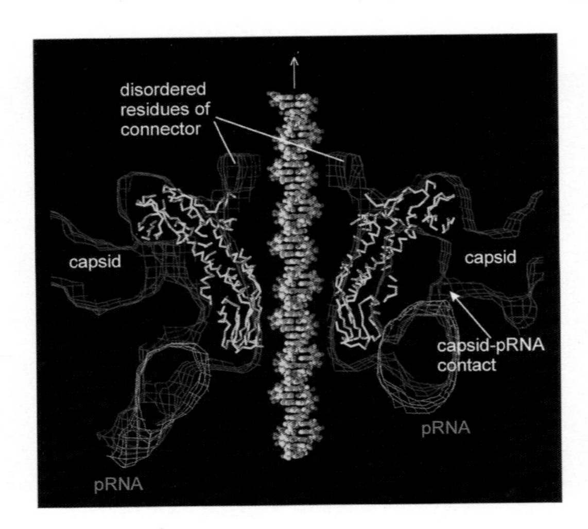

FIG 7.12 Cryo-EM density fitted with atomic structures. Cross section of the cryo-EM prohead density (*red*) fitted with the Cα backbone of the connector (*yellow*) and the cryo-EM pRNA difference map (*green*). Shown also is a DNA molecule placed through the central channel of the connector. The prohead capsid, one of the five contacts between the pRNA with the capsid, and the partially disordered residues 229–246 and 287–309 in the connector are indicated. Drawn with the programs XTALVIEW and RASTER3D. (From Simpson *et al.*, 2000, Fig. 3, p. 748.)

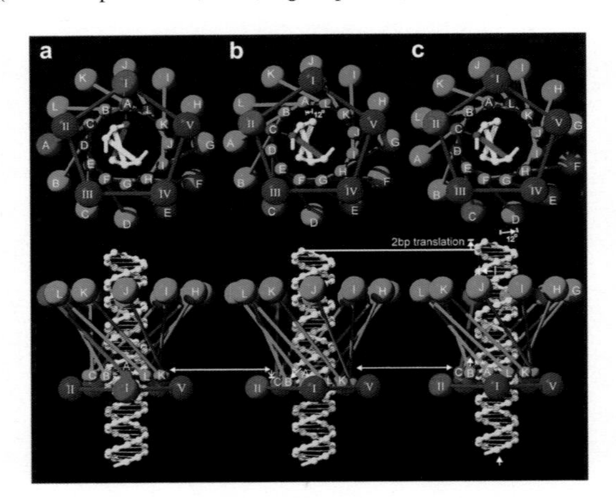

FIG 7.17 Model of the DNA-packaging mechanism. Shown is one cycle in the mechanism that rotates the connector and translates the DNA into the head. The view down the connector axis (*top*) is toward the head (as in Fig. 5a), whereas the bottom row shows side views corresponding to that seen in Fig. 5b. Eleven of the 12 subunits (A–L) of the connector are shown in green; the "active" monomer is shown in red. The connector is represented as a set of small spheres at the narrow end and a set of larger spheres at the wide end connected by a line representing the central helical region. The pRNA–ATPase complexes (I–V) surrounding the narrow end, are shown by a set of four blue spheres and one red sphere. The DNA base aligned with the active connector monomer is also shown in red. (A) The active pRNA–ATPase I interacts with the adjacent connector monomer (A), which in turn contacts the aligned DNA base. (B) The narrow end of the connector has moved counterclockwise by 12° to place the narrow end of monomer C opposite ATPase II, the next ATPase to be fired, causing the connector to expand lengthwise by slightly changing the angle of the helices in the central domain (*white arrow with asterisk*). (C) The wide end of the connector has followed the narrow end, while the connector relaxes and contracts (*white arrow with two asterisks*), thus causing the DNA to be translated into the phage head. For the next cycle, ATPase II is activated, causing the connector to be rotated another 12°, and so forth. Drawn with the program RASTER3D. (From Simpson *et al.*, 2000, Fig. 4, p. 749.)

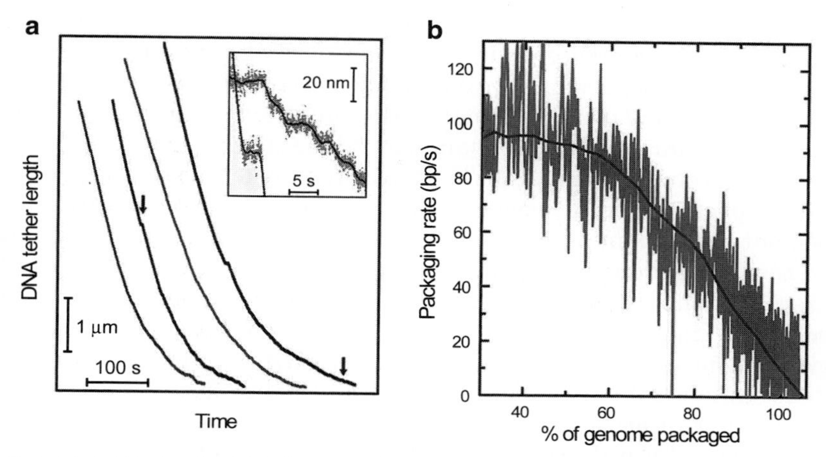

FIG. 14. (a) Plots of DNA tether length versus time for four different complexes with a constant force of ~5 pN using a 34.4-kb ϕ29–lambda DNA construct. Inset: Blowups of regions indicated by arrows, showing pauses (curves have been shifted). The solid lines are a 100-point average of the raw data. (b) Packaging rate versus the amount of DNA packaged, relative to the original 19.3-kbp ϕ29 genome. The trace for a single complex is derived from the rightmost trace in (a). Rates were obtained by linear fitting in a 1.5-sec sliding window. The thick center line is an average of eight such measurements. Here, large pauses (velocity drops >30 bp/s below local average) were removed, and the curves were horizontally shifted to account for differences in microsphere attachment points. The center line was also smoothed using a 200-nm sliding window. The standard deviation for the ensemble of measurements varies from ~20 bp/s at the beginning down to ~10 bp/s at the end. (From Smith *et al.,* 2001, Figs. 1b and 1c.)

DNA tension of 5 pN (Fig. 13, middle). As the prohead packages the DNA into the head, the tether length between the two beads is observed to decrease over time (Fig. 14a). Movement of the beads is only observed in the presence of ATP, in agreement with the ATP dependence of the packaging process. In these experiments a ϕ29–lambda hybrid DNA substrate is used to allow observation of as much of the packaging process as possible. The motor is highly processive and movements of 5 μm are commonly observed; it takes about 5.5 min to package the equivalent of the ϕ29 genome. Nuances of packaging include pauses and slips of the motor during the packaging process (Fig. 14a, inset). When the motor pauses, it remains engaged with the DNA despite the 5 pN of applied external force, and after varying duration it restarts packaging. Pauses occur more frequently as the head fills, with an average of three pauses per micrometer of DNA packaged. Slips are also seen in single-molecule packaging when the motor loses grip on the DNA, and the DNA slides out before the motor quickly recovers and resumes packaging.

Figure 14b illustrates how the packaging rate varies during the packaging process. The trace of an individual packaging event shows the complex nature of packaging; rate fluctuations up to five times the background level (4 bp/s) are observed throughout the process. For analysis, several individual events are averaged and the curve smoothed to generate the data shown in Fig. 14b. The most striking feature of the graph is that the packaging rate is not constant. Initially, the average rate is 100 bp/s, but after about 50% of the DNA is packaged, there is a dramatic and continual drop in the packaging rate. Eventually the velocity falls to zero, and the motor stalls after about 105% of a genome-equivalent is packaged. This rate decrease, which is specific to the last half of packaging, suggests the presence of an internal force that builds up, due to DNA confinement in packaging, that opposes and slows the motor.

To further interrogate the internal force, the force/velocity relationship of the motor is determined using the no-feedback mode of the trap. In the no-feedback mode of operation, the positions of the pipet and the trap are fixed (Fig. 13, right). As the DNA is reeled into the prohead during packaging, the bead is displaced in the laser trap, and the tension on the DNA molecule increases. As this force against the packaging motor increases, the motor eventually stalls. Typical force/velocity (F–V) curves for three individual complexes are shown in Fig. 15a. Although the initial packaging rate and the stall force may vary, the motor is behaving in the same fashion, as the curve shapes are similar. The mean F–V curves for one-third capsid filling (Fig. 15b, rightmost line) and two-thirds capsid filling (leftmost line) show that the two-thirds-filling curve is offset by 14 pN. At one-third filling there is assumed to be no internal force because there is no velocity decrease, whereas at two-thirds filling it is assumed there is an internal force because it takes 14 pN less force to stall the motor. Shifting of the two-thirds curve over 14 pN (in the direction of the arrows) shows that there is very good overlap with the one-third curve, supporting the supposition that the internal and external forces are additive on the motor. From this the inherent F/V curve, in the absence of any internal force, can be determined. The motor is sensitive to very low levels of applied force, indicating that the rate-limiting step is force-dependent. The total force needed to stall the ϕ29 packaging motor is on average 57 pN (Fig. 15c), making this motor one of the strongest motors ever reported! Figure 15d shows the buildup of internal force over the course of packaging. Internal force opposing the motor is generated when about 50% of the genome has been packaged, and it reaches a maximum of about 50 pN at the end of packaging. A rough estimate of the pressure inside the phage head is ~6 MPa.

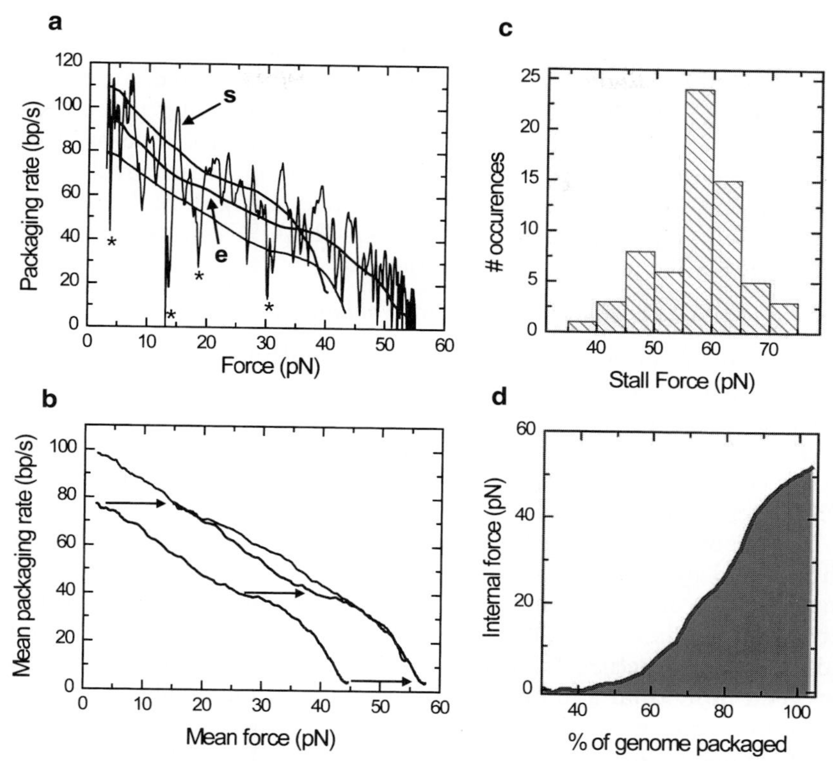

FIG. 15. Force–velocity (F–V) analysis: (a) The packaging rate for a single complex, s, was determined by linear fitting of data for packaging against increasing force (not shown) in a 0.7-s sliding window. The black line, e, is obtained by editing out large pauses (marked by asterisks; velocity drops >30 bp/s below local average) and smoothing (50-point sliding window). These long pauses were removed so as not to perturb the general trend of the F–V behavior. The other lines are data from two other complexes. (b) Mean F–V curves when ~1/3 of the genome is packaged (*rightmost line*) and when ~2/3 of the genome is packaged (*leftmost line*). These curves were obtained from 14 and 8 individual traces, respectively. Arrows show the leftmost line being shifted by +14 pN to account for the internal force (see text). (c) Histogram of total stall force measured for 65 individual complexes, indicating an average stall force of ~57 pN. DNA overstretching was not observed because linkages always broke at external forces lower than 65 pN. However, stall forces above 65 pN could be determined in cases where internal force added to the total force. (d) The internal force versus the percentage of the genome packaged. This plot is obtained by relating the packaging rate in Fig. 14b to the force acting on the motor using the F–V relationship in (b) of this figure while accounting for the 5 pN of external load in the data of Fig. 14b. (From Smith *et al.,* 2001, Fig. 3.)

Single-molecule packaging provides a window on the dynamics and energetics of the packaging process that cannot be obtained from biochemical studies and static structures. Properties of the motor that are inaccessible in bulk packaging are addressed, including precise measurements of variations in packaging rate and force. Moreover, constraints of packaging, such as pausing and slipping, are integral to eventual understanding of the mechanism of packaging. Finally, the work addresses fundamental aspects of protein–RNA–DNA interactions and chemomechanical energy transduction.

3. *Current View of the Mechanism of Packaging*

All of the observations and experiments outlined in this review are focused on a single question: How is DNA translocated from the outside of the prohead shell to the inside? The data in hand provide evidence for several possible mechanisms, which we will rationalize with respect to our current understanding of the structure and biochemistry of ϕ29 DNA packaging.

The mismatch between the 5-fold symmetry of the prohead shell and the 12-fold symmetry of the connector is the basis for the proposal that rotation of the connector within the prohead drives DNA translocation (Hendrix, 1978). The strength of the interaction between the mismatched structures is described as the sum of the bonding energy between individual subunits of the connector and capsid (Fig. 16). If the symmetries were matched, a large energy barrier would produce a rigid connector–capsid complex (Fig. 16a). On the contrary, the documented symmetry mismatch between the connector and capsid provides a loose interaction that facilitates connector rotation (Tao *et al.,* 1998) (Fig. 16b). While a connector monomer is passing through the range at which it is most attracted to the capsid, the other monomers have a mismatch. The hydrophobic character of the outside of the ϕ29 connector revealed by X-ray crystallography would facilitate rotation (Simpson *et al.,* 2001).

Hendrix (1978) proposed that the rotating connector contacts DNA as a nut on bolt to drive DNA into the prohead. However, this requires that the connector inner channel match precisely the helical grooves of the DNA, yet allow the DNA to slip axially along its length through the channel. Also, the DNA would have to be tethered inside the prohead so it would not simply rotate axially along its length with the connector. No such stringent engagement of the DNA with either the connector or the capsid has been described.

Evidence that the ϕ29 connector is capable of wrapping supercoiled DNA around the outside presents several alternative models whereby active rotation of the connector can drive DNA translocation. The

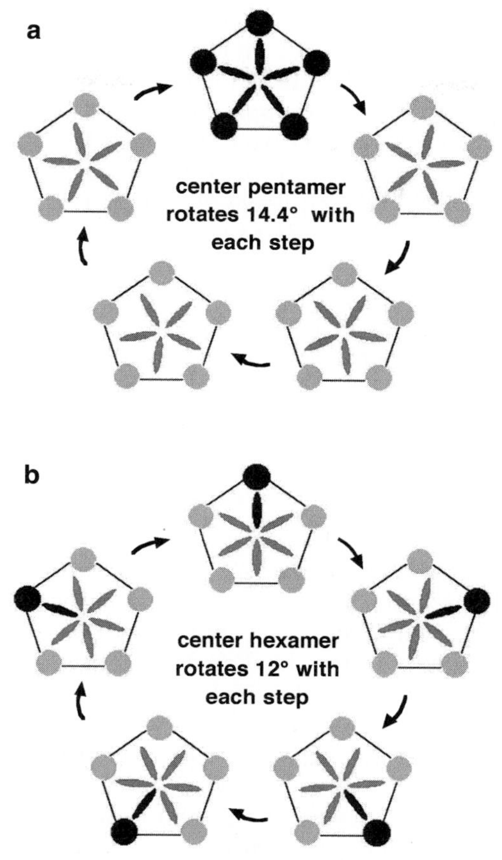

FIG. 16. Symmetry mismatch potentiates connector rotation. (a) When a structure of fivefold symmetry (*ellipses*) is rotated within another of like symmetry (*circles*), all of the inner subunits are aligned simultaneously with the outside subunits (*top orientation*). If this alignment represents the maximum binding strength between monomers (*black monomers*), the inner structure becomes trapped in the top, energetically favored orientation. (b) Conversely, if there is a mismatch, and the inner structure is sixfold (*ellipses*) compared to the outer fivefold structure (*circles*), then as the inner multimer rotates, there is only one alignment event (*dark monomers*) for any given orientation between the inside and outside structures. This distribution of the binding energy potentiates rotation of one structure with respect to the other.

simplest model is similar to a capstan of a ship: Rotation of the connector displaces DNA through a point of entry between the connector and the head shell (Turnquist *et al.*, 1992; Grimes and Anderson, 1997). A more complex idea is that rotation of the connector with bound DNA introduces superhelical stress in the DNA, which is relieved by the displacement of the chromosome through the central channel and into the head. Unlike the nut-and-bolt model, the interaction between the connector and DNA for both of these models can be reconciled with the current structural data. The wrapping of DNA around the outside of the connector, with roughly one turn of the helix binding to each monomer of the connector, suggests that the rotating connector could act as a winch to transport DNA (Turnquist *et al.,* 1992). However, cryo-EM 3D reconstructions of DNA packaging intermediates show no additional mass bound to the outside of the connector as predicted by these models. Also, there is no obvious point of entry into the head at the outside of the connector described by the capstan model. As with the nut-and-bolt mechanism, the tethering of the DNA inside the capsid required by the superhelical displacement model has not been described, nor has the feasibility of this model in terms of DNA structure been addressed in any detail.

4. Connector Rotation Provides Sequential Alignment in the Ratchet Mechanism

Solution of the structure of the ϕ29 connector by X-ray crystallography has recently provided a ratchet mechanism model for DNA packaging involving connector rotation (Simpson *et al.,* 2000) (Fig. 17; also see color insert). However, this new model refers to a passive rotation event that is driven by a more classical allosteric event. Connector conformation mediated by ATP hydrolysis serves merely to align or index the connector relative to the capsid–pRNA–ATPase for each sequential DNA translocation step.

The most compelling feature of the ϕ29 connector structure is the presence of three α alpha helices of each subunit, which traverse the height of the central region of the structure (Fig. 5). These 36 parallel α helices leave the impression that the connector can act as a spring. Compression of the connector from top to bottom would seem to result in a torsion of the connector, with the top being displaced axially from the bottom. This notion is supported by the ability to reversibly compress the connector by about 23 Å by applying pressure to the narrow end via atomic force microscopy (Muller *et al.,* 1997). It is this potential motion which is proposed by Simpson *et al.* (2000) to be the *de facto* means by which DNA is ratcheted into the prohead.

The proposed mechanism requires simple interaction of the connector with the substrate DNA. This point is most likely the inside surface of the narrow, distal end of the connector. However, it is possible that a site in the lower portion of the intact DNA packaging machine, in the pRNA and the gp16 ATPase, might be the area of contact between DNA

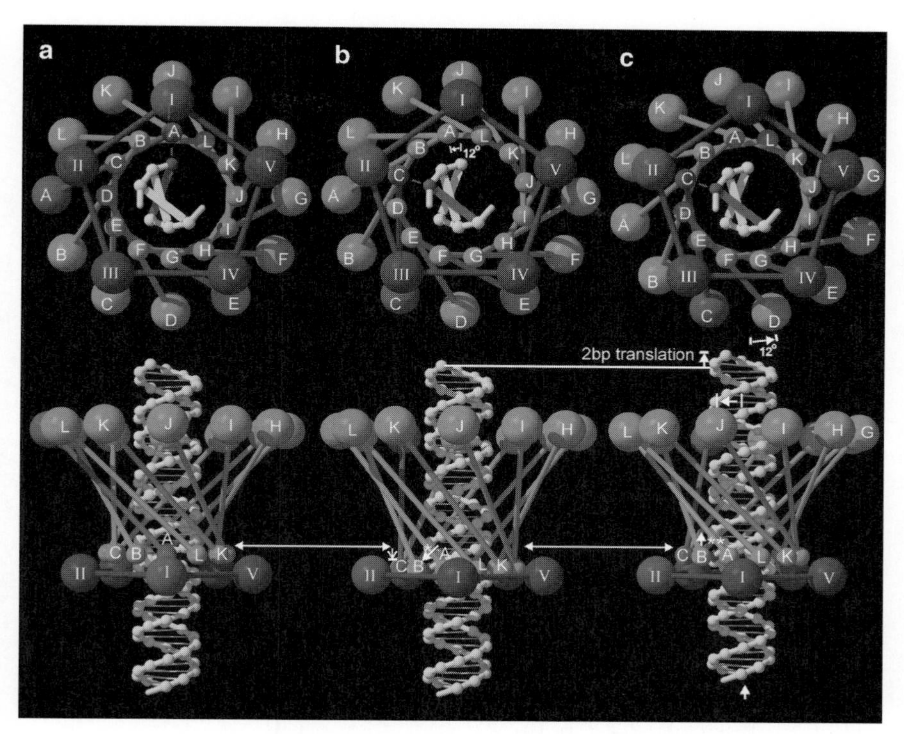

FIG. 17. Model of the DNA packaging mechanism. Shown is one cycle in the mechanism that rotates the connector and translates the DNA into the head. The view down the connector axis (*top*) is toward the head (as in Fig. 5a), whereas the bottom row shows side views corresponding to that seen in Fig. 5b. Eleven of the 12 subunits (A–L) of the connector are shown in green; the "active" monomer is shown in red. The connector is represented as a set of small spheres at the narrow end and a set of larger spheres at the wide end connected by a line representing the central helical region. The pRNA–ATPase complexes (I–V), surrounding the narrow end, are shown by a set of four blue spheres and one red sphere. The DNA base aligned with the active connector monomer is also shown in red. (a) The active pRNA–ATPase I interacts with the adjacent connector monomer (A), which in turn contacts the aligned DNA base. (b) The narrow end of the connector has moved counterclockwise by 12° to place the narrow end of monomer C opposite ATPase II, the next ATPase to be fired, causing the connector to expand lengthwise by slightly changing the angle of the helices in the central domain (*white arrow with asterisk*). (c) The wide end of the connector has followed the narrow end, while the connector relaxes and contracts (*white arrow with two asterisks*), thus causing the DNA to be translated into the phage head. For the next cycle, ATPase II is activated, causing the connector to be rotated another 12°, and so forth. Drawn with the program RASTER3D. (From Simpson *et al.,* 2000, Fig. 4, p. 749.) Also see color insert.

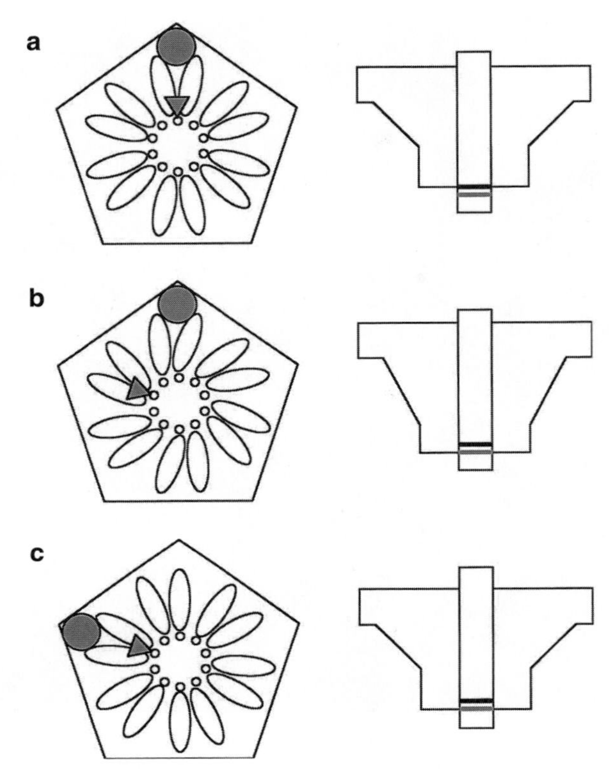

FIG. 18. Schematic of the compression ratchet mechanism. (a) A packaging stroke begins with the alignment of the outer edge of the connector with the prohead (*gray circle*) and the inner edge with the DNA helix (*gray triangle*); the connector is in a compressed form (*right*). (b) As the connector extends, the outer edge stays fixed to the prohead and the narrow end rotates with respect to the head by 12° counterclockwise, such that contact with the DNA shifts to the next pair of connector monomers and two base pairs down the DNA helix (*right*). (c) During the subsequent compression of the connector, the axially restrained DNA remains in contact with the connector and is driven two base pairs into the head (*right*). Concurrently, the outer edge of the connector rotates passively 12° counterclockwise with respect to the prohead (*left*), reestablishing contact with the head two connector monomers to the left.

and the motor (see below). The connector starts in a compressed form with an adjacent pair of connector monomers associated with the DNA (Figs. 17 and 18). In Fig. 18, pairs of connector subunits engage points of fivefold symmetry, making the effective mismatch of six within five. The DNA is released and the connector extends outward from the prohead by 7 Å. If the wide end of the connector is fixed at this point, the result of this extension would be an independent rotation of the narrow end of the connector with respect to the prohead of 12° counterclockwise when viewed from below. This would transfer the point of alignment between the narrow end of the connector and the DNA such that the tip of the connector traverses two base pairs and reestablishes its DNA contact with the next pair of connector monomers (Fig. 18b). Subsequent compression of the connector would translocate two bases of DNA into the prohead (Fig. 18c). Because the axial rotation of the DNA, and at this point the attached narrow end of the connector, are restricted, this compression would cause a counterclockwise rotation of 12° in the upper portion of the connector. This passive rotation would cause realignment of the connector and capsid such that the next vertex of the fivefold orifice of the capsid would be aligned with the next pair of connector subunits. This rotation event differs from the active rotation mechanisms of the bacterial flagellum and the F_1-ATPase and instead is the consequence of a ratchet mechanism similar to that of the myosin head, albeit in a more complex form. A different ratchet packaging mechanism for phage T3 has been described in which the large packaging protein recognizes different sites on the DNA backbone and contracts to advance the DNA; in this model connector rotation is not mentioned (Fujisawa and Morita, 1997).

The irrefutable symmetry mismatches between the connector and the prohead and between the connector and the DNA provide the necessary details for this compression-driven, passive rotation if the connector alone interacts with the substrate DNA. If the pRNA/ATPase complex has sixfold symmetry rather than fivefold symmetry and is not attached to the capsid, it is easier to rationalize mediation of connector conformational change in the mechanism described. A sixfold pRNA and ATPase would also present a potential for direct interaction with the DNA, providing more flexibility in interpretation of the dynamic events involved in the translocation process.

V. Aims and Prospects

The ability to transform free energy into motion, a ubiquitous property of biological systems, is manifest dramatically in the translocation of ϕ29 DNA into the prohead. The ϕ29 DNA packaging cascade involves protein, RNA, and DNA conformational changes and movement that are comparable to allosteric regulation in enzyme–substrate interactions. We aim to detail dynamic molecular events of packaging that will ultimately lead to the packaging mechanism. The wealth of genetic, physiological, and biochemical analyses on ϕ29 packaging will reach full consummation when the mechanism is known. For the task ahead we are blessed with the simplest and most efficient *in vitro* viral DNA packaging system known.

Three recent breakthroughs, described in detail in foregoing sections, have provided the first overall structural view of the packaging motor and demonstrated that the motor generates high force. First, the connector has been visualized for the first time *in situ* in a phage prohead (Tao *et al.,* 1998). Second, the structure of the connector has been solved to 3.2-Å resolution by X-ray crystallography and the structure fitted into a cryo-EM 3D reconstruction of the prohead (Simpson *et al.,* 2000, 2001). Third, optical tweezers have been used to pull on single ϕ29 DNA molecules as they are packaged, demonstrating that the prohead-connector–pRNA–gp16 complex is a highly processive, force-generating motor, which can work against loads of up to about 57 pN. These studies augment the substantive database on the ϕ29 *in vitro* packaging system reviewed above.

In quest of the packaging mechanism, some aims are as follows: We hope to determine atomic resolution structures of the prohead, pRNA, and gp16; obtain high-resolution cryo-EM 3D reconstructions of the partially packaged particle containing the active motor; and crystallize the partially packaged particle. In the study of dynamic motor function, we hope to observe connector rotation by polarization of single fluorophores; demonstrate conformational change in motor components during packaging by electron paramagnetic resonance (EPR) and fluorescence spectroscopy; rationalize the force/velocity relationships of DNA packaging in terms of theory on entropic, electrostatic, and bending energies of the DNA; and determine the step size and power stroke of the motor. The ϕ29 system provides a unique opportunity for determining the packaging mechanism shared by dsDNA phages and certain animal viruses.

Acknowledgments

The current ϕ29 research is supported by NIH research grants DE03606 and GM59604. We are deeply grateful for the expertise of Charlene Peterson in preparation of the manuscript.

References

Anderson, D. L., and Reilly, B. E. (1993). Morphogenesis of bacteriophage ϕ29. *In* "*Bacillus subtilis* and Other Gram Positive Bacteria: Physiology, Biochemistry and Molecular Genetics" (J. A. Hoch, R. Losick, and A. L. Sonenshein, Eds.), pp. 859–867. ASM Press, Washington, D.C.

Anderson, D. L., Hickman, D. D., and Reilly, B. E. (1966). Structure of *Bacillus subtilis* bacteriophage ϕ29 and the length of ϕ29 deoxyribonucleic acid. *J. Bacteriol.* **91,** 2081–2089.

Bailey, S., Wichitwechkarn, J., Johnson, D., Reilly, B. E., Anderson, D. L., and Bodley, J. W. (1990). Phylogenetic analysis and secondary structure of the *Bacillus subtilis* bacteriophage RNA required for DNA packaging. *J. Biol. Chem.* **265,** 22365–22370.

Bjornsti, M. A., Reilly, B. E., and Anderson, D. L. (1983). Morphogenesis of bacteriophage ϕ29 of *Bacillus subtilis:* Oriented and quantized *in vitro* packaging of DNA–gp3. *J. Virol.* **45,** 383–396.

Black, L. W. (1989). DNA packaging in dsDNA bacteriophages. *Annu. Rev. Microbiol.* **43,** 267–292.

Camacho, A., Jimenez, F., De La Torre, J., Carrascosa, J. L., Mellado, R. P., Vasquez, C., Vinuela, E., and Salas, M. (1977). Assembly of *Bacillus subtilis* phage ϕ29. 1. Mutants in the cistrons coding for the structural proteins. *Eur. J. Biochem.* **73,** 39–55.

Carlson, K., and Miller, E. S. (1994). Working with T4. *In* "Molecular Biology of Bacteriophage T4" (J. D. Karam, Ed.), pp. 421–426. ASM Press, Washington, D.C.

Carrascosa, J. L., Mendez, E., Corral, J., Rubio, V., Ramirez, G., Salas, M., and Vinuela, E. (1981). Structural organization of *Bacillus subtilis* phage ϕ29. A model. *Virology* **111,** 401–413.

Chen, C., and Guo, P. (1997). Sequential action of six virus-encoded DNA-packaging RNAs during phage ϕ29 genomic DNA translocation. *J. Virol.* **71,** 3864–3871.

Chen, C., Zhang, C., and Guo, P. (1999). Sequence requirement for hand-in-hand interaction in formation of RNA dimers and hexamers to gear ϕ29 DNA translocation motor. *RNA* **5,** 805–818.

Chen, C., Sheng, S., Shao, Z., and Guo, P. (2000). A dimer as a building block in assembling RNA. *J. Biol. Chem.* **275,** 17510–17516.

Donate, L. E., Valpuesta, J. M., Rocher, A., Mendez, E., Rojo, F., Salas, M., and Carrascosa, J. L. (1992). Role of the amino-terminal domain of bacteriophage ϕ29 connector in DNA binding and packaging. *J. Biol. Chem.* **267,** 10919–10924.

Earnshaw, W. C., and Casjens, S. R. (1980). DNA packaging by the double-stranded DNA bacteriophages. *Cell* **21,** 319–331.

Fujisawa, H., and Morita, M. (1997). Phage DNA packaging. *Genes Cells* **2,** 537–545.

Garver, K., and Guo, P. (1997). Boundary of pRNA functional domains and minimum pRNA sequence requirement for specific connector binding and DNA packaging of phage ϕ29. *RNA* **3,** 1068–1079.

Garver, K., and Guo, P. (2000). Mapping the inter-RNA interaction of bacterial virus ϕ29 packaging RNA by site-specific photoaffinity cross-linking. *J. Biol. Chem.* **275,** 2817–2824.

Grimes, S., and Anderson, D. (1989a). *In vitro* packaging of bacteriophage ϕ29 DNA restriction fragments and the role of the terminal protein gp3. *J. Mol. Biol.* **209,** 91–100.

Grimes, S., and Anderson, D. (1989b). Cleaving the prohead RNA of bacteriophage ϕ29 alters the *in vitro* packaging of restriction fragments of DNA–gp3. *J. Mol. Biol.* **209,** 101–108.

Grimes, S., and Anderson, D. (1990). RNA dependence of the bacteriophage ϕ29 DNA packaging ATPase. *J. Mol. Biol.* **215,** 559–566.

Grimes, S., and Anderson, D. (1997). The bacteriophage ϕ29 packaging proteins supercoil the DNA ends. *J. Mol. Biol.* **266,** 901–914.

Guo, P., Grimes, S., and Anderson, D. L. (1986). A defined system for *in vitro* packaging of DNA–gp3 of the *Bacillus subtilis* bacteriophage ϕ29. *Proc. Natl. Acad. Sci. USA* **83,** 3505–3509.

Guo, P., Erickson, S., and Anderson, D. L. (1987a). A small viral RNA is required for *in vitro* packaging of bacteriophage ϕ29 DNA. *Science* **236,** 690–694.

Guo, P., Peterson, C., and Anderson, D. L. (1987b). Initiation events in *in vitro* packaging of bacteriophage ϕ29 DNA–gp3. *J. Mol. Biol.* **197,** 219–228.

Guo, P., Peterson, C., and Anderson, D. L. (1987c). Prohead- and DNA–gp3-dependent ATPase activity of the DNA packaging protein gp16 of bacteriophage ϕ29. *J. Mol. Biol.* **197,** 229–236.

Guo, P., Bailey, S., Bodley, J. W., and Anderson, D. L. (1987d). Characterization of the small RNA of the bacteriophage ϕ29 DNA packaging machine. *Nucleic Acids Res.* **15,** 7081–7090.

Guo, P., Erickson, S., Xu, W., Olson, N., Baker, T. S., and Anderson, D. (1991a). Regulation of phage ϕ29 prohead shape and size by the portal vertex. *Virology* **183,** 366–373.

Guo, P., Rajagopal, B. S., Anderson, D., Erickson, S., and Lee, C. S. (1991b). sRNA of bacteriophage ϕ29 of *B. subtilis* mediates DNA packaging of ϕ29 proheads assembled in *E. coli*. *Virology* **185,** 395–400.

Guo, P., Zhang, C., Chen, C., Garver, K., and Trottier, M. (1998). Inter-RNA interaction of phage ϕ29 pRNA to form a hexameric complex for viral DNA transportation. *Mol. Cell* **2,** 149–155.

Hagen, E. W., Reilly, B. E., Tosi, M. E., and Anderson, D. L. (1976). Analysis of gene function of bacteriophage ϕ29 of *Bacillus subtilis:* Identification of cistrons essential for viral assembly. *J. Virol.* **19,** 501–517.

Harding, N. E., Ito, J., and David, G. S. (1978). Identification of the protein firmly bound to the ends of bacteriophage ϕ29 DNA. *Virology* **84,** 279–292.

Hendrix, R. W. (1978). Symmetry mismatch and DNA packaging in large bacteriophages. *Proc. Natl. Acad. Sci. USA* **75,** 4779–4783.

Hermoso, J. M., Mendez, E., Soriano, F., and Salas, M. (1985). Location of the serine residue involved in the linkage between the terminal protein and the DNA of phage ϕ29. *Nucleic Acids Res.* **13,** 7715–7728.

Ibarra, B., Caston, J. R., Llorca, O., Valle, M., Valpuesta, J. M., and Carrascosa, J. L. (2000). Topology of the components of the DNA packaging machinery in the phage ϕ29 prohead. *J. Mol. Biol.* **298,** 807–815.

Kellenberger, E., Sechaud, J., and Ryter, A. (1959). Electron microscopical studies of phage replication. IV. The establishment of the DNA pool of vegetative phage and the maturation of phage particles. *Virology* **8,** 478–498.

Lee, C. S., and Guo, P. (1994). A highly sensitive system for the *in vitro* assembly of bacteriophage ϕ29 of *Bacillus subtilis*. *Virology* **202,** 1039–1042.

Lee, C. S., and Guo, P. (1995). *In vitro* assembly of infectious virions of double-stranded DNA phage ϕ29 from cloned gene products and synthetic nucleic acids. *J. Virol.* **69,** 5018–5023.

Meijer, W. J. J., Horcajadas, J. A., and Salas, M. (2001). ϕ29 family of phages. *Microbiol. Mol. Biol. Rev.* **65,** 261–287.

Morais, M. C., Tao, Y., Olson, N. H., Grimes, S., Jardine, P. J., Anderson, D. L., Baker, T. S., and Rossmann, M. G. (2001). Cryo-EM image reconstruction of symmetry mismatches in bacteriophage ϕ29. *J. Struct. Biol.* **135,** 38–46.

Mosharrafa, E. T., Schachtele, C. F., Reilly, B. E., and Anderson, D. L. (1970). The complementary strands of bacteriophage ϕ29 DNA: Preparative separation and transcription studies. *J. Virol.* **6,** 855–864.

Muller, D. J., Engel, A., Carrascosa, J. L., and Velez, M. (1997). The bacteriophage ϕ29 head–tail connector imaged at high resolution with the atomic force microscope in buffer solution. *EMBO J.* **16,** 2547–2553.

Nelson, R. A., Reilly, B. E., and Anderson, D. L. (1976). Morphogenesis of bacteriophage ϕ29 of *Bacillus subtilis:* Preliminary isolation and characterization of intermediate particles of the assembly pathway. *J. Virol.* **19,** 518–532.

Peterson, C., Simon, M., Hodges, J., Mertens, P., Higgins, L., Egelman, E., and Anderson, D. (2001). Composition and mass of the bacteriophage ϕ29 prohead and virion. *J. Struct. Biol.* **135,** 18–25.

Rajagopal, B. S., Reilly, B., and Anderson, D. (1993). *Bacillus subtilis* mutants defective in bacteriophage ϕ29 head assembly. *J. Bacteriol.* **175,** 2357–2362.

Reid, R. J. D., Bodley, J. W., and Anderson, D. (1994a). Characterization of the prohead–pRNA interaction of bacteriophage ϕ29. *J. Biol. Chem.* **269,** 5157–5162.

Reid, R. J. D., Bodley, J. W., and Anderson, D. (1994b). Identification of bacteriophage ϕ29 prohead RNA domains necessary for *in vitro* DNA–gp3 packaging. *J. Biol. Chem.* **269,** 9084–9089.

Reid, R. J. D., Zhang, F., Benson, S., and Anderson, D. (1994c). Probing the structure of bacteriophage ϕ29 prohead RNA with specific mutations. *J. Biol. Chem.* **269,** 18656–18661.

Salas, M. (1991). Protein-priming of DNA replication. *Annu. Rev. Biochem.* **60,** 39–71.

Salas, M. (1999). Mechanisms of initiation of linear DNA replication in prokaryotes. *Genetic Eng.* **21,** 159–171.

Salas, M., Mellado, R. P., Vinuela, E., and Sogo, J. M. (1978). Characterization of a protein covalently linked to the 5′ termini of the DNA of *Bacillus subtilis* phage ϕ29. *J. Mol. Biol.* **119,** 269–291.

Schachtele, C. F., De Sain, C. V., Hawley, L. A., and Anderson, D. L. (1972). Transcription during the development of bacteriophage ϕ29: Production of host and ϕ29-specific RNA. *J. Virol.* **10,** 1170–1178.

Simpson, A. A., Tao, Y., Leiman, P. G., Badasso, M. O., He, Y., Jardine, P. J., Olson, N. H., Morais, M. C., Grimes, S., Anderson, D. L., Baker, T. S., and Rossmann, M. G. (2000). Structure of the bacteriophage ϕ29 DNA packaging motor. *Nature* **408,** 745–750.

Simpson, A. A., Leiman, P. G., Tao, Y., He, Y., Badasso, M. O., Jardine, P. J., Anderson, D. L., and Rossmann, M. G. (2001). Structure determination of the head–tail connector of bacteriophage ϕ29. *Acta Crystallog. D* **57,** 1260–1269.

Smith, D. E., Tans, S. J., Smith, S. B., Grimes, S., Anderson, D. L., and Bustamante, C. (2001). The bacteriophage ϕ29 portal motor can package DNA against a large internal force. *Nature* **413,** 748–752.

Tao, Y., Olson, N. H., Xu, W., Anderson, D. L., Rossmann, M. G., and Baker, T. S. (1998). Assembly of a tailed bacterial virus and its genome release studied in three dimensions. *Cell* **95,** 431–437.

Trottier, M., and Guo, P. (1997). Approaches to determine stoichiometry of viral assembly components. *J. Virol.* **71,** 487–494.

Tuerk, C., and Gold, L. (1990). Systematic evolution of ligands by exponential enrichment: RNA ligands to bacteriophage T4 DNA polymerase. *Science* **249,** 505–510.

Turnquist, S., Simon, M., Egelman, E., and Anderson, D. (1992). Supercoiled DNA wraps around the bacteriophage ϕ29 head–tail connector. *Proc. Natl. Acad. Sci. USA* **89,** 10479–10483.

Valle, M., Valpuesta, J. M., and Carrascosa, J. L. (1996). The interaction of DNA with bacteriophage ϕ29 connector: A study by AFM and TEM. *J. Struct. Biol.* **116,** 390–398.

Valle, M., Kremer, L., Martinez A, C., Roncal, F., Valpuesta, J. M., Albar, J. P., and Carrascosa, J. L. (1999). Domain architecture of the bacteriophage ϕ29 connector protein. *J. Mol. Biol.* **288,** 899–909.

Valpuesta, J. M., and Carrascosa, J. L. (1994). Structure of viral connectors and their function in bacteriophage assembly and DNA packaging. *Q. Rev. Biophys.* **27,** 107–155.

Valpuesta, J. M., Serrano, M., Donate, L. E., Herranz, L., and Carrascosa, J. L. (1992). DNA conformation change induced by the bacteriophage ϕ29 connector. *Nucleic Acids Res.* **20,** 5549–5554.

Wichitwechkarn, J., Bailey, S., Bodley, J. W., and Anderson, D. (1989). Prohead RNA of bacteriophage ϕ29: Size, stoichiometry, and biological activity. *Nucleic Acids Res.* **17,** 3459–3468.

Wichitwechkarn, J., Johnson, D., and Anderson, D. (1992). Mutant prohead RNAs in the *in vitro* packaging of bacteriophage ϕ29 DNA–gp3. *J. Mol. Biol.* **223,** 991–998.

Wood, W. B., and King, J. (1979). Genetic control of complex bacteriophage assembly. *In "Comprehensive Virology,"* Vol. 13 (H. Fraenkel-Conrat and R. R. Wagner, Eds.), pp. 581–633, Plenum Press, New York.

Yehle, C. O. (1978). Genome-linked protein associated with the 5′ termini of bacteriophage ϕ29 DNA. *J. Virol.* **27,** 776–783.

Yuan, R., Hamilton, D. L., and Burckhardt, J. (1980). DNA translocation by the restriction enzyme from *E. coli* K. *Cell* **20,** 237–244.

Zhang, C., Lee, C. S., and Guo, P. (1994). The proximate 5′ and 3′ ends of the 120-base viral RNA (pRNA) are crucial for the packaging of bacteriophage ϕ29 DNA. *Virology* **201,** 77–85.

Zhang, C., Tellinghuisen, T., and Guo, P. (1995). Confirmation of the helical structure of the 5′/3′ termini of the essential DNA packaging pRNA of phage ϕ29. *RNA* **1,** 1041–1050.

Zhang, C., Tellinghuisen, T., and Guo, P. (1997). Use of circular permutation to assess six bulges and four loops of DNA-packaging pRNA of bacteriophage ϕ29. *RNA* **3,** 315–323.

Zhang, F., and Anderson, D. (1998). *In vitro* selection of bacteriophage ϕ29 prohead RNA aptamers for prohead binding. *J. Biol. Chem.* **273,** 2947–2953.

Zhang, F., Lemieux, S., Wu, X., St.-Arnaud, D., McMurray, C. T., Major, F., and Anderson, D. (1998). Function of hexameric RNA in packaging of bacteriophage ϕ29 DNA *in vitro*. *Mol. Cell* **2,** 141–147.

Zucker, M. (1989). On finding all suboptimal foldings of an RNA molecule. *Science* **244,** 48–52.

ADVANCES IN VIRUS RESEARCH, VOL. 58

THE INTERACTION OF ORTHOPOXVIRUSES AND INTERFERON-TREATED CULTURED CELLS

C. Jungwirth

Institute for Virology and Immunobiology
University of Würzburg
D-97078 Würzburg, Germany

I. Introduction
II. Interaction of Poxviruses with the IFN Type I-Treated Host Cell
III. The Posttranscriptional Inhibition of Poxvirus-Specific Gene Expression in IFN Type I-Treated Chick Embryo Fibroblasts
IV. Cytotoxicity Enhancement of IFN Type I-Treated and Vaccinia Virus-Infected Cells
V. Inhibition of Vaccinia Virus Replication by Nitric Oxide Synthase Induced by IFN Type II
VI. Metabolic Activities Are Required for the Response of the IFN Type I-Treated Host Cell to Vaccinia Virus Infection
A. Induction of the dsRNA-Activated Protein Kinase PKR
B. 2′–5′ Oligo A/Ribonuclease L System
VII. Double-Stranded RNA Molecules Detectable during Vaccinia Virus Replication and Their Role in the Interaction of Vaccinia Virus with the IFN Type I-Treated Host Cell
VIII. Modulation of the Cellular IFN Type I and Type II Responses by Poxvirus-Specific Gene Products
A. Poxvirus-Specific Soluble IFN Type I and Type II Receptors and the Interception of the IFN Type II Signal Transduction Pathway
B. Poxvirus-Specific Gene Products Modulating the Activity of the 2′–5′ Oligo A/Ribonuclease L Pathway or Protein Kinase PKR
IX. Concluding Remarks
References

I. Introduction

The realization that interferons (IFNs) are a family of cytokines and the characterization of their structure and function paralleled the elucidation of the replication cycle of the poxviruses and their interaction with the host cell (Moss, 1996). An increasingly detailed dissection of the intracellular steps of the poxvirus replication cycle as affected by IFN treatment of the host cell became feasible. Many interesting observations have been collected, which are of relevance far beyond the IFN system (Goodbourn *et al.,* 2000).

0065-3527/02 $35.00

As it became clear that IFNs are proteins with pleiotropic activities, the problem that had to be solved changed (Goodbourn *et al.,* 2000; Stewart, 1979). Originally, the activity of IFN as a naturally occuring inhibitor of poxvirus replication with high specificity for viral functions and little effect on cells was studied. More recent aims are to elucidate how poxviruses interact with a host cell, with numerous properties altered due to signals activated by binding of IFNs to their cognate cellular receptors. One of the goals is to unravel the molecular mechanisms that underly the inhibition of virus replication in the intact infected cell. These mechanisms may, at least in part, contribute to the recovery of an organism from poxvirus infection (Baron and Isaacs, 1961). The physiological role of the IFN type I and type II systems in antiviral host cell defense has been proven by studying the susceptibility of IFN type I and type II receptor knockout mice to different challenge viruses (Huang *et al.,* 1993; Müller *et al.,* 1994). Compared to wild-type mice, the receptor knockout mice and mice lacking the gene for IFNβ showed a higher susceptibility to vaccinia virus infection, and both the IFN type I and the IFN type II systems were involved (Van den Broek *et al.,* 1995a, b). In a natural virus disease model IFN depletion by antibody treatment determined the outcome of mousepox (Karupiah *et al.,* 1993a). Expression of IFN type II by a recombinant vaccinia virus protects immunocompromised mice from a lethal vaccinia virus infection (Kohonen-Corish *et al.,* 1990).

II. Interaction of Poxviruses with the IFN Type I-Treated Host Cell

Early studies on the inhibition of vaccinia virus growth by pretreatment of cells with interferon or IFN-like substances were concerned with the conditions of the development of the antipoxvirus effect and the use of the virus for an accurate assay of IFNs (Isaacs *et al.,* 1958, 1961; Ho and Enders, 1959; Isaacs and Westwood, 1959; Vilcek and Rada, 1962; Gallagher and Khoobyarian, 1969). The crude or partially purified IFN preparations used were obtained from tissue cultures treated with viral inducers and were therefore most likely IFN type I. Poxvirus strains were tested for IFN sensitivity in different host cells (Stewart *et al.,* 1969).

The choice of host cells was often dictated by the availability of IFN preparations in crude or partially purified form. Later, highly purified IFNs from different species were used until purified recombinant IFNs finally became available (Weissmann and Weber, 1986). Before IFN genes were cloned, all studies were done with crude or partially purified IFN preparations. As controls, experiments were performed with mock IFN type I prepared from uninduced cells, inactivated IFN type I, or IFN type I from heterologous species. So far, no differences have been reported on the poxvirus replication cycle from studies made with IFNs from natural sources and made with recombinant IFNs.

The IFN type I concentrations used in the studies summarized in this chapter varied from 10 to several thousand international units per milliliter. In this concentration range the IFN-induced alterations were usually enhanced. Qualitatively different phenomena, although not observed, cannot be excluded because the pleiotropic alterations of the cell may depend on the IFN type I concentrations used. How critical it may be to work at a defined IFN concentration is demonstrated by the inhibition of vaccinia virus by the activity of an isoform of nitric oxide synthase (see Section V) (Harris *et al.,* 1995). Different strains of the orthopoxviruses have been used and no major difference in sensitivity has been found. However, sensitivity of virus growth to IFN type I is influenced by several parameters and differences between strains may not have been detected. Vaccinia virus strain NY 914 has, nevertheless, been reported to be 40 times more sensitive to chicken IFN type I than a vaccine strain of fowlpox virus (Asch and Gifford, 1970). Resistance to IFN type I of fowlpox virus in primary chick embryo fibroblasts (CEFs) was also observed by P. Marcus and M. Sekellik (private communication). A meaningful comparison of IFN type I sensitivity between viruses of different families is even more difficult to achieve. From many comparative studies it could, however, be concluded that orthopoxvirus replication is less sensitive to chicken IFN type I than is vesicular stomatitis virus or Sindbis virus (Stewart *et al.,* 1969; Radke *et al.,* 1974; C. Jungwirth, unpublished).

Studies on the inhibition of poxviruses by IFNs have focused on the paracrine activity of IFNs. The autocrine activity of IFNs or the activity of IFNs added to cells already infected with a poxvirus has not been analyzed in detail. To induce the refractive state IFNs were added to the medium, usually overnight, before infection. Only in some studies was it recorded whether IFNs were present in the medium after infection of cells (Jungwirth *et al.,* 1977). Once fully established, the poxvirus-refractive state induced in CEF is maintained for many hours after removal of IFN type I (C. Jungwirth, unpublished; Radke *et al.,* 1974).

An IFN type I-treated cell may respond to poxvirus infection in different ways:

1. In the virus-refractive state, progeny formation is inhibited and the host cell is protected against the viral cytopathic effect as, for example, in infected CEFs (Jungwirth *et al.,* 1977; Magee and Levine,

1970). Several assays of chicken IFN type I are based on this type of response to poxvirus infection (Gifford *et al.,* 1963; Youngner *et al.,* 1972).

2. Poxvirus replication is inhibited in the IFN type I-treated cell concomitant with an enhanced degeneration of the cell. In vaccinia WR virus-infected mouse L cells this phenomenon was first observed by Joklik and Merigan (1966) and interpreted as an example of host cell defense by self-destruction of the infected cell. This type of interaction of poxviruses with the IFN type I-treated cell has been found in poxvirus-infected cells from other species and in cells infected with several unrelated viruses. Elimination of a viral infection by IFN type I-induced suicide of the host cell (cytodestruction) is therefore a fundamental mechanism of host cell defense.

3. In several IFN type I-sensitive cells poxvirus replication is only marginally impaired even though an antiviral state directed against other viruses can be fully induced (Youngner *et al.,* 1972; Stewart *et al.,* 1969).

One strategy used to determine which step in the replication cycle of a virus is inhibited by IFN type I consists in comparing successive steps of a viral replication cycle in the infected cell with or without IFN treatment. The results can be used to explain the mechanism of inhibition of progeny virus formation in tissue culture cells. This approach is feasible if the IFN treatment inhibits virus replication with concomitant protection of the host cells. If the IFN type I-treated and -infected cells show a cytopathic effect which is different from that of infected cells, the analysis will require different strategies.

The multiple IFN-induced alterations of a eukaryotic host cell may explain how a broad spectrum of viruses with quite different replication strategies is inhibited in IFN-treated cells (Staeheli, 1990). Various targets in the viral replication cycle have been identified, and more than one step in the growth cycle of a specific virus may be affected. To unravel whether a single key step is blocked or whether the inhibitory effect is caused by additive partial effects of IFNs is a major difficulty.

Once it has been determined which step in the viral replication cycle is the primary target in the IFN-treated cells, experiments are designed to clarify the mechanism of this impairment. This requires manipulation of the system. Because IFNs alter cells in a pleiotropic way, IFN-induced alterations irrelevant for the inhibition of the growth of a virus may be studied. This pitfall becomes acute when the mechanism of inhibition of viral macromolecular synthesis in the IFN-treated cells is analyzed using transient transfection techniques and nonviral expression systems. This strategy has yielded interesting results, but it is difficult to prove their relevance to the mechanism of the antiviral effect of IFNs. More than one IFN type I-induced alteration of the cell may contribute to the impairment of a certain specific viral function or simultaneously inhibit different steps in a viral replication cycle.

III. The Posttranscriptional Inhibition of Poxvirus-Specific Gene Expression in IFN Type I-Treated Chick Embryo Fibroblasts

Infection of primary CEFs with vaccinia WR is asynchronous (Chen *et al.,* 1983). In contrast to many other host cells of vaccinia virus, the incorporation of radioactive amino acids into cellular proteins is not inhibited at early times postinfection (p.i.) (Jungwirth *et al.,* 1977; Magee and Levine, 1970). Only at late stages of virus infection is host cell protein synthesis gradually inhibited, probably as part of the death of the host cell (Jungwirth *et al.,* 1977). Incorporation of radioactive amino acids into total cellular proteins in confluent cultures is not affected by IFN type I treatment.

Pretreatment of confluent CEFs with IFN type I induces an anti-vaccinia virus effect which is thought to be the result of a specific inhibition of viral protein synthesis (Jungwirth *et al.,* 1977; Magee and Levine, 1970; Ghosh and Gifford, 1965; Ohno and Nozima, 1964). Esteban and Metz (1973) reported an inhibition of vaccinia virus protein synthesis by IFN type I in CEFs which were held in suspension at certain periods of the experiment. Under these conditions host cell protein synthesis is reduced early after vaccinia virus infection of IFN type I-treated cells.

Concomitant with the inhibition of virus growth, the CEFs are protected against the gradually developing cytopathic effect (CPE) of the virus (Magee and Levine, 1970; Osterhoff *et al.,* 1976). Even at late stages, the majority of the cells in infected cultures are protected against the CPE of the virus. After 24 hr p.i. a small number of cells in the culture seem to deteriorate. Pretreatment with IFN type I will also partially protect host cell protein synthesis in the confluent cultures (Osterhoff *et al.,* 1976). The inhibition of vaccinia virus replication by chicken IFN type I, in combination with a protective effect on the host cell, is also seen in the permanent quail cell line CEC-32 when infected at low multiplicities with the vaccinia virus (Zöller *et al.,* 1998, 2000).

In CEFs treated with low amounts of IFN type I, viral DNA synthesis still occurs (Jungwirth *et al.*, 1977; Osterhoff *et al.*, 1976). The late phase can be synchronized by a reversible treatment with cycloheximide (Chen *et al.*, 1983). Preliminary experiments attempted to study the effect of IFN type I on late viral protein synthesis under these conditions (Osterhoff *et al.*, 1976; Magee and Levine, 1970).

Studies on the synthesis of poxvirus-specific proteins have concentrated on the group of early viral proteins translated from mRNAs synthesized by enzymes brought into the cell by the infecting vaccinia virion (Jungwirth *et al.*, 1977). Because the synthesis of this group of vaccinia virus proteins shows low sensitivity to IFN type I, experimental conditions for these studies, for example, the multiplicity of infection and the age of the cells in culture, have to be controlled (Carver and Marcus, 1967; Ho, 1962; Lockart, 1968; Grün *et al.*, 1987).

Two questions were of interest: (1) What is the molecular mechanism of the inhibition of early vaccinia virus protein synthesis by IFN type I? (2) Are there structural elements in a gene which determine its sensitivity to IFN type I?

1. Early vaccinia virus protein synthesis is inhibited by a posttranscriptional mechanism. Comparable to CEFs treated with cycloheximide, early mRNA is synthesized in the IFN type I-treated cell in excess over the amount synthesized in infected cells in which protein synthesis is unimpaired. The early mRNA synthesized in the IFN type I-treated CEFs is capped and polyadenylated (Jungwirth *et al.*, 1977). A large part of the viral mRNA which is synthesized up to 6 hr p.i. is rapidly degraded. Part of the viral mRNA, corresponding approximately to the amount accumulating in the infected cell not treated with IFN type I, is resistant to degradation (Degen *et al.*, 1992). This vaccinia virus-specific mRNA can be translated in a cell-free system and seems structurally intact. Thus, besides the degradation of viral mRNA, an inhibitory system is induced in the IFN type I-treated CEF which selectively prevents the translation of vaccinia virus early mRNAs (Degen *et al.*, 1992). The translation inhibition and the degradation may be interdependent processes. A particularly fascinating facet of this regulatory mechanism is that it acts in the background of unimpaired host cell protein synthesis (Jungwirth *et al.*, 1977). In an attempt to unravel the molecular basis of this specificity, the IFN type I sensitivity of the expression of various foreign genes inserted into the vaccinia virus genome and of authentic viral genes was compared.

2. IFN type I sensitivity can be conveyed to the expression of the bacterial chloramphenicol acetyltransferase gene by inserting it

into the vaccinia virus thymidine kinase locus and expressing it under the control of an early promoter (Grün *et al.,* 1987). The IFN type I sensitivity of the expression of an influenza hemagglutinin gene integrated into the vaccinia virus genome is determined by the poxvirus vehicle in CEF and MDBK cells (Degen *et al.,* 1992). The expression of the histone H1(0) and H5 genes after integration into the vaccinia virus genome also becomes IFN type I-sensitive, albeit only to a much lower extent than the expression of vaccinia virus thymidine kinase in cells infected with wild-type vaccinia virus (Grün *et al.,* 1991; Blum *et al.,* 1993). An analysis of different histone–thymidine kinase fusion genes expressed under the control of an early vaccinia promoter revealed that about 45% or more of the 5′-amino-terminal or the 3′-carboxy-terminal parts of the histone genes is sufficient to secure reduced IFN type I sensitivity. If the cellular part of the fusion gene is further deleted to 32%, the IFN type I sensitivity of the expression of the fusion protein is comparable to that of the authentic thymidine kinase or other early viral enzymes. Thus, it seems that IFN type I sensitivity of gene expression is determined by both noncoding and coding regions, which may contain structural elements that influence the IFN type I sensitivity of gene expression (Blum *et al.,* 1993). Although a fusion gene consisting of 45% of the 5′-amino-terminal end of the histone H5 gene maintains, with respect to the IFN type I system, its identity as a cellular gene after integration into the vaccinia virus genome, it is regulated like an authentic viral gene with respect to switch-off of host cell protein synthesis by vaccinia virus (Blum *et al.,* 1993; Moss, 1968).

IV. Cytotoxicity Enhancement of IFN Type I-Treated and Vaccinia Virus-Infected Cells

In some IFN type I-treated poxvirus-infected cells an inhibition of progeny formation is observed with concomitantly enhanced destruction of the host cell (Joklik and Merigan, 1966; Horak *et al.,* 1971). This phenomenon has a striking parallel in the destruction of IFN type I-sensitized cells by toxic compounds, such as double-stranded RNA (Stewart *et al.,* 1972; Williams and Bellanti, 1983). Cytotoxicity enhancement by IFN type I-infected cells has been found in chicken cells infected with an Aujezky virus strain (Lomnici, 1974). Inhibition of Mengo virus replication without ameliorative effect on the viral cytopathic effect has been described in L cells (Levy, 1964; Gauntt and Lockart, 1966).

Premature death (cytodestruction) of IFN type I-treated L929 cells infected with poxviruses may result in a limitation of the reinfection of neighboring cells. Conditions of development of the enhanced destruction of the L929 cells indicated that IFN type I treatment induces a subtle alteration of the cell, which, together with a late viral function, leads to premature killing of the infected cell (Horak *et al.,* 1971). Besides the decreased cell viability and an often detected morphological destruction, an enhanced early switch-off of host cell protein synthesis occurred (Joklik and Merigan, 1966; Suh *et al.,* 1974). This indicates that the IFN type I-treated infected L cells may be altered before enhanced CPE can be observed by morphological inspection.

The increased cytotoxicity of vaccinia virus WR-infected, IFN type I-treated L929 cells can be followed by ^{51}Cr release into the medium (Horak *et al.,* 1971). The IFN type I-treated quail cell line CEC 32 shows cytotoxicity enhancement at high multiplicities of infection with vaccinia virus WR by morphological inspection, but an increased release of ^{51}Cr into the medium was not observed (N. Feirer and C. Jungwirth, unpublished observation). Therefore, the mechanism of destruction is probably not the same in all cases. The subtlety of the phenomenon and the multitude of different cell strains used makes the study of the mechanism additionally difficult.

28S and 18S ribosomal RNA species of IFN type I-treated L cells infected with vaccinia virus are degraded in a characteristic cleavage pattern reminiscent of the action of the 2′–5′ A oligo A/ribonuclease L pathway (see below) (Goswami and Sharma, 1984; Esteban *et al.,* 1984). A degradation of ribosomal RNA into discrete fragments has been associated with apoptotic cell death induced by different viral and nonviral stimuli (Diaz-Guerra *et al.,* 1997a; Houge *et al.,* 1995; Castelli *et al.,* 1997). Apoptotic death of the host cell would be harmful for the virus and can therefore be considered as host cell defense (Clouston and Kerr, 1985; Martz and Howell, 1989; Shen and Shenk, 1995). More recent evidence indicates that apoptotic mechanisms may also be involved in the self-destruction of the IFN type I-sensitized virus-infected cell (Tanaka *et al.,* 1998). Regardless of whether enhanced cell death of IFN type I-treated and poxvirus-infected cells is mediated by a necrotic or an apoptotic mechanism, it may limit poxvirus infection effectively. The extent of inhibition of virus growth will depend on the time that cytotoxicity enhancement develops relative to the production of progeny virus. This is perhaps in part the explanation of why enhanced cytotoxicity was not observed in some IFN type I-treated strains of L929, HeLa, and CV1 cells and why different poxvirus strains replicate unimpaired in these cell lines (Youngner *et al.,* 1972; Rice *et al.,* 1984; Paez and Esteban, 1984a).

V. Inhibition of Vaccinia Virus Replication by Nitric Oxide Synthase Induced by IFN Type II

The replication of ectromelia and vaccinia viruses is sensitive to IFN type II in primary murine macrophage cultures and macrophage-like cells (Karupiah *et al.,* 1993b; Melcova and Esteban, 1994; Reiss and Komatsu, 1994). At strictly defined IFN type II concentrations the inhibitory effect is caused by NO produced from L-arginine by a cytokine-inducible isoform of nitric oxide synthase (iNOS). The growth of vaccinia virus was severely inhibited and the restriction could be reversed by N^{ω}-monomethyl-L-arginine and other inhibitors of iNOS, such as N^{ω}-nitro-L-arginine and N-iminoethyl-L-ornithine. The iNOS-mediated antiviral effect was thereby distinguished from antiviral effects induced by higher concentrations of IFN type II (Harris *et al.,* 1995). Inhibition of vaccinia virus plaque formation in human 143B cells by pretreatment with IFN type II by an undefined mechanism has been reported (Kohonen-Corish *et al.,* 1989). The NO affected a function required for vaccinia virus-specific DNA synthesis in the cytokine-treated macrophage, but also inhibited virus growth in untreated contiguous cocultivated human renal epithelial cells 293 (Harris *et al.,* 1995). Ectromelia virus can also be inhibited in primary mouse macrophages by IFN type I by a mechanism not mediated by NO (Karupiah *et al.,* 1993b).

The developmental stages of the vaccinia virus replication affected in the cytokine-treated RAW 264.7 cells were viral DNA synthesis, late protein synthesis, and virus particle formation, but early viral protein synthesis was not affected (Melcova and Esteban, 1994; Harris *et al.,* 1995). The cellular enzymes and/or specific vaccinia viral functions that are inactivated by nitrosylation and cause the inhibition of viral DNA synthesis have not yet been identified. The production of NO by IFN type II can be synergistically enhanced by tumor necrosis factor and lipopolysaccharides. A combination of lipopolysaccharides and IFN type I also enhances NO production (Drapier and Hibbs, 1988; Ding *et al.,* 1988). This antiviral mechanism offers the possibility of cooperation of IFN type II with other agents in host cell defense (Reiss and Komatsu, 1998).

VI. Metabolic Activities Are Required for the Response of the IFN Type I-Treated Host Cell to Vaccinia Virus Infection

The cytoplasmic replication of poxviruses and the autonomy of the viral early phase with regard to the genetic information of the cell made it possible to study the effect of IFN type I on the early phase of infection in enucleated CEFs (Radke *et al.,* 1974; Moss *et al.,* 1991). An anti-vaccinia virus state could not be established in enucleated cells by pretreatment with IFN type I. If CEFs were treated with IFN type I before the removal of the nucleus, early poxvirus-specific functions and viral DNA synthesis were inhibited. The virus-refractive state was maintained in the cytoplasmic fragments over 16 hr. Even though only this indirect evidence is available, it can be assumed that cellular IFN-susceptible genes (ISGs) mediate inhibition of poxvirus replication. This is also likely because the development of the poxvirus-refractive state in CEFs requires several hours (C. Jungwirth, unpublished). Comparable studies with enucleated L929 cells have not been reported. It is not known if cellular metabolic activity is required for the development of enhanced cytotoxicity in IFN type I-treated vaccinia virus-infected cells. More recent observations support a role of ISGs in the development of the enhanced destruction of IFN type I-treated L929 cells (see below).

After binding to their cognate receptors, IFNs activate signal transduction pathways which lead to transcriptional activation of ISGs (Gilmour and Reich, 1995; De Veer *et al.,* 2001). Which of the hundreds of ISGs that have been detected by different techniques are involved in the development of the poxvirus-refractive state induced by IFN type I and what cascades lead to their activation has been and continues to be an active field of investigation. As the mechanisms that inhibit poxvirus replication vary, the effectors of the IFN type I action are also likely to be different. The dsRNA-activated protein kinase PKR and the 2′–5′ oligoA/ribonuclease L system have been implicated in the antipoxvirus effect induced by IFN type I.

A. Induction of the dsRNA-Activated Protein Kinase PKR

The protein kinase PKR is constitutively expressed in a latent form in many cells and can be transcriptionally induced by IFN type I (Hovanessian, 1993; De Haro *et al.,* 1996). Upon activation by dsRNA, one of the principal activators, protein kinase PKR is autophosphorylated and phosphorylates the alpha subunit (eIF-2α) of the eukaryotic initiation factor 2 (Farrel *et al.,* 1977). An increased level in eIF-2α phosphorylation would cause inhibition of viral and cellular protein synthesis and lead to the inhibition of virus progeny formation. Besides eIF-2α, other exogenous substrates have been identified, supporting the central role of protein kinase PKR in cell physiology (Williams, 1999; De Haro *et al.,* 1996). In addition to the putative role in the antiviral effect of IFN type I, the kinase has been implicated in regulation of cell growth and antitumoral activity. A crucial function in stress-induced signal transduction pathways has been ascertained (Der *et al.,* 1997; Yeung *et al.,* 1996; Maran *et al.,* 1994; Srivastava *et al.,* 1998). This function as an apoptotic effector may be mediated by downregulation of translation, but also by other mechanisms (Srivastava *et al.,* 1998; Balachandran *et al.,* 1998).

Based on the assumption that virus-specific protein synthesis is the primary target of the antiviral activity of IFN type I, a role of protein kinase PKR in the development of the IFN type I-induced virus-resistant state was postulated (Lebleu *et al.,* 1976; Roberts *et al.,* 1976; Zilberstein *et al.,* 1976).

Vaccinia virus protein synthesis and replication are only marginally inhibited by IFN type I in a variety of rabbit and mouse cell lines, even though the IFN type I-treated cells are refractive to vesicular stomatitis and picorna virus infection (Youngner *et al.,* 1972; Whitaker-Dowling and Youngner, 1986; Fout and Simon, 1981; Thacore and Youngner, 1973). The replication of vesicular stomatitis and picorna viruses can be rescued from the inhibitory effect of IFN type I by superinfection with vaccinia virus in some of the cell lines (Whitaker-Dowling and Youngner, 1986; Thacore and Youngner, 1973). A mechanism explaining how vaccinia virus might rescue an IFN type I-sensitive virus on the level of protein synthesis was discovered by correlating rescue experiments with the activity of protein kinase PKR in cell-free extracts from single- or double-infected L929 cells (Whitaker-Dowling and Youngner, 1983). Extracts from IFN type I-treated L929 cells and various human cell lines infected with vaccinia virus contain an inhibitor of autophosphorylation of protein kinase PKR and of the downstream phosphorylation of eIF-2α (Paez and Esteban, 1984a; Whitaker-Dowling and Youngner, 1984; Rice and Kerr, 1984). Relief of the inhibitory effect by adding excess dsRNA to the reaction mixture demonstrated that the vaccinia virus-specific inhibitor named SKIF (Specific Kinase Inhibitory Factor) interacted with the dsRNA activator noncatalytically (Whitaker-Dowling and Youngner, 1984). SKIF seems to be identical to the vaccinia virus-specific gene product pE3 identified later (see below). These studies in cell-free extracts suggested that viral protein synthesis in the intact IFN type I-treated cell might also be spared from the inhibitory activity of the protein kinase PKR by the viral dsRNA-binding protein SKIF. This possibility was supported by the observation that Poly(rI:rC)-induced autophosphorylation of PKR is inhibited by a concomitant infection of IFN type I-treated cells with the vaccinia virus (Jagus and Gray, 1994). The discovery of the vaccinia virus gene product SKIF (Whitaker-Dowling and Youngner, 1983), which could counteract an IFN type I-inducible protein kinase PKR in cell-free extracts, was seminal for an avalanche of attempts to characterize the role of protein kinase PKR as effector of the antipoxvirus effect of IFN type I and its modulation by specific viral gene products in the intact cell.

To obtain formal proof for the putative role of protein kinase PKR in the inhibition of vaccinia virus protein synthesis in the IFN type I-treated intact cell, which had been inferred from studies with cell-free extracts, protein kinase PKR was inducibly expressed by vaccinia virus-based expression-transfection systems. Indeed, expression of protein kinase PKR lead to the inhibition of vaccinia virus protein synthesis and replication in various mammalian cell lines (Lee and Esteban, 1993). The interpretation of these results is, however, complicated, as an overexpression of protein kinase PKR from a vaccinia virus vector in HeLa cells triggers pathways that lead to apoptotic cell death (Lee and Esteban, 1994). The downregulation of protein synthesis may have been in part the consequence of the premature death of the host cell. dsRNA can directly stimulate or repress the transcription of a whole set of functionally different cellular genes (Geiss *et al.,* 2001). Viral dsRNA formed in the late phase of infection with vaccinia virus recombinants may alter the pattern of gene expression drastically. The use of vaccinia recombinants expressing specific ISGs could be a promising strategy for studying the mechanism of IFN type I-induced suicide of a poxvirus-infected cell.

The role of protein kinase PKR in the development of an antiviral state has also been adressed by analyzing fibroblasts from protein kinase PKR-devoid mice (Tanaka *et al.,* 1998; Yang *et al.,* 1995; Abraham *et al.,* 1999). Growth curves of vaccinia WR and virus yield in fibroblasts from wild-type and protein kinase PKR-knockout mice showed no significant differences (Abraham *et al.,* 1999). IFN type I sensitivity of poxvirus replication in these cells was not reported.

Synthesis of early poxvirus proteins in the IFN type I-treated CEF is inhibited to various extents and concurrent host cell protein synthesis is not affected (Jungwirth *et al.,* 1977). If protein kinase PKR is a mediator in the selective inhibition of poxvirus-specific protein synthesis in CEF, additional, at present unknown, mechanisms that confer specificity to the enzyme must exist. Interestingly, poxvirus infection cannot rescue vesicular stomatitis virus in rabbit RK1337 cells and in CEFs from the inhibitory effect of chicken IFN type I (Thacore and Youngner, 1973; V. Eisert and C. Jungwirth, unpublished).

Detection of protein kinase PKR activity in extracts from CEF has been successful only recently (Martinez-Costas *et al.,* 2000). A study of the interaction of an avian protein kinase PKR with a vaccinia virus-specific gene product like SKIF/pE3 may shed additional light on the biological meaning of the results obtained with mammalian protein kinase PKR. Should it turn out that protein kinase PKR from CEF is pivotal for the observed selective inhibition of viral protein synthesis, the elucidation of the mechanism of discrimination between poxvirus-specific and cellular protein synthesis will be a challenging problem.

B. 2′–5′ Oligo A/Ribonuclease L System

The properties of the enzymes of the 2′–5′ oligo A/ribonuclease L pathway have been determined by extensive enzymatic studies (Kerr and Brown, 1978; Hovanessian, 1991). Upon activation by dsRNA, the 2′–5′ oligo A synthetases catalyze the conversion of ATP into a series of 5′-triphosphorylated 2′–5′ oligo adenylates of the general formula ppp5′(A2′p5′A)n (Kerr and Brown, 1978). Most types of the oligo adenylates (2′–5′ A) activate a latent endonuclease, ribonuclease L, which degrades ribosomal RNA and viral and cellular mRNAs (Hovanessian, 1991; Clemens and Williams, 1978). Enzymes of the 2′–5′ oligo A pathway are constitutively expressed at a basal level in many cells and are induced by IFN type I and other compounds (Hovanessian, 1991). Several degradative enzyme activities have been detected in normal and vaccinia virus-infected cells which may be regulating the intracellular concentrations of 2′–5′ oligo adenylates (Schmidt *et al.,* 1979; Minks *et al.,* 1979; Paez and Esteban, 1984b; Williams *et al.,* 1978). Susceptibility to viruses does not correlate with the basal or induced level of the 2′–5′ oligo A synthetases in different murine cell lines (Hovanessian *et al.,* 1981). This emphasizes that mechanisms of antiviral activity of IFNs may exist which are not mediated by the 2′–5′ oligo A system. The constitutive expression of the enzymes of the 2′–5′ oligo A pathway indicates that the 2′–5′ oligo A synthetases also have functions in other physiological processes besides their role in the IFN system (Stark *et al.,* 1979).

The ribosomal RNA from cells of different mammalian species dying by an apoptotic mechanism is degraded and in some cases the cleavage pattern indicates an involvement of ribonuclease L (Houge *et al.,* 1995; Castelli *et al.,* 1997; Silverman and Cirino, 1997; Wreschner *et al.,* 1981; Delic *et al.,* 1993). Degradation of ribosomal RNA into discrete products characteristic for ribonuclease L activity was detected in IFN type I-treated L cells infected with vaccinia virus (Goswami and Sharma, 1984; Esteban *et al.,* 1984; Rice *et al.,* 1984). Simultaneously, an increased synthesis of 2′–5′ A was detected (Goswami and Sharma, 1984; Rice *et al.,* 1984). Expression of 2′–5′ oligo A synthetase and ribonuclease L in L cells by using a recombinant vaccinia virus vector also indicates that the degradation of RNA observed in IFN type I-treated and vaccinia virus-infected monkey BSC40 and in certain murine cells is mediated by the 2′–5′ A/ribonuclease L pathway (Diaz-Guerra *et al.*, 1997a, b). The 2′–5′ oligo A/ribonuclease L pathway and protein kinase PKR may be mediating apoptotic death independently and the inhibition of vaccinia virus growth by cytodestruction (Diaz-Guerra *et al.*, 1997b).

Why is it that in some IFN type I-treated mammalian cell lines the 2′–5′ oligo A/ribonuclease L pathway is induced, but ribosomal RNA is not degraded and the vaccinia virus replicates unimpaired? To explain this, it has been postulated that in these cells the poxviruses may activate mechanisms that subvert the activity of the induced protein kinase PKR and the 2′–5′ oligo/ribonuclease L pathway (Paez and Esteban, 1984a; Whitaker-Dowling and Youngner, 1984; Rice and Kerr, 1984).

The existence of an IFN type I-induced RNA degradative enzyme system in chicken cells can be inferred from an enzymatic activity synthesizing an oligonucleotide with quite similar properties to 2′–5′ oligo A from mammalian cells (Ball and White, 1978). Furthermore, the inhibition of protein synthesis by dsRNA in a cell-free system from IFN type I-treated CEF could be traced to a nuclease activated by 2′–5′ oligo adenylate (Ball and White, 1979). However, these observations do not provide an explanation for the selectivity of inhibition of poxvirus protein synthesis by IFN type I in CEF.

VII. Double-Stranded RNA Molecules Detectable during Vaccinia Virus Replication and Their Role in the Interaction of Vaccinia Virus with the IFN Type I-Treated Host Cell

Complementary viral transcripts which can form duplex structures are synthesized in the cytoplasm during replication of poxviruses (Colby *et al.*, 1971; Boone *et al.*, 1979; Varich *et al.*, 1979). Symmetrical transcription occurs preferentially in the late phase of infection, whereas the amount of viral transcripts having the potential to form dsRNA structures is very low at early times after infection and in cycloheximide-treated infected cells. The formation of 2′–5′ oligo A during unimpaired replication of vaccinia virus in certain strains of HeLa and L cells also supports the existence of viral dsRNA in infected cells (Rice *et al.*, 1984).

Vaccinia virus-specific dsRNA that can induce interference against a Sindbis virus challenge but is inefficient in inducing IFN type I has been extracted from infected CEFs (Colby and Duesberg, 1969; Bakay and Burke, 1972). Only very low amounts of IFN type I were produced by CEFs infected for many hours with fowlpox or vaccinia virus (Asch and Gifford, 1970). In mouse embryo cells held in tissue culture, vaccinia virus is able to induce IFN-like activity (Lowell *et al.,* 1962). The material has been detected because of its ability to induce a refractive state against different viruses and has properties comparable to crude IFN preparations induced by other viruses.

The product of the A18R open reading frame of the vaccinia virus genome is a negative transcription elongation factor (Xiang *et al.,* 1998). A mutation of the A18R gene will cause readthrough on converging transcription units resulting in the transcription of extended regions of the viral genome late in infection (Bayliss and Condit, 1993). As a consequence, an increased amount of self-complementary RNA accumulates in the infected monkey BSC40 cells (Bayliss and Condit, 1993). A temperature-sensitive mutant of the A18R gene displays an abortive late phenotype at the nonpermissive temperature in which late viral protein synthesis ceases. Depending on the host all viral mRNA and ribosomal RNA may be degraded in the absence of IFN type I treatment (Pacha and Condit, 1985). Isatin β-thiosemicarbazone (IBT) treatment also limits wild-type vaccinia virus infection by causing a loss of control of late viral transcription (Bayliss and Condit, 1993). The pattern of ribosomal RNA degradation in the A18R gene mutant-infected monkey BSC40 cells at the nonpermissive temperature and in IBT-treated, wild-type vaccinia virus-infected monkey BSC40 cells is similar and indicates an involvement of the 2′–5′ oligo A synthetase/ribonuclease L pathway secondarily activated by aberrant viral mRNA transcription (Pacha and Condit, 1985; Cohrs *et al.,* 1989).

The identification of poxvirus-coded inhibitors of dsRNA-activated enzymes (see below) presents further indirect support for the existence of poxviral duplex RNA structures in the infected cell and their eventual importance in determining the fate of the IFN type I-treated infected cell (Jacobs and Langland, 1996). The observation that IFN type I-treated cells are sensitized to destruction by Poly(rI:rC) and that dsRNA triggers apoptosis of vaccinia virus-infected cells is also relevant for a critical role of viral dsRNA in mobilizing the IFN type I host cell defense response (Stewart *et al.,* 1972; Kibler *et al.,* 1997).

VIII. Modulation of the Cellular IFN Type I and Type II Responses by Poxvirus-Specific Gene Products

The pathogenicity and virulence of poxviruses is modulated by a set of viral gene products that subvert host cell defense mechanisms. Several poxvirus genes have been identified that are able to counteract the interferon type I and type II systems (Buller and Palumbo, 1991).

A. Poxvirus-Specific Soluble IFN Type I and Type II Receptors and the Interception of the IFN Type II Signal Transduction Pathway

Poxviruses encode secreted proteins that are homologous to IFN type I and IFN type II receptor-binding domains which bind IFNs and thereby prevent the interaction with their cognate receptors. The IFN signal is neutralized and the induction of the pleiotropic alteration of the cells is prevented. The activity of paracrine- and autocrine-acting IFNs is inhibited. The infected and neighboring cells remain in a permissive state for virus growth. The prevention of a virus-resistant state is only one manifestation of a more general phenomenon.

The genes for the soluble viral counterparts of cellular IFN type I and IFN type II receptors have been mapped, like those for other host cell defense modifiers, toward the termini of the poxvirus genomes (Troktmen, 1990.) The myxoma T7 gene encodes a soluble IFN type II receptor which was first isolated from supernatants of myxoma virus-infected BGMK cells (Upton *et al.,* 1992). A soluble receptor for IFN type I, a product of the B18R open reading frame of the vaccinia WR strain, could be detected in the medium and on the surface of cells infected with most orthopoxviruses. (Symons *et al.,* 1995; Ueda *et al.,* 1969; Colamonici *et al.,* 1995). The interaction of soluble receptor proteins with IFNs in comparison to the corresponding cellular receptors, as well as the functional consequences of this interaction for the development of the IFN-induced pleiotropic alterations, are well understood and the observations have been summarized in several extensive reviews (McFadden and Graham, 1994; Smith, 1994; Spriggs, 1994).

It is likely, but has not been directly demonstrated, that the poxvirus-refractive state induced by IFN type I or type II is activated by the Jak/STAT signal transduction pathway (Darnell *et al.,* 1994). A critical step in this pathway is a tyrosine phosphorylation of a signal-transducing and transcription-activating protein, STAT1, and its translocation to the nucleus (Schindler *et al.,* 1992; Shuai *et al.,* 1992). The vaccinia virus H1 open reading frame codes for a tyrosine/serine protein phosphatase (VH1), which is expressed late in the infectious cycle (Guan *et al.,* 1991). This virus-encoded enzyme activity specifically dephosphorylates STAT1 in IFN type II-treated infected HeLa cells (Najarro *et al.,* 2001). Other upstream-signaling molecules of the Jak/STAT pathway are not affected. STAT1 is not translocated into the nucleus and transcriptional activation of IFN type II-inducible genes is prevented.

The interruption of the IFN type II signal cascade by the activity of VH1 seems to be another strategy the virus has developed to subvert the host cell defense response. Support of this interpretation might be derived from a comparable study in IFN type II-treated macrophages (see above). It will also be of interest to see whether vaccinia virus infection is able to intercept other STAT1-mediated signal transduction pathways and how signal transduction by autocrine IFN type II activity is influenced by VH1.

B. Poxvirus-Specific Gene Products Modulating the Activity of the 2′–5′ Oligo A/Ribonuclease L Pathway or Protein Kinase PKR

Several proteins have been described which are thought to account for the ability of poxviruses to replicate in the presence of dsRNA, the activated protein kinase PKR, or the 2′–5′ oligo A/ribonuclease L pathway.

1. Enzymes Modulating the Oligo A Synthetase/Ribonuclease L Pathway

Enzymes have been identified which, by degradation of 2′–5′A, may modulate the function of this RNA-degradative pathway. An IFN type I-inducible phosphodiesterase (2′-phosphodiesterase-interferon) was purified from mouse L cells and reticulocytes which is able to cleave ppp(2′–5′) ApApA to 5′-AMP and ATP (Schmidt *et al.,* 1979). In extracts from vaccinia virus-infected cells and in purified virion preparations, an ATPase and a phosphatase have been detected which degrade ATP and dephosphorylate 2′–5′ oligo A (Paez and Esteban, 1984b). There is, however, no proof that any of these 2′–5′ A-degrading enzymatic activities contribute to the IFN type I resistance during poxvirus infection. The product of the A18R gene and a nucleoside triphosphate phosphohydrolase I (NPH-I), both of which control viral transcription, may indirectly affect the IFN type I sensitivity of vaccinia virus by modulating the intracellular amount of dsRNA (see above and Diaz-Guerra *et al.,* 1993).

2. The E3L/SKIF and the K3L Gene Products of Vaccinia Virus

Biochemical and genetic analysis identified poxvirus-specific proteins which are thought to overcome the IFN type I effect by downregulating the activity of the enzymes of the 2′–5′ oligo A pathway or of protein kinase PKR (Jagus and Gray, 1994).

The poxvirus-specific E3L gene codes for two proteins, of 19 and 25 kDa, which bind to various dsRNAs in cell-free extracts (Yuwen *et al.,* 1993; Chang *et al.,* 1992). The 25-kDa protein corresponds to the complete E3L reading frame and has properties similar to SKIF isolated from vaccinia virus-infected cells (Akkaraju *et al.,* 1989; Watson *et al.,* 1991). It is expressed in the cytoplasm and the nucleus early in the replication cycle and in the absence of viral DNA synthesis (Yuwen *et al.,* 1993; Hruby *et al.,* 1980; Beattie *et al.,* 1995a). Because viral dsRNA is formed in the cytoplasm of the infected cell, it was suggested that the pE3 may have several functions (Yuwen *et al.,* 1993). Recently, such functions have been found (Smith *et al.,* 2001; Liu *et al.,* 2001). Physical and functional studies have determined the requirements for the interaction of recombinant pE3, dsRNA, and protein kinase PKR (Ho and Shuman, 1996; Chang and Jacobs, 1993; Romano *et al.,* 1998; Sharp *et al.,* 1998). pE3 can prevent the activation of protein kinase PKR presumably by sequestering the dsRNA activator in mammalian cells (Chang *et al.,* 1992). In addition, an inhibitory interaction of the pE3 with protein kinase PKR has been identified by binding assays using culture cells and by studying protein–protein interaction in yeast cells (Romano *et al.,* 1998; Sharp *et al.,* 1998). Circumstantial evidence has been presented that pE3 can prevent the inhibition of protein synthesis and induction of apoptosis mediated by the action of the 2′–5′ oligo A/ribonuclease L pathway (Rivas *et al.,* 1998).

In a transient expression system the products of the E3L and K3L gene (see below) alleviate the inhibitory activity of protein kinase PKR as assayed by stimulation of translation of a reporter plasmid (Davies *et al.,* 1992, 1993). This system offered the possibility of studying the modulation of the activity of protein kinase PKR by pE3 and pK3 and how the expression of the two vaccinia virus proteins affects the status of phosphorylation of eIF2.

The K3L gene of vaccinia virus is also expressed early in infection, in the absence of DNA synthesis, and codes for a 10-kDa protein that shares a 28% identity with the amino-terminal end of the α subunit (eIF-2α) of the eukaryotic initiation factor 2 (Beattie *et al.,* 1991; Goebel *et al.,* 1990). This amino acid sequence is highly conserved in different orthopox viruses (Kawagishi-Kobayashi *et al.,* 2000). pK3 physically associates with protein kinase PKR as shown in a cell-free system and in the yeast two-hybrid assay (Carrol *et al.,* 1993; Sharp *et al.,* 1997; Craig *et al.,* 1996; Gale *et al.,* 1996; Kawagishi-Kobayashi *et al.,* 1997). Autophosphorylation is inhibited, and as a pseudosubstrate inhibitor it competitively spares the α subunit of eIF-2. The inhibitory effect of protein kinase PKR on translation of viral and cellular mRNAs is thereby prevented. Binding of endogenous protein kinase PKR to pK3 in intact vaccinia virus-infected cells was proven by coimmunoprecipitation, using a monoclonal antibody against protein kinase PKR (Jagus and Gray, 1994). On the basis of the interaction of pK3 with protein kinase PKR expressed in yeast an elegant system for characterizing a spectrum of regulators of mammalian protein kinase PKR has been developed (Kawagishi-Kobayashi *et al.,* 1997).

3. *How Do pK3 and pE3/SKIF Modulate the IFN Type I Sensitivity of Poxvirus Replication in the Infected Cell?*

The concept that the poxvirus-specific anti-antiviral gene products pK3 and pE3/SKIF abrogate the IFN type I-induced anti-poxvirus effect was based on the assumption that viral translation inhibition is the crucial target of the antipoxvirus activity of IFN type I and that protein kinase PKR and the 2′–5′ oligo A/ribonuclease L pathway are central players in this mechanism.

The IFN type I sensitivity of replication of wild-type vaccinia virus and a K3L-minus mutant (vP872) derived from it were compared in mouse L929 cells (Beattie *et al.,* 1991). The wild-type virus yield and viral protein synthesis were only slightly inhibited by IFN type I under experimental conditions. Incorporation of [^{35}S]methionine into cellular proteins was inhibited as the infection proceeded. In cells infected with the K3L-minus mutant, virus yield and [^{35}S]methionine incorporation into viral proteins were sensitive to IFN type I (Beattie *et al.,* 1991). A quantitative comparison of the extent of the switch-off of host cell protein synthesis in IFN type I-treated cells infected with wild-type virus or K3L-minus mutant and observations on the development of CPE were not reported. The results indicated that pK3 inhibits the activity of the IFN type I-induced PKR in the wild-type vaccinia virus-infected L929 cells, thereby sparing viral protein synthesis and progeny formation (Beattie *et al.,* 1991). In IFN type I-treated K3L-minus-mutant-infected L929 cells degradation of 18S ribosomal RNA in the pattern indicative of ribonuclease L activity was detectible, indicating an activation of the 2′–5′ oligo A synthetase/ribonuclease L pathway (Beattie *et al.,* 1995a). The degradation of ribosomal RNA indicated apoptotic death of the K3L-minus-mutant-infected L929 cells which may be responsible for the IFN type I sensitivity of the K3L-minus virus mutant (see above; Goswami and Sharma, 1984; Beattie *et al.,* 1991, 1995a). IFN type I-treated L929 cells infected with wild-type vaccinia virus WR or the K3L-minus mutant (constructed by Drs. R. Blasco and B. Moss, National Institutes of Health) used in the author's laboratory showed an enhanced cytopathic effect compared to infected L cells not treated with IFN type I (Eisert, 1995).

IFN type I sensitivity of wild-type vaccinia virus WR and K3L-minus mutant was also compared in primary CEFs. No robust difference in

IFN type I sensitivity of virus yield or viral protein synthesis was detected. Wild-type and K3L-minus-mutant-infected CEFs pretreated with IFN type I (500–1000 units/ml) showed a reduced viral cytopathic effect up to 24 hr p.i. (Eisert *et al.,* 1997; Eisert, 1995).

The deletion of the E3L gene restricted the viral host range and rendered the vaccinia virus in L929 and RK-13 cells sensitive to IFN type I (Beattie *et al.,* 1995b; Chang *et al.,* 1995). The insensitivity to IFN type I could be restored by either reinserting the E3L gene or expressing the gene of the reovirus dsRNA-binding protein sigma 3 in the E3L-minus background (Beattie *et al.,* 1995b; Shors *et al.,* 1997). Degradation of 18S ribosomal RNA indicating ribonuclease L activity was again only observed in the IFN type I-treated L cells infected with the E3L-minus virus mutant, but not in infected cells where pE3 or reovirus sigma 3 protein were expressed (Beattie *et al.,* 1995b). In HeLa cells the E3L gene product prevents death of the infected cell by an apoptotic mechanism (Lee and Esteban, 1994; Kibler *et al.,* 1997). This ability to act as an antiapoptotic gene correlates with the ability to bind dsRNA. dsRNA-binding proteins from reovirus or porcine rotavirus group C can substitute for the E3L gene product. The IFN type I-treated (500–1000 units/ml) L cell strain used in the author's laboratory developed, after infection with the E3L-minus mutant (vP1080), a CPE which differed clearly from the enhanced cytotoxicity shown by the IFN type I-treated cells infected with wild-type vaccinia WR. Similar observations were also made with a vaccinia virus recombinant in which the E3L gene defect had been compensated by inserting the S4 reovirus gene encoding the sigma 3 protein (vP1112) (Eisert, 1995). (Vaccinia virus recombinants vP1080 and vP1112 were kindly provided by Drs. B. Jacobs and E. Paoletti.) The interaction of the vaccinia virus mutants and the wild-type virus with the IFN type I-treated L cell strain used were so divergent that no conclusion as to the function of the viral dsRNA-binding protein pE3 was attempted.

Wild-type and E3L-minus-mutant (vP1080) yield were inhibited to the same extent in the IFN type I-treated CEF. Protection of the host cell against viral CPE could only be detected up to 5 hr p.i. (Eisert, 1995; Eisert *et al.,* 1997). So far, there is no evidence that pE3 and pK3 modulate IFN type I sensitivity of vaccinia virus replication in CEF. The orthopox viruses used in these studies are mammalian viruses and it may be that their anti-IFN gene products are not able to interfere with the IFN system of chicken. It will be of interest to carry out a comparative study with fowlpox virus, which has been reported to be less sensitive to chicken IFN type than a vaccinia strain (Asch and Gifford, 1970).

When it was discovered that PKR and the 2′–5′ oligo A synthetase/ribonuclease L pathway may be involved in the development of apoptotic cell death induced by different stimuli, the analysis of the function of dsRNA, pE3, and pK3 also took a new twist (Lee and Esteban, 1994; Diaz-Guerra *et al.*, 1997a; Houge *et al.*, 1995; Castelli *et al.*, 1997; Der *et al.*, 1997; Yeung *et al.*, 1996; Srivastava *et al.*, 1998). The two putative mediators and any viral and cellular proteins interacting with them might determine the antipoxvirus state, not by a specific inhibition of viral mRNA translation, but by programming cytodestruction of the IFN type I-treated poxvirus-infected host cell. Replication of poxviruses may be possible only in a tightly regulated equilibrium between antiviral proteins, such as the 2′–5′ oligo A/ribonuclease L system, the protein kinase PKR, and other IFN type I-inducible genes and activators or inhibitors interacting with them. By unbalancing this sensitive scenario the fate of the IFN type I-treated infected cell would be programmed and the mechanism and the extent of the impediment of poxvirus progeny formation determined.

Youngner and coworkers suggested the importance of studying the interaction of SKIF with the homologous PKR in CEFs (Whitaker-Dowling and Youngner, 1984). The recent detection of protein kinase PKR activity in CEFs makes it feasible to study the interaction of pK3 and pE3 with the avian homologues of IFN type I-induced enzymes believed to be resposible for the IFN type I-induced poxvirus-resistant state in mammalian cells (Martinez-Costas *et al.*, 2000). Comparisons of the observations made in cells, in which the replication of vaccinia virus is either sensitive or resistant to IFN type I, may lead to a better understanding of the physiological role and the mechanisms underlying the modulating effect of the K3L and E3L gene products. Fowlpox and vaccinia virus differ in sensitivity to IFN type I in CEF (Asch and Gifford, 1970). Perhaps this is caused by subtle differences in the interaction of poxvirus gene products, with the avian IFN type I system. These studies may also reveal the role of protein kinase PKR and the 2′–5′ oligo A pathway in the antiviral host cell defense system of Aves.

IX. Concluding Remarks

Work on the mechanism of inhibition of poxvirus replication has focused on the paracrine mode of action of IFN type I or type II. In IFN type I-pretreated vertebrate cells, poxvirus replication can be limited by different mechanisms. One mechanism of IFN type II-induced abrogation of orthopox virus replication in macrophages is mediated by NO radicals produced by the enzyme iNOS.

In CEF, poxvirus-specific macromolecular synthesis is inhibited by IFN type I leading to an impairment of viral replication. An involvement of IFN type I-induced proteins in this effect is inferred from the pioneering work with vaccinia virus infection in enucleated CEF. It is not known which specific IFN type I-induced genes are involved in the classic antipox effect of IFN type I and therefore we have no knowledge of the principles underlying the mechanisms for discrimination between pox viral and cellular protein synthesis. An ever-increasing number of cellular genes transcriptionally activated by IFN type I and type II is being identified and may contribute to the antipoxvirus effect. Our lack of sufficient knowledge of the IFN type I-induced genes involved in the development of the antipoxvirus effect—in particular the role of PKR is not clearly defined—also hampers the understanding of the function of the intracellular-acting anti-antiviral genes. In this respect, it will be of interest to compare the interaction of the poxvirus-specific proteins pK3 and pE3 with IFN type I-inducible protein kinase PKR, 2′–5′ oligo A synthetase, and other dsRNA-activated proteins from mammalian and avian cells. This approach might contribute to our understanding of how these enzymes function in the IFN type I-induced host cell defense system in mammals and Aves.

Abrogation of virus replication by cytodestruction of the infected cell is increasingly considered as a general mechanism of host cell defense. Several observations support a role of the 2′–5′ oligo A/ribonuclease L pathway and protein kinase PKR in this phenomenon. It will be important to ascertain to what extent apoptotic mechanisms contribute to the enhanced destruction of IFN type I-treated and -infected cells, as it has been reported that the suicidal death of the cells may be caused by a combination of apoptotic and necrotic mechanisms. Some IFN type I-treated cells can be conditioned to show alternatively both types of response to poxvirus infection. The IFN type I-treated, poxvirus-infected host cell may turn out to be a particularly suitable model in which to study the mechanisms involved. IFN-induced specific inhibition of virus replication with a concomitant protection of the host cell, and inhibition of virus replication by cytodestruction of the IFN-treated cells, are also mechanisms limiting virus infections in the animal (Guidotti and Chisari, 2001).

ACKNOWLEDGMENTS

The help of Drs. R. Drillien, W. K. Joklik, G. Bodo, K. Koschel, M. Buller, J. Youngner, and R. Condit is gratefully acknowledged. Their constructive and critical comments based on their own studies were an important help in the presentation of the different and often controversial observations taken from the various realms in which IFNs play a role. I am also especially grateful to Dr. T. Hünig for financial support and C. Hofmann for help with the English language and the preparation of the manuscript.

References

Abraham, N., Stojdl, D. F., Duncan, P. I., Methot, N., Ishii, T., Dube, M., Vanderhyden, B. C., Atkins, H. L., Gray, D. A., McBurney, M. W., Koromilas, A. E., Brown, E. G., Sonenberg, N., and Bell, J. C. (1999). Characterization of transgenic mice with targeted disruption of the catalytic domain of the double-stranded RNA-dependent protein kinase, PKR. *J. Biol. Chem.* **274,** 5953–5962.

Akkaraju, G. R., Whitaker-Dowling, P., Youngner, J. S., and Jagus, R. (1989). Vaccinia specific kinase inhibitory factor prevents translational inhibition by double-stranded RNA in rabbit reticulocyte lysate. *J. Biol. Chem.* **264,** 10321–10325.

Asch, B. B., and Gifford, G. E. (1970). Fowlpox virus: Interferon production and sensitivity. *Proc. Soc. Exp. Biol. Med.* **135,** 177–179.

Bakay, M., and Burke, D. (1972). The production of interferon in chick cells infected with DNA viruses: A search for double-stranded RNA. *J. Gen. Virol.* **16,** 399–403.

Balachandran, S., Kim, C. N., Yeh, W. C., Mak, T. W., Bhalla, K., and Barber, G. (1998). Activation of PKR induces apoptosis through FADD. *EMBO J.* **17,** 6888–6902.

Ball, L. A., and White, C. N. (1978). Oligonucleotide inhibitor of protein synthesis made in extracts of interferon-treated chick embryo cells: Comparison with the mouse low molecular weight inhibitor. *Proc. Natl. Acad. Sci. USA* **75,** 1167–1171.

Ball, L. A., and White, C. N. (1979). Nuclease activation by double-stranded RNA and by 2′, 5′ oligo-adenylate in extracts of interferon-treated chick cells. *Virology* **93,** 348–356.

Baron, S., and Isaacs, A. (1961). Interferon and natural recovery from viral diseases. *New Sci.* **243,** 81–82.

Bayliss, C. D., and Condit, R. C. (1993). Temperature-sensitive mutants in the vaccinia virus A18R gene increase double-stranded RNA synthesis as a result of aberrant viral transcription. *Virology* **194,** 254–262.

Beattie, E., Tartaglia, J., and Paoletti, E. (1991). Vaccinia virus-encoded eIF-2 alpha homolog abrogates the antiviral effects of interferon. *Virology* **183,** 419–422.

Beattie, E., Paoletti, E., and Tartaglia, J. (1995a). Distinct patterns of IFN sensitivity observed in cells infected with vaccinia K3L- and E3L-mutant viruses. *Virology* **210,** 254–263.

Beattie, E., Denzler, K. L., Tartaglia, J., Perkus, M.E., Paoletti, E., and Jacobs, B. L. (1995b). Reversal of the interferon-sensitive phenotype a vaccinia virus lacking E3L by expression of the reovirus S4 gene. *J. Virol.* **69,** 499–505.

Blum, D., Degen, H.-J., Redmann-Müller, I., Campos, M., and Jungwirth, C. (1993). Interferon sensitivity of expression of histone H5/H1(0)–vaccinia thymidine kinase fusion genes expressed by recombinant vaccinia viruses is enhanced by shortening the histone sequences. *Virology* **196,** 419–426.

Boone, R. F., Parr, R. P., and Moss, B. (1979). Intermolecular duplexes formed from polyadenylated vaccinia virus RNA. *J. Virol.* **30,** 365–374.

Buller, M., and Palumbo, G. J. (1991). Poxvirus pathogenesis. *Microbiol. Rev.* **55,** 80–122.

Carroll, K., Elroy-Stein, O., Moss, B., and Jagus, R. (1993). Recombinant vaccinia virus K3L gene product prevents activation of double-stranded RNA-dependent initiation factor 2 alpha specific protein kinase. *J. Biol. Chem.* **268,** 12837–12842.

Carver, D. H., and Marcus, P. I. (1967). Enhanced interferon production from chick embryo cells aged *in vitro*. *Virology* **32,** 247–257.

Castelli, J. C., Hassel, B. A., Wood, K. A., Li, X.-L., Amemiya, K., Dalakas, M. C., Torrence, P. F., and Youle, R. J. (1997). A study of the interferon antiviral mechanism: Apoptosis activation by the 2–5 A system. *J. Exp. Med.* **186,** 967–972.

Chang, H. W., and Jacobs, B. L. (1993). Identification of a conserved motif that is necessary for binding of vaccinia virus E3L gene products to a double-stranded RNA. *Virology* **194,** 537–547.

Chang, H.-W., Watson, J. C., and Jacobs, B. L. (1992). The EL3 gene of vaccinia virus encodes an inhibitor of the interferon-induced, double-stranded RNA-dependent protein kinase. *Proc. Natl. Acad. Sci. USA* **89,** 4825–4829.

Chang, H.-W., Uribe, L. H., and Jacobs, B. L. (1995). Rescue of vaccinia virus deleted for the E3L gene by mutants of E3L. *J. Virol.* **69,** 6605–6608.

Chen, X., Rösel, J., Dunker, R., Goldschmidt, R., Maurer-Schultze, B., and Jungwirth, C. (1983). Reversible inhibition of poxvirus replication by cycloheximide during the early phase of infection. *Virology* **124,** 308–317.

Clemens, M. J., and Williams, B. R. G. (1978). Inhibition of cell-free protein synthesis by pppA 2′p5′A2′p5′A: A novel oligonucleotide synthesized by interferon-treated L cell extracts. *Cell* **13,** 565–572.

Clouston, W. M., and Kerr, J. F. R. (1985). Apoptosis, lympho-toxicity and the containment of viral infections. *Med. Hypotheses* **18,** 399–404.

Cohrs, R. J., Condit, R. C., Pacha, R. F., Thompson, C. L., and Sharma, O. K. (1989). Modulation of ppp(A2′p)nA-dependent RNase by a temperature-sensitive mutant of vaccinia virus. *J. Virol.* **63,** 948–951.

Colamonici, O. R., Domanski, P., Sweitzer, S. M., Larner, A., and Buller, R. M. (1995). Vaccinia virus B18R gene encodes a type I interferon-binding protein that blocks interferon alpha transmembrane signaling. *J. Biol. Chem.* **270,** 15974–15978.

Colby, C., and Duesberg, P. H. (1969). Double-stranded RNA in vaccinia virus-infected cells. *Nature* **222,** 940–944.

Colby, C., Jurale, C., and Kates, J. R. (1971). Mechanism of synthesis of vaccinia virus double-stranded ribonucleic acid *in vivo* and *in vitro*. *J. Virol.* **7,** 71–76.

Craig, A. W., Cosentino, G. P., Donze, O., and Sonenberg, N. (1996). The kinase insert domain of interferon-induced protein kinase PKR is required for activity but not for interaction with the pseudosubstrate K3L. *J. Biol. Chem.* **271,** 24526–24533.

Darnell, J. E., Jr., Kerr, I. M., and Stark, G. R. (1994). Jak-STAT Pathways and transcriptional activation in response to IFNs and other extracellular signaling proteins. *Science* **264,** 1415–1421.

Davies, M. V., Elroy-Stein, O., Jagus, R., Moss, B., and Kaufmann, R. J. (1992). The vaccinia virus K3L gene product potentiates translation by inhibiting double-stranded RNA-activated protein kinase and phosphorylation of the alpha subunit of eukaryotic initiation factor 2. *J. Virol.* **66,** 1943–1950.

Davies, M. V., Chang, H. W., Jacobs, B. L., and Kaufman, R. J. (1993). The E3L and K3L vaccinia virus gene products stimulate translation through inhibition of the double-stranded RNA-dependent protein kinase by different mechanisms. *J. Virol.* **67,** 1688–1692.

Degen, H.-J., Blum, D., Grün, J., and Jungwirth, C. (1992). Expression of authentic vaccinia virus-specific and inserted viral and cellular genes under control of an early vaccinia virus promoter is regulated post-transcriptionally in interferon-treated chick embryo fibroblast. *Virology* **188,** 114–121.

De Haro, C., Mendez, R., and Santoyo, J. (1996). The eIF-2alpha kinases and the control of protein synthesis. *FASEB J.* **10,** 1378–1387.

Delic, J., Coppey-Moisan, M., and Magdelenat, H. (1993). Gamma-ray-induced transcription and apoptosis-associated loss of 28S rRNA in interphase human lymphocytes. *Int. J. Radiat. Biol.* **64,** 39–46.

Der, S. D., Yang, Y. L., Weissmann, C., and Williams, B. R. (1997). A double-stranded RNA-activated protein kinase-dependent pathway mediating stress-induced apoptosis. *Proc. Natl. Acad. Sci. USA* **94,** 3279–3283.

De Veer, M. J., Holko, M., Frevel, M., Walker, E., Der, S., Paranjape, J. M., Silverman, R. H., and Williams, B. R. (2001). Functional classification of interferon-stimulated genes identified using microarrays. *J. Leukocyte Biol.* **69,** 912–9264.

Diaz-Guerra, M., Kahn, J. S., and Esteban, M. (1993). A mutation of the nucleoside triphosphate phosphohydrolase I (NPH-I) gene confers sensitivity of vaccinia virus to interferon. *Virology* **197,** 485–491.

Diaz-Guerra, M., Rivas, C., and Esteban, M. (1997a). Activation of the IFN-inducible enzyme RNase L causes apoptosis of animal cells. *Virology* **236,** 354–363.

Diaz-Guerra, M., Rivas, C., and Esteban, M. (1997b). Inducible expression of the 2-5A synthetase/RNase L system results in inhibition of vaccinia virus replication. *Virology* **227,** 220–228.

Ding, A. H., Nathan, C. F., and Stuehr, D. J. (1988). Release of reactive nitrogen intermediates and reactive oxygen intermediates from mouse peritoneal macrophages. Comparison of activating cytokines and evidence for independent production. *J. Immunol.* **141,** 2407–2412.

Drapier, J. C., and Hibbs, J. B., Jr. (1988). Differentiation of murine macrophages to express nonspecific cytotoxicity for tumor cells results in L-arginine-dependent inhibition of mitochondrial iron-sulfur enzymes in the macrophage effector cells. *J. Immunol.* Apr 15; **140**(8), 2829–38.

Eisert, V. (1995). Diplomarbeit, Medical Faculty, University of Würzburg, Würzburg, Germany.

Eisert, V., Jungwirth, C., Feirer, N., and Kroon, E. (1997). The interaction of vaccinia virus WR; K3L minus and E3L minus mutants with recombinant interferon type I (Ch-IFN-I)-treated primary and permanent chicken fibroblasts (CEF). *J. Interferon Cytokine Res.* **17** [Abstract M11], 567.

Esteban, M., and Metz, D. H. (1973). Inhibition of early vaccinia virus protein synthesis in interferon-treated chicken embryo fibroblasts. *J. Gen. Virol.* **20,** 111–115.

Esteban, M., Benavente, J., and Paez, E. (1984). Effect of interferon on integrity of vaccinia virus and ribosomal RNA in infected cells. *Virology* **134,** 40–51.

Farrel, P. J., Balkow, K., Hunt, T., Jackson, R. J., and Trachsel, H. (1977). Phosphorylation of initiation factor eIF-2 and the control of reticulocyte protein synthesis. *Cell* **11,** 187–200.

Fout, G. S., and Simon, E. H. (1981). Studies on an interferon-sensitive mutant of Mengo virus: Effects on host RNA and protein synthesis. *J. Gen. Virol.* **52,** 391–394.

Gale, M., Jr., Tan, S. L., Wambach, M., and Katze, M. G. (1996). Interaction of the interferon-induced PKR protein kinase with inhibitory proteins p581 PK and vaccinia virus K3L is mediated by unique domains: Implications for kinase regulation. *Mol. Cell. Biol.* **16,** 4172–4181.

Gallagher, J. G., and Khoobyarian, N. (1969). Adenovirus susceptibility to interferon: Sensitivity of types 2, 7, and 12 to human interferon. *Proc. Soc. Exp. Biol. Med.* **130,** 137–142.

Gauntt, C. J., and Lockart, R. Z., Jr. (1966). Inhibition of Mengo virus by interferon. *J. Bacteriol.* **91,** 176–182.6.

Geiss, G., Jin, G., Guo, J., Bumgarner, R., Katze, M. G., and Sen, G. S. (2001). A comprehensive view of regulation of gene expression by double-stranded RNA-mediated cell signaling. *J. Biol. Chem.* **276,** 30178–30182.

Ghosh, S. N., and Gifford, G. E. (1965). Effect of interferon on the dynamics of H3-thymidine incorporation and thymidine kinase induction in chick fibroblast cultures infected with vaccinia virus. *Virology* **27,** 186–192.

Gifford, G. E., Toy, S. T., and Lindenmann, J. (1963). Studies on vaccinia virus plaque formation and its inhibition by interferon. II. Dynamics of plaque formation by vaccinia virus in the presence of interferon. *Virology* **19,** 294–301.

Gilmour, K. C., and Reich, N. C. (1995). Signal transduction and activation of gene transcription by interferons. *Gene Expression* **5,** 1–18.

Goebel, S. J., Johnson, G. P., Perkus, M. E., Davis, S. W., Winslow, J. P., and Paoletti, E. (1990). The complete DNA sequence of vaccinia virus. *Virology* **179,** 517–563.

Goodbourn, S., Didcock, L., and Randall, R. E. (2000). Interferons: Cell signaling, immune modulation, antiviral responses and virus countermeasures. *J. Gen. Virol.* **81,** 2341–2364.

Goswami, B. B., and Sharma, O. K. (1984). Degradation of rRNA interferon-treated vaccinia virus-infected cells. *J. Biol. Chem.* **259,** 1371–1374.

Grün, J., Kroon, E., Zöller, B., Krempien, U., and Jungwirth, C. (1987). Reduced steady-state levels of vaccinia virus-specific early mRNAs in interferon-treated chick embryo fibroblasts. *Virology* **158,** 28–33.

Grün, J., Redmann-Müller, I., Blum, D., Degen, H. J., Doenecke, D., Zentgraf, H. W., and Jungwirth, C. (1991). Regulation of histone H5 and H1 zero gene expression under the control of vaccinia virus-specific sequences in interferon-treated chick embryo fibroblasts. *Virology* **180,** 535–542.

Guan, K. L., Broyles, S. S., and Dixon, J. E. (1991). A Tyr/Ser protein phosphatase encoded by vaccinia virus. *Nature* **350,** 359–362.

Guidotti, L. G., and Chisari, F. V. (2001). Noncytolytic control of viral infections by the innate and adaptive immune response. *Annu. Rev. Immunol.* **19,** 65–91.

Harris, N., Buller, R. M., and Karupiah, G. (1995). Gamma interferon-induced, nitric oxide-mediated inhibition of vaccinia virus replication. *J. Virol.* **69,** 910–915.

Ho, M. (1962). Kinetic considerations of the inhibitory action of an interferon produced in chick cultures infected with Sindbis virus. *Virology* **17,** 262–275.

Ho, M., and Enders, J. F. (1959). Further studies on an inhibitor of viral activity appearing in infected cell cultures and its role in chronic viral infections. *Virology* **9,** 446–477.

Ho, C. K., and Shuman, S. (1996). Physical and functional characterization of the double-stranded RNA binding protein encoded by the vaccinia virus E3 gene. *Virology* **217,** 272–284.

Horak, I., Jungwirth, C., and Bodo, G. (1971). Poxvirus-specific cytopathic effect in interferon-treated L cells. *Virology* **45,** 456–462.

Houge, G., Robaye, B., Eikhom, T. S., Golstein, J., Mellgren, G., Gjertsen, B. T., Lanotte, M., and Doskeland, S. O. (1995). Fine mapping of 28S rRNA sites specifically cleaved in cells undergoing apoptosis. *Mol. Cell. Biol.* **15,** 2051–2062.

Hovanessian, A. G. (1991). Interferon-induced and double-stranded RNA-activated enzymes: A specific protein kinase and 2′, 5′ oligoadenylate synthetase. *J. Interferon Res.* **11,** 199–205.

Hovanessian, A. G. (1993). Interferon-induced dsRNA-activated protein kinase(PKR): Anti-proliferative, antiviral, antitumoral functions. *Semin. Virol.* **4,** 237–245.

Hovanessian, A. G., Meurs, E., and Montagnier, L. (1981). Lack of systematic correlation between the interferon-mediated antiviral state and the levels of 2–5 A synthetase and protein kinase in three different types of murine cells. *J. Interferon Res.* **1,** 179–190.

Hruby, D. E., Lynn, D. L., and Kates, J. R. (1980). Identification of a virus-specified protein in the nucleus of vaccinia virus-infected cells. *J. Gen. Virol.* **47,** 293–299.

Huang, S., Hendriks, W., Althage, A., Hemmi, S., Bluethmann, H., Kamijo, R., Vilcek, J., Zinkernagel, R. M., and Aguet, M. (1993). Immune response in mice that lack the interferon-gamma receptor. *Science* **259,** 1742–1745.

Isaacs, A., and Westwood, M. A. (1959). Inhibition by interferon of the growth of vaccinia virus in the rabbit skin. *Lancet* **ii,** 324–325.

Isaacs, A., Burke, D. C., and Fadeeva, L. (1958). Effect of interferon on the growth of viruses on the chicken chorion. *Br. J. Exp. Pathol.* **39,** 447–451.

Isaacs, A., Porterfield, J. S., and Baron, S. (1961). The influence of oxygenation on virus growth. II. Effect on the antiviral action of interferon. *Virology* **14,** 450–455.

Jacobs, B. L., and Langland, J. O. (1996). When two strands are better than one: The mediators and modulators of the cellular responses to double-stranded RNA. *Virology* **219,** 339–349.

Jagus, R., and Gray, M. M. (1994). Proteins that interact with PKR. *Biochimie* **76,** 779–791.

Joklik, W. K., and Merigan, T. C. (1966). Concerning the mechanism of action of interferon. *Proc. Natl. Acad. Sci. USA* **56,** 558–565.

Jungwirth, C., Kroath, H., and Bodo, G. (1977). Effect of interferon on poxvirus replication. *Tex. Rep. Biol. Med.* **35,** 247–259.

Karupiah, G., Frederickson, T. N., Holmes, K. L., Khairallah, L. H., and Buller, R. M. (1993a). Importance of interferons in recovery from mousepox. *J. Virol.* **67,** 4214–4226.

Karupiah, G., Xie, Q. W., Buller, R. M., Nathan, C., Duarte, C., and Macmicking, J. D. (1993b). Inhibition of viral replication by interferon-gamma-induced nitric oxide synthase. *Science* **261,** 1445–1448.

Kawagishi-Kobayashi, M., Silverman, J. B., Ung, T. L., and Dever, T. E. (1997). Regulation of protein kinase PKR by the vaccinia virus pseudosubstrate inhibitor K3L is dependent on residues conserved between the K3L protein and the PKR substrate eIF2alpha. *Mol. Cell. Biol.* **17,** 4146–4158.

Kawagishi-Kobayashi, M., Cao, C., Lu, J., Ozato, K., and Dever, T. E. (2000). Pseudosubstrate inhibition of protein kinase PKR by swine pox virus C8L gene product. *Virology* **276,** 424–434.

Kerr, I. M., and Brown, R. E. (1978). pppA2′p5A: An inhibitor of protein synthesis synthesized with an enzyme fraction from interferon-treated cells. *Proc. Natl. Acad. Sci. USA* **75,** 256–260.

Kibler, K. V., Shors, T., Perkins, K. B., Zeman, C. C., Banaszak, M. P., Biesterfeldt, J., Langland, J. O., and Jacobs, B. L. (1997). Double-stranded RNA is a trigger for apoptosis in vaccinia virus-infected cells. *J. Virol.* **71,** 1992–2003.

Kohonen-Corish, M. R., Blanden, R. V., and King, N. J. C. (1989). Induction of cell surface expression of HLA antigens by human IFN-γ encoded by recombinant vaccinia virus. *J. Immunol.* **143,** 623–627.

Kohonen-Corish, M. R. J., King, N. J. C., Woodhams, C. E., and Ramshaw, I. A. (1990). Immunodeficient mice recover from infection with vaccinia virus expressing interferon-γ. *Eur. J. Immunol.* **20,** 157–161.

Lebleu, B., Sen, G. C., Shaila, S., Cabrer, B., and Lengyel, P. (1976). Interferon, double-stranded RNA, and protein phosphorylation. *Proc. Natl. Acad. Sci. USA* **73,** 3107–3111.

Lee, S. B., and Esteban, M. (1993). The interferon-induced double-stranded RNA-activated human p68 protein kinase inhibits the replication of vaccinia virus. *Virology* **193,** 1037–1041.

Lee, S. B., and Esteban, M. (1994). The interferon-induced double-stranded RNA-activated kinase induces apoptosis. *Virology* **199,** 491–496.

Levy, B. H. (1964). Studies on the mechanism of interferon action. II. The effect of interferon on some early events in Mengo virus infection in L cells. *Virology* **22,** 575–579.

Liu, Y., Wolff, K. C., Jacobs, B. L., and Samuel, C. E. (2001). Vaccinia virus E3L interferon resistance protein inhibits the interferon-induced adenosine deaminase A-I editing activity. *Virology* **289,** 378–387.

Lockart, R. Z., Jr. (1968). Viral interference in aged cultures of chick embryo cells. In "Medical and Applied Virology" (M. Sanders and E. H. Lennette, Eds.), pp. 45–55. Warren H. Green, St. Louis, Missouri.

Lomniczi, B. (1974). Biological properties of Aujeszky's disease (pseudorabies) virus strains with special regard to interferon production and interferon sensitivity. *Arch. Ges. Virusforsch.* **44,** 205–214.

Lowell, A., Glasgow, L. A., and Habel, K. (1962). The role of interferon in vaccinia virus infection of mouse embryo tissue culture. *J. Exp. Med.* **115,** 503–512.

Magee, W. E., and Levine, S. (1970). The effects of interferon on vaccinia virus infection in tissue culture. *Ann. N.Y. Acad. Sci.* **173,** 362–378.

Maran, A., Maitra, R. K., Kumar, A., Dong, B., Xiau, W., Li, G., Williams, B. R., Torrence, P. F., and Silverman, R. H. (1994). Blockage of NF-kappa-B signaling by selective ablation of an mRNA target by 2′–5′A antisense chimeras. *Science* **265,** 789–792.

Martinez-Costas, J., Gonzalez-Lopez, C., Vakharia, V. N., and Benavente, J. (2000). Possible involvement of the double-stranded RNA-binding core protein sigmaA in the resistance of avian reovirus to interferon. *J. Virol.* **74,** 1124–1131.

Martz, E., and Howell, D. M. (1989). CTL: Virus control cells first and cytolytic cells second. DNA fragmentation, apoptosis and the prelytic halt hypothesis. *Immunol. Today* **10,** 79–86.

McFadden, G., and Graham, K. (1994). Modulation of cytokine networks by pox-virus: The myxoma model. *Semin. Virol.* **5,** 421–429.

Melcova, Z., and Esteban, M. (1994). Interferon-gamma severely inhibits DNA synthesis of vaccinia virus in a macrophage cell line. *Virology* **198,** 731–735.

Minks, M. A., Benvin, S., Maroney, P. A., and Baglioni, C. (1979). Metabolic stability of 2′ 5′ oligo (A) and activity of 2′ 5′ oligo (A)-dependent endonuclease in extracts of interferon-treated and control HeLa cells. *Nucleic Acids Res.* Feb; **6**(2), 767–80.

Moss, B. (1968). Inhibition of HeLa cell protein synthesis by the vaccinia virion. *J. Virol.* **2,** 1028–1037.

Moss, B. (1996). Poxviridae: The Viruses and Their Replication, in "*Fields Virology,*" 4th ed. (D. M. Knipe and P. W. Howley, Eds.), Vol. 2, pp. 2649–2883. Lippincott Williams and Wilkins, New York.

Moss, B., Ahn, B. Y., Amegadzie, B., Gershon, P. D., and Keck, J. G. (1991). Cytoplasmic transcription system encoded by vaccinia virus. *J. Biol. Chem.* **266,** 1355–1358.

Müller, U., Steinhoff, U., Reis, L. F., Hemmi, S., Pavlovic, J., Zinkernagel, R. M., and Aguet, M. (1994). Functional role of type I and type II interferons in antiviral defense. *Science* **264,** 1918–1921.

Najarro, P., Traktman, P., and Lewis, J. A. (2001). Vaccinia virus blocks gamma interferon signal transduction: Viral VH1 phosphatase reverses Stat1 activation. *J. Virol.* **75,** 3185–3196.

Ohno, S., and Nozima, T. (1964). Inhibitory effect of interferon on the induction of thymidine kinase in vaccinia virus-infected chick embryo fibroblasts. *Acta Virol.* **8,** 479.

Osterhoff, J., Jäger, M., Jungwirth, C., and Bodo, G. (1976). Inhibition of poxvirus-specific functions induced in chick-embryo fibroblasts by treatment with homologous interferon. *Eur. J. Biochem.* **69,** 535–543.

Pacha, R. F., and Condit, R. C. (1985). Characterization of a temperature-sensitive mutant of a vaccinia virus reveals a novel function that prevents virus-induced breakdown of RNA. *J. Virol.* **56,** 395–403.

Paez, E., and Esteban, M. (1984a). Resistance of vaccinia virus to interferon is related to an interference phenomenon between the virus and the interferon system. *Virology* **134,** 12–28.

Paez, E., and Esteban, M. (1984b). Nature and mode of action of vaccinia virus products that block activation of the interferon-mediated ppp(A2′p)nA-synthetase. *Virology* **134,** 29–39.

Radke, K. L., Colby, C., Kates, J. R., Krider, H. M., and Prescott, D. M. (1974). Establishment and maintenance of the interferon-induced antiviral state: Studies in enucleated cells. *J. Virol.* **13,** 623–630.

Reiss, C. S., and Komatsu, T. (1998). Does nitric oxide play a critical role in viral infections? *J. Virol.* **72,** 4547–4551.

Rice, A. P., and Kerr, I. M. (1984). Interferon-mediated, double-stranded RNA-dependent protein kinase is inhibited in extracts from vaccinia virus-infected cells. *J. Virol.* **50,** 229–236.

Rice, A. P., Roberts, W. K., and Kerr, I. M. (1984). 2–5 A accumulates to high levels in interferon-treated, vaccinia virus-infected cells in the absence of any inhibition of virus replication. *J. Virol.* **50,** 220–228.

Rivas, C., Gil, J., Melkova, Z., Esteban, M., and Diaz-Guerra, M. (1998). Vaccinia virus E3L protein is an inhibitor of the interferon (IFN)-induced 2–5 A synthetase enzyme. *Virology* **243,** 406–414.

Roberts, W. K., Hovanessian, A., Brown, R. E., Clemens, M. J., and Kerr, I. M. (1976). Interferon-mediated protein kinase and low molecular weight inhibitor of protein synthesis. *Nature* **264,** 477–480.

Romano, P. R., Zhang, F., Tan, S. L., Garcia-Barrio, M. T., Katze, M. G., Dever, T. E., and Hinnebusch, A. G. (1998). Inhibition of double-stranded RNA-dependent protein kinase PKR by vaccinia virus E3. Role of complex formation and the E3N-terminal domain. *Mol. Cell. Biol.* **18,** 7304–7316.

Schindler, C., Shuai, K., Prezioso, V. R., and Darnell, J. E., Jr. (1992). Interferon-dependent tyrosin phosphorylation of a latent cytoplasmic transcription factor. *Science* **257,** 809–815.

Schmidt, A., Chernajovsky, Y., Shulman, L., Federman, P., Berissi, H., and Revel, M. (1979). An interferon-induced phosphodiesterase degrading (2′–5′) oligoisoadenylate and the C–C–A terminus of tRNA. *Proc. Natl. Acad. Sci. USA* **76,** 4788–4792.

Sharp, T. V., Witzel, J. E., and Jagus, R. (1997). Homologous regions of the alpha subunit of eukaryotic translational initiation factor 2 (eIF2 alpha) and vaccinia virus K3L gene product interact with the same domain within the dsRNA-activated proteinkinase (PKR). *Eur. J. Biochem.* **250,** 85–91.

Sharp, T. V., Moonan, F., Romashko, A., Joshi, B., Barber, G. N., and Jagus, R. (1998). The vaccinia virus E3L gene product interacts with both the regulatory and substrate binding regions of PKR: Implication for PKR autoregulation. *Virology* **250,** 302–315.

Shen, Y., and Shenk, T. E. (1995). Viruses and apoptosis. *Curr. Opin. Genet. Dev.* **5,** 105–111.

Shors, T., Kibler, K., Perkins, K. B., Seidler-Wulff, R., Banaszak, M. P., and Jacobs, B. L. (1997). Complementation of vaccinia virus lacking of the E3L gene by mutants of E3L. *Virology* **239,** 269–276.

Shuai, K., Schindler, C., Precioso, V. R., and Darnell, J. E., Jr. (1992). Activation of transcription by IFN-gamma: Tyrosin phosphorylation of 91kD binding protein. *Science* **258,** 1808–1812.

Silverman, R. H., and Cirino, N. M. (1997). RNA decay by the interferon-regulated 2-5A system as a host defense against viruses. *In* "mRNA Metabolism and Post-Transcriptional Gene Regulation," pp. 295–309. Wiley-Liss, New York.

Smith, G. L. (1994). Virus strategies for evasion of the host response to infection. *Trends Microbiol.* **2,** 81–88.

Smith, E. J., Marie, I., Prakash, A., Garcia-Sastre, A., and Levy, D. E. (2001). IRF3 and IRF7 phosphorylation in virus-infected cells does not require double-stranded RNA-dependent protein kinase R or I kappa B kinase but is blocked by vaccinia virus E3L protein. *J. Biol. Chem.* **276,** 8951–8957.

Spriggs, M. K. (1994). Poxvirus-encoded soluble cytokine receptors. *Virus Res.* **33,** 1–10.

Srivastava, S. P., Kumar, K. U., and Kaufman, R. J. (1998). Phosphorylation of eukaryotic translation initiation factor 2 mediates apoptosis in response to activation of the double-stranded RNA-dependent protein kinase. *J. Biol. Chem.* **273,** 2416–2423.

Staeheli, P. (1990). Interferon-induced proteins and the antiviral state. *Adv. Virus Res.* **38,** 147–200.

Stark, G. R., Dower, W. J., Schimke, R. T., Brown, R. E., and Kerr, I. M. (1979). 2–5 A synthetase: Assay, distribution and variation with growth or hormone status. *Nature* **278,** 471–473.

Stewart II, W. E. (1979). "The Interferon System." Springer-Verlag, New York.

Stewart II, W. E., Scott, W. D., and Sulkin, S. E. (1969). Relative sensitivities of viruses to different species of interferon. *J. Virol.* **4,** 147–153.

Stewart II, W. E., De Clercq, E., Billiau, A., Desmyter, J., and De Somer, P. (1972). Increased susceptibility of cells treated with interferon to the toxicity of polyriboinosinic–polyribocytidylic acid. *Proc. Nat. Acad. Sci. USA* **69,** 1851–1854.

Suh, M., Bodo, G., Wolf, W., Viehhauser, G., and Jungwirth, C. (1974). Interferon effect on cellular functions: Enhancement of virus-induced inhibition of host cell protein synthesis in interferon-treated cells. *Z. Naturforsch.* **29C,** 623–629.

Symons, J. A., Alcami, A., and Smith, G. L. (1995). Vaccinia virus encodes a soluble type I interferon receptor of novel structure and broad species specificity. *Cell* **81,** 551–560.

Tanaka, N., Sato, M., Lamphier, M. S., Nozawa, H., Oda, E., Noguchi, S., Schreiber, R. D., Tsujimoto, Y., and Taniguchi, T. (1998). Type I interferons are essential mediators of apoptotic death in virally infected cells. *Genes Cells* **3,** 29–37.

Thacore, H. R., and Youngner, J. S. (1973). Rescue of vesicular stomatitis virus from interferon-induced resistance by superinfection with vaccinia virus I. Rescue in cell cultures from different species. *Virology* **56,** 505–511.

Traktman, P. (1990). Poxviruses: an emerging portrait of biological strategy. *Cell.* Aug 24; **62**(4), 621–6. Review.

Ueda, Y., Ito, M., and Tagaya, I. (1969). A specific surface antigen induced by poxvirus. *Virology* **38,** 180–182.

Upton, C., Mossman, K., and McFadden, G. (1992). Encoding of a homolog of the IFN-gamma receptor by myxoma virus. *Science* **258,** 1369–1372.

Van den Broek, M. F., Müller, U., Huang, S., Zinkernagel, R. M., and Aguet, M. (1995a). Immunodefense in mice lacking type I and/or type II interferon receptors. *Immunol. Rev.* **148,** 5–18.

Van den Broek, M. F., Müller, U., Huang, S., Aguet, M., and Zinkernagel, R. M. (1995b). Antiviral defense in mice lacking both alpha/beta and gamma interferon receptors. *J. Virol.* **69,** 4792–4796.

Varich, N. L., Sychova, I. V., Kaverin, N. V., Antonova, T. P., and Chernos, V. I. (1979). Transcription of both DNA strands of vaccinia virus genome *in vivo*. *Virology* **96,** 412–430.

Vilcek, J., and Rada, B. (1962). Studies on an interferon from tick-borne encephalitis virus-infected cells. III. Antiviral action of IF. *Acta Virol.* **6,** 9–16.

Watson, J. C., Chang, H. W., and Jacobs, B. L. (1991). Characterization of a vaccinia virus-encoded double-stranded RNA-binding protein that may be involved in inhibition of the double-stranded RNA-dependent protein kinase. *Virology* **185,** 206–216.

Weissmann, C., and Weber, H. (1986). The interferon genes. *Prog. Nucleic Acid Res. Mol. Biol.* **33,** 251–300.

Whitaker-Dowling, P., and Youngner, J. S. (1983). Vaccinia rescue of VSV from interferon-induced resistance: Reversal of translation block and inhibition of protein kinase activity. *Virology* **131,** 128–136.

Whitaker-Dowling, P., and Youngner, J. S. (1984). Characterization of a specific kinase inhibitory factor produced by vaccinia virus which inhibits the interferon-induced protein kinase. *Virology* **137,** 171–181.

Whitaker-Dowling, P., and Youngner, J. S. (1986). Vaccinia-mediated rescue of encephalomyocarditis virus from the inhibitory effects of interferon. *Virology* **152,** 50–57.

Williams, B. R. G. (1999). PKR: A sentinel kinase for cellular stress. *Oncogene* **18,** 6112–6120.

Williams, T. W., and Bellanti, J. A. (1983). *In vitro* synergism between interferons and human lymphotoxins: Enhancement of lymphotoxin-induced target cell killing. *J. Immunol.* **130,** 518–520.

Williams, B. R., Kerr, I. M., Gilbert, C. S., White, C. N., and Ball, L. A. (1978). Synthesis and breakdown of pppA2′p5′A2′p5′A and transient inhibition of protein synthesis in extracts from interferon-treated and control cells. *Eur J. Biochem.* Dec; **92**(2), 455–62.

Wreschner, D. H., James, T. C., Silverman, R. H., and Kerr, J. M. (1981). Ribosomal RNA cleavage, nuclease activation, and 2–5 A (ppp(A2′p)nA) in interferon-treated cells. *Nucleic Acids Res.* **9,** 1571–1581.

Xiang, Y., Simpson, D. A., Spiegel, J., Zhou, A., Silverman, R. H., and Condit, R. C. (1998). The vaccinia virus A18R DNA helicase is a postreplicative negative transcription elongation factor. *J. Virol.* **72,** 7012–7023.

Yang, Y. L., Reis, L. F., Pavlovic, J., Aguzzi, A., Schafer, R., Kumar, A., Williams, B. R., Aguet, M., and Weissmann, C. (1995). Deficient signaling in mice devoid of double-stranded RNA-dependent protein kinase PKR. *EMBO J.* **14,** 6095–6106.

Yeung, M. C., Liu, J., and Lau, A. S. (1996). An essential role for the interferon-inducible, double-stranded RNA-activated protein kinase PKR in the tumor necrosis factor-induced apoptosis in U937 cells. *Proc. Natl. Acad. Sci. USA* **93,** 12451–12455.

Youngner, J. S., Thacore, H. R., and Kelly, M. E. (1972). Sensitivity of ribonucleic acid and deoxyribonucleic acid viruses to different species of interferon in cell cultures. *J. Virol.* **10,** 171–178.

Yuwen, H., Cox, J. H., Yewdell, J. W., Bennink, J. R., and Moss, B. (1993). Nuclear localization of a double-stranded RNA-binding protein encoded by the vaccinia virus E3L gene. *Virology* **195,** 732–744.

Zilberstein, A., Federman, P., Shulman, L., and Revel, M. (1976). Specific phosphorylation *in vitro* of a protein associated with ribosomes of interferon-treated mouse L cells. *FEBS Lett.* **68,** 119–124.

Zöller, B., Popp, M., Walter, A., Redmann-Müller, I., Lodemann, E., and Jungwirth, C. (1998). Overexpression of chicken interferon regulatory factor-1 (Ch-IRF-1) induces constitutive expression of MHC class I antigens but does not confer virus resistance to a permanent chicken fibroblast cell line. *Gene* **222,** 269–278.

Zöller, B., Redmann-Müller, I., Nanda, I., Guttenbach, M., Dosch, E., Schmid, M., Zoorob, R., and Jungwirth, C. (2000). Sequence comparison of avian interferon regulatory factors and identification of the avian CEC-32 cell as a quail cell line. *J. Interferon Cytokine Res.* **20,** 711–717.

INDEX

A

AcMNPV, *see Autographica californica* M nucleopolyhedrovirus
Active immunization, subunit vaccines, 44
Adenoviruses, live vaccines, 48–49
Agrobacterium tumefaciens vaccines, 88
AHF, *see* Argentine hemorrhagic fever
Alfalfa mosaic virus, coat proteins, 101–102
Aluminum hydroxide, vaccine formulation, 58
Aluminum phosphate, vaccine formulation, 58
Annexin XIIIb, lipid rafts, 6
Antibodies
- live attenuated vaccines, 37
- vaccination function, 34–35
- vaccines, 107–108

Antigens
- hepatitis B virus surface antigen, 15, 110
- immune responses, 110–111
- tumor-assisted antigens, 44–45
- viral antigens, 42

Antiviral drug resistance, HIV mechanism
- fusion and integrase inhibitors, 172
- general aspects, 168–169
- protease inhibitors, 171–172
- RT inhibitors, 169–171

Antiviral drug therapy, HIV
- envelope glycoprotein, 166–167
- history, 158
- integrase, 168
- protease, 164–166
- reverse transcriptase, 159–160
- reverse transcriptase inhibition, 162–164
- reverse transcriptase structure, 160–162
- targets, 159

Antiviral response
- bracovirus, 235
- ichnovirus, 225

Arenaviruses
- agents of emerging disease, 128–131
- assembly and budding, 142–143
- attachment, 133–135
- characteristics, 131–132
- classification, 126–128
- host cell infection, 132–133
- human disease, 143–145
- internalization and uncoating, 135–137
- protein maturation and exocytic transport, 140–142
- RNA transcription and replication, 137–140

Argentine hemorrhagic fever, 144–145
Assembly pathway, bacteriophage ϕ29, 257
ATPase gp16, bacteriophage ϕ29 packaging, 272
Autographica californica, 42
Autographica californica M nucleopolyhedrovirus, 225, 235
Avridine, vaccine formulation, 58
AZT, HIV antiviral drug resistance mechanism, 169–170

B

Bacteria
- live vaccines, 48–49
- products as vaccine formulation, 62–64
- subunit vaccines, 82

Bacteriophage ϕ29
- DNA packaging
  - assembly pathway, 257
  - definition, 257, 259
  - DNA–gp3, 260
  - DNA maturation, 273–275
  - efficiency, 256
  - gp16, 272
  - head–tail connector, 263–264

Bacteriophage ϕ29 *(continued)*
initiation, 259–260, 275–276
intermolecular pseudoknot, 268–269
multiple functions, 271–272
pRNA mutants, 266–268
pRNA properties, 264–266
prohead, 261–263
SELEX studies, 269–271
DNA packaging, DNA–gp3 translocation
mechanism, 284–286
motor structure and symmetry, 277–279
motor studies, 279–284
overview, 276–277
ratchet mechanism, 286–289
dsDNA, overview, 255–256
Braconid wasp, *see Cardiochiles nigriceps*
Bracovirus
antiviral response, 235
cellular response, 235–237
developmental alterations, 237–239
gene families, 233–234
genome organization, 233
history, 228
humoral response, 237
pathogenic host, 231–233
replicative host, 229–231
BV, *see* Bracovirus

C

Campoletis sonorensis ichnovirus
antiviral response, 225
capsid morphology and composition, 214
cellular response inhibition, 225–226
Cys-motif genes, 218–221
humoral response inhibition, 226–227
infection, 215
molecular characteristics, 217
rep gene, 222
segment types, 211
vinnexin gene, 223
CaMV, *see* Cauliflower mosaic virus
Cancer vaccines, subunit vaccines, 44–45
Capsids
bracovirus, 230–231
ichnovirus, 213–214
Cardiochiles nigriceps
bracovirus associations, 228–229
bracovirus history, 228
ecology, 229
Cardiochiles nigriceps bracovirus
capsids, 230–231
developmental alterations, 238–239
gene families, 234
genome organization, 232
Castration, subunit vaccines, 45–46
Cauliflower mosaic virus
subunit vaccination, 93–94
vaccines, 86
Caveolin-1/VIP21, lipid rafts, 6
*Cc*BV, *see Cotesia congregata* bracovirus
CEFs, *see* Chick embryo fibroblasts
Cell entry, bracovirus, 231
Cell-mediated immunity, vaccines, 106–108
Cellular response
bracovirus, 235–237
ichnovirus, 225–226
Chelonus inanitus bracovirus, 238–239
Chick embryo fibroblasts, IFN type I-treated
poxvirus-specific gene expression, 299–301
protein kinase PKR, 306–307
sensitivity of poxvirus, 313–315
Chloroplast genome, vaccines, 89
Cholera toxin
subunit vaccines, 103
vaccine delivery, 55–56
vaccine formulation, 64
*Ci*BV, *see Chelonus inanitus* bracovirus
*Cm*BV, *see Cotesia melanoscela* bracovirus
CMI, *see* Cell-mediated immunity
*Cn*BV, *see Cardiochiles nigriceps* bracovirus
Coat proteins
alfalfa mosaic virus, 101–102
icosahedral plant viruses, 102
tobacco mosaic virus, 101–102
Conventional vaccines
killed vaccines, 37–39
live attenuated vaccines, 36–37
Cotesia congregata bracovirus
immune system, 235–237
replication, 229–230
Cotesia melanoscela bracovirus

capsids, 230–231
genome organization, 233
Cotesia rubecula bracovirus
gene families, 234
immune system, 237
Cowpea mosaic virus
subunit vaccination, 91–93
vaccines, 86
CpG motifs, vaccine formulation, 62–63
CPMV, *see* Cowpea mosaic virus
*Cr*BV, *see Cotesia rubecula* bracovirus
*Cs*IV, *see Campoletis sonorensis* ichnovirus
CT, *see* Cholera toxin
CTP synthetase, arenaviruses, 140
Cyclopentenylcytosine, arenaviruses, 140
Cyclopentylcytosine, arenaviruses, 140
Cys-motif, ichnovirus, 211, 218–221
Cytokines
live vaccines, 49
vaccine formulation, 64
Cytoskeleton, lipid raft in budding process, 18
Cytotoxicity, IFN type I-treated and vaccinia virus-infected cells, 301–302

D

DDA, vaccine formulation, 58–59
Deletion, live vaccines, 46–47
Deoxyribonucleoside derivatives, HIV reverse transcriptase inhibition, 162–163
Detergent-insoluble glycolipid-enriched complexes, Triton X-100, 3–5
Development
bracovirus effect, 237–239
ichnovirus effects, 227–228
α-DG, *see* α-Dystroglycan
Diadromus pulchrellus ascovirus, evolutionary model, 241
Differentiate Infected from Vaccinated Animals, 52–53
DIGs, *see* Detergent-insoluble glycolipid-enriched complexes
Diseases
emerging diseases, 128–131
human, arenaviruses, 143–145
DIVA vaccines, *see* Differentiate Infected from Vaccinated Animals
DNA
bacterial, vaccine formulation, 62–63
bacteriophage ϕ29 dsDNA, 255–256
bracovirus, 232–233
HIV reverse transcriptase, therapy, 159–160
ichnovirus, molecular characteristics, 216–217
subunit vaccine production, 40
T-DNA, vaccines, 88
DNA–gp3 translocation
mechanism, 284–286
motor structure and symmetry, 277–279
motor studies, 279–284
overview, 276–277
ratchet mechanism, 286–289
DNA immunization, polynucleotide vaccines, 50–52
DNA maturation, bacteriophage ϕ29 DNA packaging, 273–275
DNA packaging, bacteriophage ϕ29
assembly pathway, 257
definition, 257, 259
DNA–gp3, 260
DNA–gp3 translocation
mechanism, 284–286
motor structure and symmetry, 277–279
motor studies, 279–284
overview, 276–277
ratchet mechanism, 286–289
DNA maturation, 273–275
efficiency, 256
gp16, 272
head–tail connector, 263–264
initiation, 259–260, 275–276
intermolecular pseudoknot, 268–269
multiple functions, 271–272
pRNA mutants, 266–268
pRNA properties, 264–266
prohead, 261–263
SELEX studies, 269–271
DNA-translocating ATPase gp16, bacteriophage ϕ29 packaging, 272
DNA vaccines
gene gun, 55
polynucleotides, 51–52

Double-stranded DNA, bacteriophage ϕ29, 255–256
Double stranded RNA
activated protein kinase PKR, 304–307
vaccinia virus–IFN type I-treated host cell interaction, 308–309
*Dp*AV, *see Diadromus pulchrellus* ascovirus
Drugs, *see* Antiviral drug therapy
Drug sensitivity, HIV
assay interpretation, 183–185
clinical use, 185–187
enzymatic assays, 180
genotyping, 180–183
overview, 173
phenotypic testing
primary HIV isolates, 173–174
recombinant HIV, 174–177
replication-incompetent HIV particles, 177–179
virus entry and integrase, 183
dsDNA, *see* Double-stranded DNA
dsRNA, *see* Double stranded RNA
α-Dystroglycan, viral attachment protein, 133–134

E

E3L/specific kinase inhibitory factor, 311–313
Ecology, *Cardiochiles nigriceps,* 229
Edible plant vaccines, human health, 112–113
Electroporation, vaccine delivery, 56
Emerging diseases, arenaviruses as agents, 128–131
Emulsions, oil, vaccine formulation, 59
Endocytosis, lipid rafts, 8
Envelope glycoprotein, HIV therapy, 166–167
Enzymatic assays, HIV drug resistance, 180
Enzymes, oligo A synthetase/ribonuclease L pathway, 311
Episomes, wasp-encoded, polydnavirus evolution, 241–242
Epitopes, small, expression, 100–102
Escherichia coli heat-labile toxin, vaccine delivery, 55–56
Eukaryotic expression system, subunit vaccine production, 40–41
Evolutionary model, polydnavirus
ancestor, 239–241
wasp-encoded episome, 241–242
Exocytic transport, arenaviruses, 140–142

F

FMDV, *see* Foot-and-mouth disease virus
Foot-and-mouth disease virus
icosahedral plant viruses, 92
immune responses, 111
marker vaccines, 53
Foot-and-mouth disease virus VP1, subunit vaccine production, 40
F protein, lipid raft role at budding site, 14–15
Fusion inhibitors, HIV drug resistance, 172

G

GAD, *see* Glutamic acid decarboxylase
GALT, *see* Gut-associated lymphoid tissue
Gastroenteritis, viruses causing, 104, 106
Gene gun, vaccine delivery, 55
Genes
bracovirus, 233–234
E3L/SKIF and K3L products of vaccinia virus, 311–313
M24, 224
M27, 224
M40, 224
modification for live vaccines, 46–47
poxvirus expression in IFN type I-treated fibroblasts, 299–301
Rep, 221–222
vinnexin, 222–224
Genetically engineered vaccines
live vaccines, 46–49
polynucleotide vaccines, 50–52
subunit vaccines, 39–46
Geno2pheno, HIV drug sensitivity, 184
Genome organization
bracovirus, 233
chloroplast, vaccines, 89

HIV reverse transcriptase, therapy, 159–160
ichnovirus
genomics, 217–218
molecular characteristics, 216–217
overview, 215–216
polydnaviruses, 205–206
Genotyping, HIV drug resistance, 180–183
*Gi*BV, *see Glyptapanteles indiensis* baculovirus
GLD, *see* Glucose dehydrogenase
Glucose dehydrogenase, bracovirus and immune system, 236
Glutamic acid decarboxylase, oral delivery, 110
Glycoproteins
arenaviruses, 131
envelope glycoprotein, 166–167
viral attachment, 134
Glyptapanteles indiensis baculovirus
cell entry, 231
genome organization, 233
gp3, bacteriophage ϕ29 packaging, DNA–gp3, 260
gp16, bacteriophage ϕ29 packaging, 272
GPC, arenaviruses, 140–141
Gut-associated lymphoid tissue, 109

H

HBsAg, *see* Hepatitis B virus surface antigen
Head–tail connector, bacteriophage ϕ29 packaging, 263–264
Heat-labile enterotoxin, subunit vaccines, 103
Helical plant viruses, subunit vaccines, 94–97
Hemolymph, ichnovirus effect, 227–228
Hepatitis B virus surface antigen
immune responses, 110
lipid raft role at budding site, 15
Herpesvirus, live vaccines, 48–49
HIV, *see* Human immunodeficiency virus
Hormones
ichnovirus effect, 227
subunit vaccines, 45
Host cells
arenavirus infection, 132–133
IFN type I-treated
cytotoxicity enhancement, 301–302
dsRNA-activated protein kinase PKR, 304–307
2′–5′ oligo A/ribonuclease L system, 307–308
overview, 303–304
poxvirus interaction, 296–299
vaccinia virus interaction, dsRNA role, 308–309
poxvirus replication, 313–315
H protein, lipid raft role at budding site, 14–15
Human disease, arenaviruses, 143–145
Human health, edible vaccines, 112–113
Human immunodeficiency virus
antiviral drug resistance mechanism
fusion and integrase inhibitors, 172
general aspects, 168–169
protease inhibitors, 171–172
RT inhibitors, 169–171
antiviral therapy history, 158
cowpea mosaic virus, 92
drug sensitivity
assay interpretation, 183–185
clinical use, 185–187
enzymatic assays, 180
genotyping, 180–183
overview, 173
virus entry and integrase, 183
infection characteristics, 157–158
lipid raft role at budding site, 16
phenotypic sensitivity testing
primary HIV isolates, 173–174
recombinant HIV, 174–177
replication-incompetent HIV particles, 177–179
therapy
envelope glycoprotein, 166–167
genome reverse transcription, 159–160
inhibition mechanisms, 162–164
integrase, 168
protease, 164–166
structure and function, 160–162
tomato bushy stunt virus, 93
Humoral response
bracovirus, 237
ichnovirus, 226–227

Hyposoter spp. ichnovirus
capsid morphology and composition, 213–214
infection, 215
Hyposoter virescens ichnovirus
M24, M27, M40 genes, 224
rep gene, 222
vinnexin gene, 222–224

I

Ichnovirus
antiviral response inhibition, 225
cellular response inhibition, 225–226
Cys-motif genes, 218–221
developmental regulation, 227–228
genome organization
genomics, 217–218
molecular characteristics, 216–217
overview, 215–216
humoral response inhibition, 226–227
M24, M27, M40 genes, 224
pathogenic host, 215
phylogeny, 208–209
rep gene, 221–222
replicative host, 210–215
capsid morphology and composition, 213–214
infection, 215
segment types, 211–213
vinnexin gene, 222–224
Icosahedral plant viruses
coat proteins, 102
subunit vaccines, 91–94
Ideal vaccine, characteristics, 30–32
IFN type I, *see* Interferon type I
IFN type II, *see* Interferon type II
Immune response
alteration by ichnovirus, 225–228
infection, 33
oral delivery of protein, 108–110
plant-derived antigens, 110–111
vaccination, 31
Immune-stimulating complexes, vaccine formulation, 60–61
Immune system
bracovirus, 235–237
vaccines, 106–108
Immunity
vaccination induction, 31–32
vaccines, 106–108
Immunization
polynucleotide vaccines, 50–52
subunit vaccines, 44
Immunogenic carrier molecules, 102–103
Immunogenicity, killed vaccines, 38–39
Immunogenic molecules, small epitope expression, 100–102
Infection
body response, 33–34
HIV therapy
envelope glycoprotein, 166–167
integrase, 168
protease, 164–166
reverse transcription, 159–164
host cells with arenaviruses, 132–133
ichnovirus, 215
NK cell role, 35
recovery after vaccination, 32–33
vaccinia virus, IFN type I-treated host cell response
dsRNA-activated protein kinase PKR, 304–307
2′–5′ oligo A/ribonuclease L system, 307–308
overview, 303–304
Influenza viral proteins, lipid raft role, 9–10
Influenza virus
assembly, 19–22
lipid raft role at budding site, 13–14
Influenza virus HA
lipid raft role, 9–10
lipid rafts, 7–8
Influenza virus neuraminidase
lipid raft role, 10–11
lipid rafts, 7
Integrase, HIV
drug sensitivity, 183
therapy, 168
Integrase inhibitors, HIV drug resistance, 172
Interferon type I
cytotoxicity enhancement, 301–302
poxvirus interaction, 295–299
poxvirus sensitivity, 313–315
poxvirus-specific gene expression, 299–301
response to vaccinia virus infection

dsRNA-activated protein kinase PKR, 304–307
2′–5′ oligo A/ribonuclease L system, 307–308
overview, 303–304
vaccinia virus interaction, 308–309
Interferon type II
signal transduction pathway, poxvirus, 310–311
vaccinia virus replication, 303
Interferon type I receptor, 310–311
Interferon type II receptor, 310–311
Intermolecular pseudoknot, bacteriophage ϕ29 pRNA, 268–269
Internalization, arenaviruses, 135–137
ISCOMs, *see* Immune-stimulating complexes
IV, *see* Ichnovirus

J

Jet propulsion, vaccine delivery, 54–55
JGMV, *see* Johnsongrass mosaic virus
JH, *see* Juvenile hormone
JHE, *see* Juvenile hormone esterase
Johnsongrass mosaic virus, subunit vaccination, 96–97
Junín virus
emerging disease agent, 130
inhibition, 141–142
internalization and uncoating, 135–137
RNA transcription and replication, 140
JUNV, *see* Junín virus
Juvenile hormone, bracovirus effect, 238
Juvenile hormone esterase, ichnovirus effect, 227

K

K3L, vaccinia virus gene products, 311–313
Killed vaccines, characteristics, 37–39

L

L929 cells, IFN type I-treated and vaccinia virus-infected, 301–302
Lassa fever, present treatment, 143–144
Lassa virus
emerging disease agent, 130
present treatment, 143–144
LASV, *see* Lassa virus
LCMV, *see* Lymphocytic choriomeningitis virus
Life cycle
bracovirus pathogenic host, 231–233
bracovirus replicative host, 229–231
polydnavirus, 203–205
Lipid amines, vaccine formulation, 58–59
Lipid carriers, vaccine formulation, 60–62
Lipid rafts
endocytosis, 8
membrane signaling, 8
overview, 2–5
protein sorting, 5–8
protein targeting, 5–8
protein transport, 5–8
role in budding process, 17–19
viral budding site role, 12–17
viral components, 8–12
Lipid vehicles, vaccine formulation, 60–62
Lipopolysaccharides
vaccination reaction, 31
vaccine formulation, 62
Liposomes, vaccine formulation, 61–62
Live attenuated vaccines
characteristics, 36–37
gene modification, 46–47
Live vaccines
gene deletion, 46–47
protective protein carrying, 48–49
LPS, *see* Lipopolysaccharides
LT, *see* Heat-labile enterotoxin
Lymphocytic choriomeningitis virus
attachment, 134
emerging disease agent, 129–130
GPC, 141
host cells infection, 132
internalization and uncoating, 135–136

M

M1, *see* Matrix protein
M24 gene, ichnovirus, 224
M27 gene, ichnovirus, 224
M40 gene, ichnovirus, 224

Machupo virus, emerging disease agent, 130
MACV, *see* Machupo virus
Madin–Darby kidney cells, lipid raft
  function, 5–6
  overview, 4
  role at budding site, 12–13
  role in influenza virus HA, 9–10
  role in influenza virus NA, 10
Major histocompatibility complex I proteins, 106–107
MAL/VIP17, lipid rafts, 6–7
Mammalian cells, subunit vaccine production, 41
Marker vaccines, 52–53
Matrix protein, influenza virus assembly, 19–22
Maturation
  bacteriophage ϕ29 DNA packaging, 273–275
  protein, arenaviruses, 140–142
*Md*BV, *see Microplitis demolitor* bracovirus
MDCK cells, *see* Madin–Darby kidney cells
MDP, *see* Muramyl dipeptide
Measles virus, lipid raft role at budding site, 14–15
Membrane signaling, lipid rafts, 8
MHC, *see* Major histocompatibility complex I proteins
Microplitis demolitor bracovirus
  DNA, 232
  gene families, 233–234
  immune response, 237
Microspheres, vaccine formulation, 64–65
Mineral salts, vaccine formulation, 58
Models, polydnavirus
  ancestor, 239–241
  wasp-encoded episome, 241–242
Monophosphoryl lipid A, vaccine formulation, 62
Morphology, ichnovirus capsids, 213–214
Motors, bacteriophage ϕ29 DNA packaging
  single-molecule studies, 279–284
  structure and symmetry, 277–279
MPL, *see* Monophosphoryl lipid A
Mucosal immune responses, oral delivery of protein, 108–110
Muramyl dipeptide, vaccine formulation, 63
Mutations
  bacteriophage ϕ29 pRNA, 266–268
  HIV protease, 171–172
  HIV reverse transcriptase, 170
MV, *see* Measles virus

N

Natural killer cells, infection role, 35
Needle-free delivery devices, vaccines, 54–56
Neonates, vaccination, 32
New World arenaviruses, lineages, 126–128
Nitric oxide synthase, vaccinia virus replication, 303
NK cells, *see* Natural killer cells
NNRTIs, *see* Nonnucleoside reverse transcriptase inhibitors
Nonnucleoside reverse transcriptase inhibitors
  HIV drug resistance, 170–171
  HIV reverse transcriptase inhibition, 163
Norwalk virus, oral vaccination, 104
NP protein, arenavirus
  assembly and budding, 142–143
  overview, 137–139
NTRIs, *see* Nucleotidic reverse transcriptase inhibitors
Nucleotidic reverse transcriptase inhibitors, HIV drug resistance, 169–171

O

Oil emulsions, vaccine formulation, 59
Old World arenaviruses, lineages, 126–128
2′–5′ Oligo A/ribonuclease L system
  IFN type I-treated host cell, 307–308
  modulating enzymes, 311
Oral delivery, protein for mucosal immune responses, 108–110
Oral vaccination
  plant proteins, 87
  virus-like particles, 104–106

P

Paramyxoviruses, budding, 12
Particles
  replication-incompetent HIV, 177–179
  virus-like particles, 41–44, 104–106
Pathogenic host
  bracovirus, 231–233
  ichnovirus, 215
Pathogens, vaccination, 32
PBMCs, *see* Peripheral blood mononuclear cells
PDV, *see* Polydnavirus
pE3/specific kinase inhibitory factor, 313–315
Peripheral blood mononuclear cells, HIV drug sensitivity, 174
Phenosense, HIV drug sensitivity, 184–185
Phenotypic assay, HIV drug sensitivity
  clinical use, 186–187
  interpretation, 183–185
  primary HIV isolates, 173–174
  recombinant HIV, 174–177
Phospholipase D2, lipid rafts, 7–8
Phylogeny
  bracovirus associations, 228–229
  ichnovirus, 208–209
PIs, *see* Protease inhibitors
pK3, IFN type I sensitivity of poxvirus, 313–315
PKI, *see* Protein kinase inhibitors
Plant-derived antigens, immune responses, 110–111
Plant products, vaccine formulation, 59–60
Plant proteins
  expression systems, 86–87
  recombinant expression, 83–86
  yield, 98–99
Plant vaccines
  human health, 112–113
  overview, 82
  production, 42–43
  transgenic plants, 87–89
Plant viruses
  expression system overview, 89–91
  helical viruses, 94–97
  icosahedral viruses, 91–94
  protein expression advantages, 97–98
  virus-like particle expression, 104–106
Plasma membranes, lipid rafts, 2–3
Plux pox potyvirus, subunit vaccines, 97
Polio virus vaccination, reaction, 30–31
Polydnavirus
  future research, 242–244
  genome organization, 205–206
  life cycle and transmission, 203–205
  origins, 207–208, 239
  viral ancestor model, 239–241
  virus comparison, 206–207
  wasp-encoded episome, 241–242
Polymerase chain reaction, HIV drug sensitivity, 174–176
Polynucleotide vaccines, DNA immunization, 50–52
Polypeptides, plant expression, 85
Potato virus X
  coat proteins, 102
  subunit vaccination, 96
  vaccines, 86
Potexvirus-based vectors, subunit vaccination, 96
Potyvirus-based vectors, subunit vaccination, 96–97
Poxviruses
  E3L/SKIF and K3L gene products, 311–313
  gene expression in IFN type I-treated fibroblasts, 299–301
  interferon interaction, type I-treated host cell, 296–299
  interferon interaction overview, 295–296
  live vaccines, 48–49
  2′–5′ oligo A/ribonuclease L system, 311
  replication, IFN type sensitivity, 313–315
  specific IFN type I and II receptors, 310–311
PPV, *see* Plux pox potyvirus
PR, *see* Protease
pRNA, *see* Prohead RNA
Prohead RNA, bacteriophage ϕ29 packaging
  DNA–gp3, 261–263
  general properties, 264–266
  intermolecular pseudoknot, 268–269

Prohead RNA, bacteriophage *(continued)*
multiple functions, 271–272
mutants, 266–268
SELEX studies, 269–271
Protease, HIV
antiviral therapy target, 159
drug resistance enzymatic assays, 180
drug resistance genotyping, 181–183
drug sensitivity, 174–176
inhibition mechanism, 165–166
proteolysis in replication, 164–165
structure and function, 165
Protease inhibitors, HIV drug resistance, 171–172
Protective proteins, live vaccines, 48–49
Protein kinase inhibitors, bracovirus, 238
Protein kinase PKR
dsRNA-activated, induction, 304–307
poxvirus-specifc gene products, 311–315
Proteins
coat proteins, 101–102
expression in plant viruses, 97–98
F protein, 14–15
H protein, 14–15
influenza viral proteins, 9–10
lipid rafts, 5–8
matrix protein, 19–22
maturation, arenaviruses, 140–142
MHC I proteins, 106–107
mucosal immune responses, 108–110
NP protein, 137–139, 142–143
plant proteins, 83–87
protective proteins, 48–49
recombinant proteins, 83–86
surface proteins, 39–40, 106
viral attachment protein, 133–134
yield in plants, 98–99
Proteolysis, HIV replication, 164–165
PVX, *see* Potato virus X
Pyrophosphate, HIV reverse transcriptase inhibition, 163–164

R

Rabies virus, live vaccines, 49
Ratchet mechanism, bacteriophage ϕ29 DNA packaging, 286–289
Recombinant DNA, subunit vaccine production, 40
Recombinant proteins, plant expression, 83–86
Rep gene, ichnovirus, 221–222
Replication
arenavirus RNA, 137–140
bracovirus, 229–230
HIV, 164–165
ichnovirus, 210–211
poxvirus, 313–315
vaccinia virus, 303
vaccinia virus dsRNA, 308–309
Replicative host
bracovirus, 229–231
ichnovirus capsid morphology, 213–214
ichnovirus infection, 215
ichnovirus segment types, 211–213
Reverse transcriptase, HIV
antiviral therapy target, 159
drug resistance, 180–183
drug sensitivity, 174–176
inhibition mechanisms, 162–164
reverse transcription, 159–160
structure and function, 160–162
Reverse transcriptase inhibitors, HIV drug resistance, 169–171
Reverse transcription polymerase chain reaction, HIV
drug resistance, 180
drug sensitivity, 174–176
RNA
activated protein kinase PKR, 304–307
arenaviruses, 131–132
cowpea mosaic virus-based vectors, 91–92
plant virus overview, 90–91
prohead, *see* Prohead RNA
transcription in arenaviruses, 137–140
vaccinia virus–IFN type I-treated host cell interaction, 308–309
Rotavirus, oral vaccination, 104, 106
RT, *see* Reverse transcriptase
RT-PCR, *see* Reverse transcription polymerase chain reaction
Rubella virus, lipid raft role at budding site, 15

S

Sabiá virus, emerging disease agent, 131
SABV, *see* Sabiá virus
Saponin, vaccine formulation, 60
SEAP, *see* Secretable alkaline phosphatase
Secretable alkaline phosphatase, HIV drug sensitivity, 175
Segments, ichnovirus, 211–213
SELEX, *see* Systematic Evolution of Ligands by Exponential Enrichment
Signal transduction, IFN type II, poxvirus, 310–311
Simian immunodeficiency virus, lipid raft role at budding site, 16
SIV, *see* Simian immunodeficiency virus
SKIF, *see* Specific kinase inhibitory factor
Small epitopes, expression, 100–102
Smallpox vaccination, reaction, 30
Specific kinase inhibitory factor
  dsRNA, 305–306
  E3L/SKIF, 311–313
  pE3/SKIF, 313–315
Sphingolipid–cholesterol clusters, 3–5
STAT1, poxvirus-specific IFN type I and II receptors, 310–311
Subunit vaccines
  advantages, 43
  bacteria and yeast, 82
  castration, 45–46
  caulimovirus, 93–94
  cowpea mosaic virus-based vectors, 91–93
  definition, 39
  immunogenic carrier molecules, 102–103
  plant production, 42–43
  plant virus overview, 89–91
  potexvirus-based vectors, 96
  potyvirus-based vectors, 96–97
  production, 40–41
  specific epitope production, 43–44
  surface proteins as vaccines, 39–40
  tobamoviridae-based vectors, 94–96
  tomato bushy stunt virus, 93
  tumor-assisted antigens, 44–45
  turkey egg production, 46
Surface proteins
  oral vaccination, 106
  use as subunit vaccines, 39–40
Surfactants, oil emulsion vaccine formulation, 59
Synthetic adjuvant, vaccine formulation, 58–59
Systematic Evolution of Ligands by Exponential Enrichment, 269–271

T

TAAs, *see* Tumor-assisted antigens
Tamiami virus, emerging disease agent, 129
TAMV, *see* Tamiami virus
TBSV, *see* Tomato bushy stunt virus
TCRV
  inhibition, 142
  RNA transcription and replication, 140
T-DNA, vaccines, 88
TEV, *see* Tobacco etch virus
TGEV, *see* Transmissible gastroenteritis virus
TMD, *see* Transmembrane domain
TMV, *see* Tobacco mosaic virus
Tobacco etch virus
  subunit vaccines, 97
  vaccines, 86
Tobacco mosaic virus
  coat protein, 101–102
  subunit vaccines, 94–96
  vaccines, 86
Tobamoviridae-based vectors, subunit vaccination, 94–96
Tomato bushy stunt virus, subunit vaccination, 93
Transcription, arenavirus RNA, 137–140
Transgenic plants, vaccines, 87–89
Transmembrane domain
  influenza virus HA, 7
  lipid raft in influenza virus HA, 9–10
  lipid raft in NA, 10–11
  lipid rafts, 3
Transmissible gastroenteritis virus, immune responses, 110–111
Transmission, polydnavirus, 203–205
Transosema rostrale ichnovirus
  antiviral response, 225
  developmental regulation, 227–228

Transosema rostrale (continued)
humoral response inhibition, 227
Triton X-100
influenza virus assembly, 19–20
influenza virus HA, 9–10
sphingolipid–cholesterol clusters, 3–5
vesicular stomatitis virus, 17
*Tr*IV, *see Transosema rostrale* ichnovirus
Tumor-assisted antigens, subunit vaccines, 44–45
Turkeys, egg production, subunit vaccines, 46
TX-100, *see* Triton X-100

U

Ubiquitin, lipid raft in budding process, 18
Uncoating, arenaviruses, 135–137
5′ Untranslated regions, plant protein yield, 99
UTRs, *see* 5′ Untranslated regions

V

Vaccines
basic principles, 32–35
cancer vaccines, 44–45
conventional vaccines, 36–39
developmental trends, 81–83
DNA vaccines, 51–52, 55
edible, human health, 112–113
formulation, 56–58, 60–65
genetically engineered vaccines, 39–52
ideal, characteristics, 30–32
immune system, 106–108
killed vaccines, 37–39
live attenuated vaccines, 36–37, 46–47
marker vaccines, 52–53
mineral salt formulation, 58
needle-free delivery devices, 54–56
oil emulsion formulation, 59
oral vaccination, 87, 104–106
plant product formulations, 59–60
plant protein-expression systems, 86–87
polio virus vaccination, 30–31
polynucleotide vaccines, 50–52
smallpox vaccination, 30
subunit, *see* Subunit vaccines
synthetic adjuvant formulation, 58–59
transgenic plants, 87–89
Vaccinia virus
E3L/SKIF and K3L gene products, 311–313
IFN type I-treated host cell interaction, 308–309
IFN type I-treated host cell response
dsRNA-activated protein kinase PKR, 304–307
2′–5′ oligo A/ribonuclease L system, 307–308
overview, 303–304
infected cells, cytotoxicity enhancement, 301–302
live vaccines, 49
replication inhibition, 303
VAP, *see* Viral attachment protein
Venturia canescens ichnovirus
Cys-motif genes, 220
infection, 215
Vesicular stomatitis virus, lipid raft role at budding site, 16–17
Vinnexin gene, ichnovirus, 222–224
Viral ancestor model, polydnavirus, 239–241
Viral antigens, *Autographica californica,* 42
Viral assembly
arenaviruses, 142–143
influenza virus, 19–22
lipid raft role, 8–19
Viral attachment protein, arenaviruses, 133–134
Viral budding
arenaviruses, 142–143
lipid raft at assembly site, 8–12
lipid raft at budding site, 12–17
lipid raft in budding process, 17–19
Viral delivery, lipid raft role, 8–12
Viral entry
arenavirus attachment, 133–135
arenavirus internalization and uncoating, 135–137
HIV, 166–167
HIV drug sensitivity, 183

Viral envelope, 1–2
Viral sorting, lipid raft role, 8–12
Viral targeting, lipid raft role, 8–12
VirtualPhenotype, HIV drug sensitivity, 184
Virus-like particle
  oral vaccination, 104–106
  subunit vaccines, 41–44
Virus–polydnavirus comparison, 206–207
VLP, *see* Virus-like particle
VP6, oral vaccination, 106
VSV, *see* Vesicular stomatitis virus

W

Wasp-encoded episome, polydnavirus evolution, 241–242
Whitewater Arroyo virus, emerging disease agent, 129
WWAV, *see* Whitewater Arroyo virus

Y

Yeast, subunit vaccines, 82

ISBN 0-12-039858-3